Wireless Communications and Networking

Jon W. Mark and Weihua Zhuang
Centre for Wireless Communications
Department of Electrical and Computer Engineering
University of Waterloo
Waterloo, Ontario, Canada

Pearson Education, Inc.
Upper Saddle River, New Jersey 07458

Library of Congress Cataloging-in-Publication Data on file

Vice President and Editorial Director, ECS: *Marcia J. Horton*
Publisher: *Tom Robbins*
Vice President and Director of Production and Manufacturing, ESM: *David W. Riccardi*
Executive Managing Editor: *Vince O'Brien*
Managing Editor: *David A. George*
Production Editor: *Scott Disanno*
Director of Creative Services: *Paul Belfanti*
Creative Director: *Carole Anson*
Art Director: *Jayne Conte*
Cover Designer: *Bruce Kenselaar*
Art Editor: *Greg Dulles*
Manufacturing Manager: *Trudy Pisciotti*
Manufacturing Buyer: *Lisa McDowell*
Marketing Manager: *Holly Stark*

Pearson Education, Inc.
Upper Saddle River, NJ 07458

Printed in the United States of America
10 9 8 7 6 5 4 3 2 1

ISBN 0-13-040905-7

Pearson Education Ltd., *London*
Pearson Education Australia Pty. Ltd., *Sydney*
Pearson Education Singapore, Pte. Ltd.
Pearson Education North Asia Ltd., *Hong Kong*
Pearson Education Canada Inc., *Toronto*
Pearson Educación de Mexico, S.A. de C.V.
Pearson Education—Japan, *Tokyo*
Pearson Education Malaysia, Pte. Ltd.
Pearson Education, Inc., *Upper Saddle River, New Jersey*

To my wife, Betty—J.W.M.

To my husband, Xuemin Shen and our sons, Alan and Alvin—W.Z.

Contents

Preface xi

1 Overview of Wireless Communications and Networking 1

1.1 Historical Overview of Wireless Communications 1

1.2 Challenges in Wireless Communication Networking 2

1.2.1 The Wireless Channel 3

1.2.2 User Mobility 3

1.3 Wireless Communications Standards 4

1.3.1 First Generation Cellular Systems 4

1.3.2 Second Generation Cellular Systems 5

1.3.3 Third Generation Wireless Communications Networks 7

1.3.4 Coverage Extension 8

1.3.5 Types of Wireless Communication Networks 8

1.4 Organization of This Text 9

Summary 10

Endnotes 10

Problems 11

2 Characterization of the Wireless Channel 12

2.1 Multipath Propagation Environment 12

2.2 Linear Time-Variant Channel Model 16

2.2.1 Channel Impulse Response 16

2.2.2 Time-Variant Transfer Function 18

2.2.3 Doppler Spread Function and Delay-Doppler Spread Function 19

2.2.4 Example on the Channel Functions 22

2.3 Channel Correlation Functions 25
2.3.1 Delay Power Spectral Density 25
2.3.2 Frequency and Time Correlation Functions 27
2.3.3 Doppler Power Spectral Density 29
2.3.4 Examples on the Channel Correlation Functions 31
2.4 Large-Scale Path Loss and Shadowing 35
2.4.1 Free Space Propagation Model 36
2.4.2 Propagation Over Smooth Plane 38
2.4.3 Log-Distance Path Loss with Shadowing 40
2.4.4 Okumura–Hara Path Loss Model 41
2.4.5 Lee's Path Loss Model 43
2.4.6 Radio Cell Coverage 45
2.5 Small-Scale Multipath Fading 48
2.5.1 First-Order Statistics 48
2.5.2 Second-Order Statistics 51
Summary 57
Endnotes 57
Problems 58

3 Bandpass Transmission Techniques for Mobile Radio **63**
3.1 Introduction 63
3.2 Signal Space and Decision Regions 68
3.2.1 Vector-Space Representation of M-ary Signals 68
3.2.2 Signal Detection and Optimal Receiver 70
3.3 Digital Modulation 76
3.3.1 M-ary Phase Shift Keying (MPSK) 76
3.3.2 Minimum Shift Keying (MSK) 84
3.3.3 Gaussian MSK (GMSK) 88
3.3.4 Orthogonal Frequency Division Multiplexing (OFDM) 90
3.4 Power Spectral Density 93
3.5 Probability of Transmission Error 100
3.5.1 Coherent Reception in an AWGN Channel 100
3.5.2 Coherent Reception in a Flat Slow Rayleigh Fading Channel 108
Summary 112
Endnotes 112
Problems 112

4 Receiver Techniques for Fading Dispersive Channels **117**
4.1 Overview of Channel Impairment Mitigation Techniques 117
4.2 Diversity 121
4.2.1 Diversity Mechanisms 121
4.2.2 Linear Combining 122
4.2.3 Performance Improvement 126
4.3 Channel Equalization 134

4.3.1 Linear Equalization 136
4.3.2 Decision Feedback Equalization 146
Summary 155
Endnotes 155
Problems 156

5 Fundamentals of Cellular Communications 160
5.1 Introduction 160
5.2 Frequency Reuse and Mobility Management 161
5.2.1 Cellular Communications and Frequency Reuse 161
5.2.2 Mobility Management 163
5.3 Cell Cluster Concept 163
5.3.1 Capacity Expansion by Frequency Reuse 164
5.3.2 Cellular Layout for Frequency Reuse 165
5.3.3 Geometry of Hexagonal Cells 168
5.3.4 Frequency Reuse Ratio 171
5.4 Cochannel and Adjacent Channel Interference 172
5.4.1 Cochannel Interference 172
5.4.2 Adjacent Channel Interference 175
5.5 Call Blocking and Delay at the Cell-Site 176
5.6 Other Mechanisms for Capacity Increase 180
5.6.1 Cell Splitting 180
5.6.2 Directional Antennas (Sectoring) 182
5.7 Channel Assignment Strategies 184
Summary 185
Endnotes 185
Problems 186

6 Multiple Access Techniques 189
6.1 Multiple Access in a Radio Cell 189
6.2 Random Access 191
6.2.1 Aloha Systems 191
6.2.2 Carrier Sense Multiple Access (CSMA) 195
6.3 Conflict-Free Multiple Access Technologies 197
6.3.1 FDMA 197
6.3.2 TDMA 198
6.3.3 CDMA 201
6.4 Spectral Efficiency 212
6.4.1 FDMA Systems 213
6.4.2 TDMA Systems 218
6.4.3 DS-CDMA Systems 222
Summary 229
Endnotes 229
Problems 230

7 Mobility Management in Wireless Networks **235**
7.1 Introduction 235
7.2 Call Admission Control (CAC) 237
7.2.1 Prioritized Call Admission 238
7.3 Handoff Management 238
7.3.1 Handoff Strategies 239
7.3.2 Types of Handoff 239
7.3.3 Design Issues 240
7.3.4 Feedback-Based MAHO Strategy 241
7.3.5 AP/MSC Identification 242
7.3.6 Profile 244
7.3.7 Capability of the Mobile 244
7.3.8 Mobility Model 245
7.3.9 Intraswitch Handoff Algorithm 247
7.4 Location Management for Cellular Networks 250
7.4.1 Two-Tiered Architecture of IS-41 252
7.4.2 SS7 Network and Common Channel Signaling 253
7.4.3 Location Update, Call Setup, and Paging 255
7.5 Location Management for PCS Networks 257
7.5.1 Overlay Approach 258
7.5.2 Local Anchor Approach 259
7.6 Traffic Calculation 261
7.6.1 System and Traffic Parameters 262
7.6.2 Handoff Rate Calculation 263
Summary 266
Endnotes 266
Problems 267

8 Wireless/Wireline Interworking **272**
8.1 Background 272
8.2 Mobile IP 274
8.2.1 Operation of Mobile IP 276
8.2.2 Local Anchor for Mobile IP 281
8.2.3 Hierarchical Routing 282
8.3 Internet Protocol (IP) 284
8.3.1 IPv6 versus IPv4 284
8.3.2 Mobile IPv6 287
8.4 Transmission Control Protocol (TCP) 289
8.4.1 Flow Control 289
8.4.2 Modified TCP 290
8.4.3 Modified UDP 293
8.5 Network Performance 294
8.5.1 Network Model 294

8.5.2 Mobility Model 3 295
8.5.3 Handoff Delay with Local Anchor 295
8.6 Wireless Application Protocol (WAP) 300
8.6.1 Wireless Application Environment 300
8.6.2 WAP Protocol Stack 301
8.6.3 WAP Gateway 302
8.7 Mobile AD HOC Networks 303
8.7.1 Ad Hoc Routing Protocols 303
8.7.2 Comments 305
Summary 305
Endnotes 306
Problems 306

Appendix **311**
A Gram–Schmidt Orthogonalization 311
B Maximum Likelihood Detection 313
C MSK Signal Representation 314
D Derivation of the Rician Distribution 315
E Pseudorandom Noise Sequences 317
F The Erlang-B and Erlang-C Tables 319

Abbreviations and Acronyms **327**

Bibliography **333**

Answers to Selected Problems **345**

Index **350**

Preface

Signal propagation through a guided wire (e.g., a coaxial cable or an optical fiber) is relatively free of interference. With a wireless channel, the impairments are much more severe. A signal propagating through the wireless channel will be subject to additive background noise, and will experience signal fading, multipath spread, cochannel interference, adjacent channel interference, etc. However, a wireless system has the flexibility to support user roaming, while a wired system lacks this flexibility. Because of its ability to support user mobility, the wireless system has emerged as the key information transport platform to meet the demands of modern society. However, wireless systems suffer a number of shortcomings, the most important being the severe channel impairments that limit the usable spectral width. Also, a wireless system has either a limited geographical coverage (e.g., ground radio) or a long propagation delay (e.g., geostationary satellite). Wireless cellular communication based on radio propagation has been evolving from narrowband (i.e., the first generation (1G) and second generation (2G) wireless systems) to wideband (i.e., the third generation (3G) wireless systems). With their geographical coverage limitation, wireless systems need a backbone network to extend their geographical coverage to enable global communications. A wireline network such as the Internet has universal appeal. The interworking of a wireless network as the front-end and the Internet as the backbone has been receiving much attention in recent years. With a hybrid wireless/Internet network, the wireless front-end supports user roaming while the Internet backbone offers global coverage.

The severe impairments in the wireless channel introduce a set of challenging problems for the network provider. When the demand for system bandwidth is not very high, the problems of spectral limitation and transmission errors associated with conventional modulation and coding methods may not be so intolerable. This is the case with narrowband cellular communications systems (e.g., 1G and 2G). To support multimedia communications (e.g., in 3G) mitigation of channel dispersive fading and multiple access interference effects are critically important.

The present text is aimed at providing the fundamentals of wireless communications and networking to senior undergraduate students. The materials in the text have been given as a one-semester course to fourth year students several times at the University of Waterloo. The student who takes this course would have in prior semesters already had courses in the principles of analog and digital communications, probability theory, and signal analysis methods.

The book begins with an overview of wireless communications and networking by providing a road map on these topics. The salient features of first and second generation wireless cellular systems, and those perceived for the third generation wireless systems, are highlighted in this overview. The focus of the text is on fundamentals at the senior undergraduate level. The materials on wireless communications and networking are organized into seven chapters, Chapters 2 through 8. Chapter 2 aims at providing an informative exposition and an understanding of the characteristics of the wireless channel, which form the basis for the development of bandpass transmission techniques in Chapter 3 and reception techniques in Chapter 4. These three chapters focus on the properties of the physical transmission layer, with the objective of providing an understanding of the characteristics of the various types of interference and the methods used at both the transmitter and receiver for mitigating these types of interference.

The fundamentals of cellular communications, including the rationales for cellular systems and the properties of frequency reuse to enlarge system capacity, are described and discussed in Chapter 5. One of the main benefits from wireless cellular communications is the ability to support multiple users who are on the move. Simultaneous transmissions by multiple users can lead to conflict, resulting in destructive interference. Techniques to resolve conflicts, and hence permit multiple access of the common resources, are described and the ramifications are discussed in Chapter 6.

Although wireless systems offer the flexibility for users to roam, user mobility nevertheless creates a set of challenging problems for the network provider. Mobility management that takes care of handoff and locating the whereabouts of the mobile unit to facilitate information delivery are discussed in Chapter 7. On the other hand, the geographical coverage of a wireless system is limited. Global communications that permit anyone to communicate from anywhere at anytime require global geographical coverage. The Internet is the most pervasive global network that has been enjoying universal acceptance. The ramifications of, and the interworking of, a wireless front-end and an Internet-Protocol-based backbone as the global communications platform for supporting user mobility are described and discussed in Chapter 8.

Solutions to selected end of chapter problems are available to instructors who have adopted the text for classroom use in academic instituitions. They can be obtained by contacting the publisher. Also available from the publisher is set of MATLAB® simulations that may be used with various exercises throughout the book. The simulations have a user-friendly interface and were created using MATLAB version 6.0. They can be used for demos in the classroom.

Societal demand for services from anywhere at anytime will likely escalate with time. It is hoped that the fundamentals of wireless communications and networking presented in this text provide the right foundation and appetite for individuals graduating from the bachelor degree program in engineering to pursue further learning in the area.

We would like to thank Ka Chun (Kyle) Chan for the development of the simulation programs. We would also like to thank Professor Vijay Bhargava and the anonymous reviewers for the valuable comments on, and suggestions for improving, the flow of the manuscript.

JON W. MARK AND WEIHUA ZHUANG
Waterloo, Ontario

1

Overview of Wireless Communications and Networking

This book aims at providing students at the senior level an in-depth understanding of end-to-end communications over an information transport platform consisting of a wireless segment and a wireline segment. It points out the obstacles that cause spectral limitation in the wireless propagation channel, the difficulty in maintaining service continuity as the user roams, the problems in bridging the wireless and wireline domains, and the techniques to overcome the difficulties in handling the end-to-end information transfer.

This chapter gives the reader an overview of the salient features of wireless communications and networking and the organization of the materials presented in the ensuing seven chapters.

1.1 HISTORICAL OVERVIEW OF WIRELESS COMMUNICATIONS

Long distance wireless transmission has a century-old history, dating from the time Guglielmo Marconi sent telegraphic signals over a distance of approximately 1800 miles from Cornwall, across the Atlantic Ocean, to St. John's, Newfoundland in 1901. Over the past century, wireless transmission has progressed through the development and deployment of radio, radar, television, satellite and mobile telephone technologies.

In the early years of wireless communications, radio was the most intensively deployed wireless technology, both in the public domain and by law enforcement establishments (e.g., police forces at all levels). Amplitude modulation (AM) of a radio frequency carrier was the transmission technique used until Edwin Armstrong demonstrated the feasibility of frequency modulation (FM) in 1935. Since the late 1930s, FM has been the main modulation method deployed for mobile communication systems worldwide.

The development of radar technology was escalated and matured during World War II when practically every university radar research laboratory in the United States, notably the University of Michigan at Ann Arbor and the Lincoln Laboratories at Massachusetts Institute of Technology (MIT), were converted to concentrate mainly on military applications, such as surveillance of the battle front in Europe. In the late 1940s and 1950s, commercial deployment of one-way and two-way radio and television systems flourished.

The growth of cellular radio and personal communication systems began to accelerate in the late 1970s. This growth was spurred on with the successive introduction of the first generation (1G), the second generation (2G), and the third generation (3G) cellular systems.

1.2 CHALLENGES IN WIRELESS COMMUNICATION NETWORKING

A wireless communication network offers a flexible information transport platform that allows mobile users to roam without suffering intolerable performance degradation. A wireless communication networking scenario is depicted in Figure 1.1. Here, a base station (BS) is the information distribution center for all mobile stations (MSs) within its signaling coverage area. The signal propagation medium between the MSs and the BS is wireless. The radio channel from an MS to its serving BS is called the uplink or reverse channel, and the radio channel from the BS to the MSs is called the downlink or forward channel. Base stations are connected to Mobile Switching

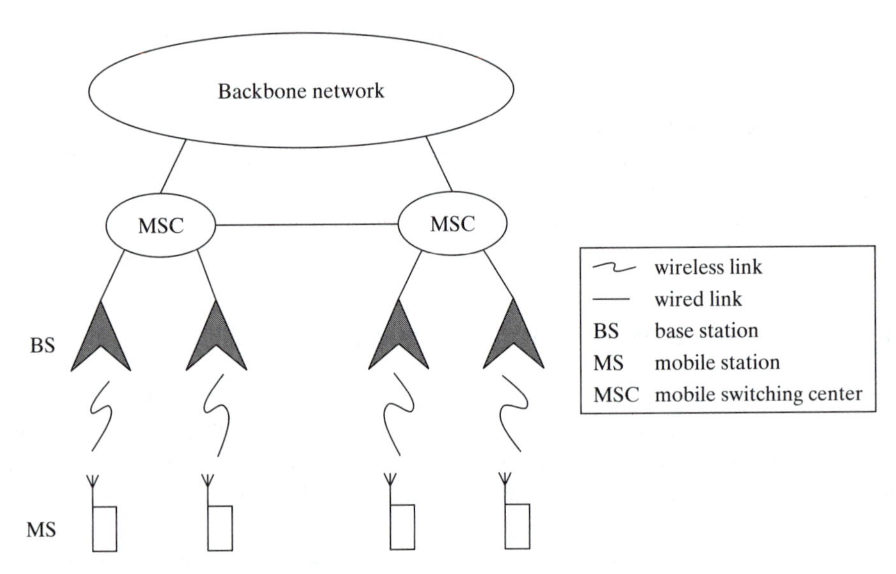

Figure 1.1 An illustration of wireless communications network.

Centers (MSCs) by wirelines to extend the geographical coverage of a single BS. Also, an MSC may be connected by wirelines to other MSCs and/or to a wired backbone network, e.g., an ATM (Asynchronous Transfer Mode) network or the Internet. In this way, a pair of communicating mobile users can be separated by a large distance. The key challenge is the creation of a viable information transport platform for the support of communications between mobile users. This platform is a hybrid connection of a front-end wireless network and a backbone wireline network.

Definition of Radio Cell A radio cell is a geographical area served by a single base station supporting the services of many mobile stations.

Depending on the size of the area covered, radio cells are categorized into picocells, microcells, and macrocells. The main problems in wireless communications come from (*a*) the hostile wireless propagation medium and (*b*) user mobility.

1.2.1 The Wireless Channel

The radio propagation channel exhibits many different forms of channel impairments, notably multipath delay spread, Doppler spread, intracell interference, intercell interference, fading, ambient noise, etc. Interference, distortion and noise can be differentiated into multiplicative and additive types as follows:

a. Multiplicative interference and distortion are normally signal-dependent, and include fading, intersymbol interference, etc. It causes attenuation, mutilation, etc., of the transmitted signal. The net result is a reduction in usable frequency spectrum. This form of disturbance cannot be suppressed by using filtering.
b. Additive noise is not as severe as multiplicative noise, but it still reduces signal detectability. Out-of-band noise can be suppressed by filtering, but in-band noise will still penetrate through the filter.

Effective and efficient transmitters and receivers are needed to combat interference and distortion. Transmitter/receiver design requires a good knowledge of the channel characteristics. Thus, a good understanding of the characteristics of the propagation channel is essential to allow for the design of effective transmitters, receivers, and communications protocols.

1.2.2 User Mobility

To provide communication services to a larger number of users using the limited radio spectrum, wireless systems are designed based on the cellular concept for frequency reuse. By dividing a large service area into small nonoverlapping (in an ideal case) cells and letting each base station communicate with all the mobile stations in the cell with low transmitter power, radio frequency spectrum can be reused in different cells subject to transmission quality satisfaction. As a mobile station moves from cell to cell, the serving base station changes. The process in which a mobile station switches its serving base station while crossing the cell boundary is referred to as *handoff management*. The process that tracks the user's movement, supports user roaming on a large scale, and delivers calls to the user at its current location is referred to as *location management*. As a result, the operation of a wireless network requires proper *mobility management* functions, which include both handoff management and location management.

While the information is represented and transmitted in the form of signal waveforms, communications between the sender and the receiver can be conveniently viewed as taking place in three conceptual layers: *physical, link* and *networking*. The physical layer involves the actual signal transmission and reception over the propagation channel. The link layer deals with the signals at the output of the cell-site (base station) receiver. Radio resource management functions such as power control, rate allocation and error control, and network resource management functions such as service scheduling and call admission control, can be exercised at the link layer. The networking layer comprises a protocol stack that includes handoff management, location management, traffic management and traffic control. As reflected by its title, the focus of this text is on elucidating and understanding the physical layer and the networking layer properties. The rationale behind this approach is that resource management and service scheduling strategies at the link layer are issues that aim at maximizing resource utilization and enhancing quality of service. Works in these areas are currently under intensive research.

1.3 WIRELESS COMMUNICATIONS STANDARDS

Wireless communications systems that have been in deployment for sometime are those of the first generation and second generation. Third generation systems are also currently under deployment, but continue to evolve. The first generation (1G) wireless communications systems use *frequency division multiple access* (FDMA) as the multiple access technology. FDMA is an analog transmission technique that is inherently narrowband. The second generation (2G) wireless systems use digital transmission. The multiple access technology is both *time division multiple access* (TDMA) and *code division multiple access* (CDMA). Although the second generation wireless systems offer higher transmission rates with greater flexibility than the first generation systems, they are nevertheless narrowband systems. The service offered by both 1G and 2G systems is predominantly voice. The third generation (3G) standard is based on CDMA as the multiple access technology. With a transmission rate of up to 2 megabits per second (Mbps), 3G systems are wideband, and are expected to support multimedia services.

For the first and second generations, the main initiatives have been originated in North America, Europe and Japan. Although all regions use similar technology, the systems differ in the location of the frequency band in the radio spectrum, the channel spacing, and the data transmission rate. ITU (International Telecommunications Union) has now adopted both the European IMT-2000 (International Mobile Telecommunications by the year 2000) and the North American cdma-2000 as third generation standards. IMT-2000 is a wideband direct-sequence CDMA (DS-CDMA) and cdma-2000 is a multicarrier CDMA (MC-CDMA) technology.

1.3.1 First Generation Cellular Systems

The first generation cellular systems use analog FM for speech transmission. The individual calls use different frequencies and share the available spectrum through FDMA.

AMPS in America and Australia. America and Australia use the Advanced Mobile Phone System (AMPS) with a 25 MHz band in each uplink, from 824 to 849 MHz, and a 25 MHz band in each downlink, from 869 to 894 MHz. AMPS uses a channel spacing of 30 kHz, with a total capacity of 832 channels, and supports a data transmission rate of 10 kilobits per second (kbps).

ETACS in Europe. Europe uses the European Total Access Communications System (ETACS) with a 25 MHz band in the uplink, from 890 to 915 MHz, and a 25 MHz band in the downlink, from 935 to 960 MHz. ETACS uses a channel spacing of 25 kHz, with a total capacity of 1000 channels, and supports a data transmission rate of 8 kbps.

NTT in Japan. The first generation cellular system is the Nippon Telephone and Telegraph (NTT) system which employs a 15 MHz band in the uplink, from 925 to 940 MHz, and a 15 MHz band in the downlink, from 870 to 885 MHz, with a channel spacing of 25 kHz. The NTT system has a total capacity of 600 channels and supports a data transmission rate of 0.3 kbps.

The NTT system was later modified to increase its capacity from 600 to 2400 channels. This increase was achieved by decreasing the channel spacing from 25 kHz to 6.25 kHz. In addition, the data transmission rate of each channel was increased from 0.3 kbps to 2.40 kbps.

The radio interface technology of the first generation wireless cellular systems (AMPS in America, ETACS in Europe, and NTT in Japan) is tabulated in Table 1.1. As can be seen from this table, the systems differ in the location of the frequency bands, channel spacing, data rate, spectral efficiency and system capacity.

1.3.2 Second Generation Cellular Systems

The second generation cellular systems are completely digital, employing either TDMA or CDMA as the multiple access technology. The digital technology allows greater sharing of the radio hardware in the base station among the multiple users, and provides a larger capacity to support more users per base station per MHz of spectrum than analog systems. Digital systems offer a number of advantages over analog systems, including

Table 1.1 First Generation Analog Cellular Systems

Region	America	Europe	Japan
Parameter	AMPS	ETACS	NTT
Multiple access	FDMA	FDMA	FDMA
Duplexing	FDD	FDD	FDD
Forward channel	869–894 MHz	935–960 MHz	870–885 MHz
Reverse channel	824–849 MHz	890–915 MHz	925–940 MHz
Channel spacing	30 kHz	25 kHz	25 kHz
Data rate	10 kbps	8 kbps	0.3 kbps
Spectral efficiency	0.33 bps/Hz	0.33 bps/Hz	0.012 bps/Hz
Capacity	832 channels	1000 channels	600 channels

a. natural integration with the evolving digital wireline network,
b. flexibility for supporting multimedia services,
c. flexibility for capacity expansion,
d. reduction in RF (radio frequency) transmit power,
e. encryption for communication privacy, and
f. reduction in system complexity.

As stated previously, the second generation cellular systems use either TDMA or CDMA access technology. While the CDMA standard is strictly North American, in the form of IS-95, TDMA deployment is regionally based (i.e., Europe, North America, or Japan). In Europe, it is the Pan-European GSM (Groupe Special Mobile or Global System for Mobile Communications) and DCS 1800 (Digital Cellular System). In North America, it is IS-54 and IS-136, and in Japan it is PDC (Personal Digital Cellular) systems. The different second generation standards have the following specifications.

GSM in Europe. The channel time in TDMA is partitioned into frames, each containing eight time-slots. Each time-slot is of 0.57 ms duration. Each user transmits periodically in every eighth slot and receives in the corresponding slot. (Each slot is a channel.) The modulation scheme is Gaussian filtered minimum shift keying (GMSK).

IS-54 in North America. In the frequency domain, the channel spacing is 30 kHz, the same as that of AMPS. The modulation scheme is $\pi/4$ shifted differential quadrature phase shift keying (DQPSK) with a channel rate of 48.6 kbps. In the time domain, one TDMA frame contains six time slots supporting three full-rate users or six half-rate users, each slot having a duration of approximately 6.67 ms. The speech codec rate is 7.95 kbps or 13 kbps with error protection. The capacity of each frequency channel is three times that of AMPS.

PDC in Japan. The system is TDMA-based with three slots multiplexed onto each carrier (similar to IS-54). In the frequency domain, the channel spacing is 25 kHz. The modulation scheme is $\pi/4$-shifted DQPSK with a transmission rate of 42 kbps. The speech codec operates at a full rate of 6.7 kbps or 11.2 kbps with error protection.

IS-95 in North America. IS-95 is a CDMA-based standard. Users share a common channel for transmission within a cell. Users in adjacent cells also use the same radio channel. In other words, the frequency spectrum is reused from cell to cell. The system is designed to be compatible with the existing analog system AMPS; the allocated frequency band is 824-849 MHz for the uplink using offset quadrature phase shift keying (OQPSK) and 869-894 MHz for the downlink using QPSK. The spreading code chip rate is 1.2288 megachips per second (Mcps). The spreading factor is 128, with the maximum user data rate of 9.6 kbps. Forward and reverse links use different spreading processes. Rake receivers are used at both the base station and mobile station to resolve and to combine multipath components.

The radio interface technology of the second generation digital systems is tabulated in Table 1.2.

In cellular communications, location management is critically important. IS-41, with a two-tiered network architecture for location management, is a companion standard to IS-54 in North America. GSM in Europe performs location management using MAP (Mobile Applications Part), which is also a two-tiered network architecture.

Table 1.2 Second Generation Digital Cellular Systems

Region	U.S.	Europe	Japan	U.S.
Parameter	IS-54	GSM	PDC	IS-95
Multiple access	TDMA/FDD	TDMA/FDD	TDMA/FDD	CDMA
Modulation	$\pi/4$ DQPSK	GMSK	$\pi/4$ DQPSK	QPSK/OQPSK
Forward channel	869-894 MHz	935-960 MHz	810-826 MHz	869-894 MHz
Reverse channel	824-849 MHz	890-915 MHz	940-956 MHz	824-849 MHz
Channel spacing	30 kHz	200 kHz	25 kHz	1,250 kHz
Data/chip rate	48.6 kbps	270.833 kbps	42 kbps	1.2288 Mcps
Speech codec rate	7.95 kbps	13.4 kbps	6.7 kbps	1.2/2.4/4.8/9.6 kbps

1.3.3 Third Generation Wireless Communications Networks

Third generation standardization activities were initiated in Europe and in North America under the respective names IMT-2000 and cdma-2000. IMT-2000 is wideband direct-sequence code division multiple access (DS-CDMA), while cdma-2000 is multicarrier code division multiple access (MC-CDMA). ITU has adopted the recommendations of both IMT-2000 and cdma-2000 under the banner of Harmonized Global 3G (G3G). Both IMT-2000 and cdma-2000 use frequency division duplex (FDD) to support two-way transmissions with frequency isolation. TDD (time division duplex) has also been suggested as a third 3G mode. The gist of harmonization is that both ANSI-41 and GSM MAP based services should be fully supported in the Radio Access Network with all 3G CDMA modes.

It is likely that the third generation cellular systems will be equipped with the infrastructure to support Personal Communications Systems (PCS). The network infrastructure support will likely include

a. public land mobile networks (PLMNs),
b. Mobile Internet Protocol (Mobile IP),
c. wireless asynchronous transfer mode (WATM) networks, and
d. low earth orbit (LEO) satellite networks.

Impact of High Transmission Rate. The most prominent features of 3G, compared with 2G, are the higher transmission rate and the support of multimedia services. The higher transmission rate means that the bandwidth of the signal will be large compared with the coherence bandwidth of the propagation channel. When the signal bandwidth is large compared with the coherence bandwidth of the channel, different frequency components of the signal will experience different fading characteristics (i.e., different parts of the signal spectrum will be affected differently by channel fading). Techniques to combat frequency selective fading need to be used in order to attain an acceptable error rate at the output of the cell-site receiver. A basic approach to combat frequency selective fading is to partition the signal into contiguous frequency bands, each of which is narrow compared with the coherence bandwidth of the channel. Each of the signal components is then modulated onto a different subcarrier and the signal components are sent over the channel in parallel. In this way, each of the signal components will experience non-frequency-selective fading, i.e., the fading is uniform across a given component's frequency band. This can be achieved by converting the high rate serial data sequence into a number of lower rate parallel sequences, and then modulating each onto a subcarrier. An effective method to achieve this is orthogonal frequency division multiplexing (OFDM). OFDM, as a signaling method, will be described in detail later in the text.

Mobiles can be located anywhere within the footprint of a base station. If every mobile transmits at the same power level, the signal received at the cell-site receiver from the mobile(s) closest to it will be the strongest. This is referred to as the near–far problem, and the power levels need to be controlled to smooth out the near–far effect. Power control, rate allocation and service scheduling are radio resource management functions. The primary purpose in managing the radio resources is to maximize system utilization. Strategies to effectively mechanize these functions are challenging research issues, which are not addressed in this text.

System Capacity and Impact of User Mobility. The capacity of cellular systems is enlarged through efficient employment of the available bandwidth (i.e., frequency reuse). To support higher transmission rates in 3G systems, the limited bandwidth of the systems needs to be reused more often.

Although decreasing the cell size allows for a higher degree of frequency reuse to increase system capacity, the tradeoff for this benefit is that mobile users tend to move in and out of cells much more frequently. To maintain service continuity, the connection of a mobile user must be handed off from the serving base station to base station of the target cell. Also, once handoff is complete, the mobile needs to identify its current location within the cellular array so that messages can be delivered to it in its new location. As a result, a reduction in the cell size translates to a larger overhead for mobility management (i.e., handoff management and location management) in the network. This text provides a lucid discussion of handoff and location management issues, and approaches to construct handoff and location management algorithms.

1.3.4 Coverage Extension

Wireless systems have limited geographical coverage. A backbone network such as the Internet is needed to extend the geographical coverage. Interworking of a wireless domain with a wireline domain introduces many challenging issues, chief among which is the problem of delivering information to the mobile as it roams around the extended network. These issues and the methods to handle information delivery in an efficient manner are discussed and treated later in the text.

1.3.5 Types of Wireless Communication Networks

Depending on whether or not there is a fixed infrastructure, wireless systems can be categorized as cellular systems or ad hoc networks. A cellular system has a fixed infrastructure in the form of a base station, which performs central administration for the system. Cellular networks provide the information transport platform for wireless local area networks (WLANs) and wireless wide area networks (WWANs). Wireless LAN standards activities have been spearheaded by IEEE 802.11, while wireless WAN standards activities have been led by ANSI (American National Standards Institute) and ITU (International Telecommunications Union).

Ad hoc networks have no fixed infrastructure and the network architecture is configurable. Every node (mobile) in the ad hoc network can set up as, and play the role of, a base station in that it can directly transmit to and receive from other nodes in the network. The main focus of this book is with cellular networks, but a bird's eye view into the operational features of ad hoc networks is also given.

1.4 ORGANIZATION OF THIS TEXT

With the overview of challenging issues in wireless communications and networking presented in this chapter, in-depth discussions of channel characterization, transmitter/receiver design, cellular network architecture, multiple access technologies, mobility management, and traffic management in extended networks are provided in Chapters 2 through 8. The rationale behind the material presented in each of these chapters is concisely summarized next.

Characterization of the Wireless Channel (Chapter 2). Because of impairments in the wireless channel, signal transmission through the wireless propagation medium presents a set of very challenging problems. Impairments in the wireless channel include multipath delay spread, Doppler spread, fading, path loss, shadowing, interference and ambient noise, etc. Knowledge of the channel characteristics will facilitate transmitter and receiver design. In the absence of this knowledge, ways to characterize the propagation channel are critically important.

Bandpass Transmission Techniques for Mobile Radio (Chapter 3). To minimize the effects of signal attenuation and signal distortion, the information-bearing baseband signal needs to be suitably modulated onto an appropriate carrier frequency, or a set of subcarrier frequencies. With high transmission rates, the bandwidth of the transmitted signal will likely exceed the coherence bandwidth of the wireless channel, resulting in frequency-selective fading. Transmission techniques such as OFDM, having the ability to render the fading channel non-frequency-selective, are important to mobile radio communications. With severe channel impairments, residual errors are unavoidable. Error probabilities of different modulation schemes that provide appropriate measures of system performance need to be analyzed and evaluated.

Receiver Techniques for Fading Dispersive Channels (Chapter 4). Optimum reception of transmitted signals over fading dispersive channels subject to a suitably chosen criterion is key to optimum receiver design. With the wireless channel exhibiting impairments such as intersymbol interference, Doppler spread and multipath delay spread, techniques such as diversity reception, linear combining, channel equalization, etc., are crucial to optimal signal reception at the cell-site receiver.

Fundamentals of Cellular Communications (Chapter 5). The area covered by a single base station is limited. The area coverage can be extended by (a) linking cells in a two-dimensional array and (b) wiring the cell-sites to an MSC and then to a backbone network as illustrated in Figure 1.1.

An array structure offers reuse of the same frequency in neighboring cells. However, frequency reuse can cause interference from adjacent cells. Therefore, effective strategies to manage the available resources are needed.

Multiple Access Techniques (Chapter 6). There is no interference if only one mobile transmits to the base station (cell-site receiver). If two or more mobiles transmit to the same cell-site receiver simultaneously, collision(s) can occur leading to destructive interference. Therefore an effective multiple access control scheme to enhance spectral efficiency and capacity is needed.

Mobility Management in Wireless Networks (Chapter 7). A wireless network has the capability to support user roaming. To achieve this roaming feature, it is necessary to introduce efficient and effective methods for (*a*) call admission control, (*b*) handoff, and (*c*) location management.

Wireless/Wireline Interworking (Chapter 8). While the wireless system offers the flexibility for user roaming, a backbone network such as the Internet is needed to support global communications. The interworking of a wireless segment with a wireline segment to extend the geographical coverage is imperative for future personal communication service (PCS) networks. The interworking of wireless/IP-based networks introduces some challenging issues in terms of maintaining network integrity, maximization of network utilization, and provision of quality of service in end-to-end network information delivery.

SUMMARY

The flexibility offered by a wireless communications system, especially the inherent capability in supporting user roaming, has escalated wireless communications as an information exchange platform to the forefront of society. With the inherent severe impairments, wireless channels suffer greatly from spectral limitations and relatively high and nonconstant residual error profiles. The challenge in the transmission of signals over wireless channels lies in seeking transmission and reception methods to mitigate channel impairments in an attempt to broaden the spectral width and to make the residual error profile as low and flat as possible.

With narrowband voice service, channel impairments pose a lesser problem than the perceived high transmission rates associated with the third generation and beyond systems, which are expected to support multimedia services. In this overview chapter, we have presented an account of the first and second generation wireless technologies, and gazed into third generation standardization activities.

The intent of this text is to provide a good understanding of the fundamentals for the design and analysis of wireless communications systems and networks that could serve as an effective and efficient information transport platform for information delivery to/from mobile users.

ENDNOTES

1. For details of the first generation analog cellular systems, see the special issue on Advanced Mobile Phone Service (AMPS) of *Bell System Technical Journal* [19], and Chapter 10 of the book by Rappaport [128].
2. For details of the second generation cellular systems, see Chapter 10 of the book by Rappaport [128].
3. For an overview of wireless personal communications, see the papers by Cox [32], Li and Qiu [85], and by Padgett, Günther and Hattori [107].

4. For an overview of the third generation wireless systems, see the paper by Chaudhury, Mohr, and Onoe [27].

PROBLEMS

P1-1 True or false.

a. An FM signal offers better receptive quality compared with AM because it has a narrower signal bandwidth than that of AM.
b. A wireless channel is said to exhibit flat fading when different components of the signal spectrum experience similar fading characteristics.
c. Mobiles in a cellular network can directly communicate with each other.
d. The term reverse channel transmission is used to represent the transmissions from the base station to the mobiles located within coverage area of the base station.
e. AMPS is a digital communications technology deployed in North America.

P1-2
a. In wireless mobile communications, what is meant by *near–far problem*?
b. What step(s) should be taken to compensate for the near–far problem?

P1-3 Consider a mobile cellular system.

a. A cellular network consists of radio cells. What is meant by a radio cell?
b. How can the capacity of a cellular network be enlarged? Explain.
c. Associated with capacity enlargement is a price tag. What is the price to be paid for the capacity enlargement attained using the method in part (*b*)?

P1-4 OFDM is a technique that can be used to compensate for frequency selective fading. Explain how OFDM achieves mitigation of frequency selective fading.

P1-5
a. What is the basic difference between a mobile cellular network and an ad hoc network?
b. Describe the characteristics of the two types of wireless networks.

2

Characterization of the Wireless Channel

Impairments in the propagation channel have the effect of disturbing the information carried by the transmitted signal. Channel disturbance can be a combination of additive noise, multiplicative fading, and distortion due to time dispersion. The focus of this chapter is to characterize the wireless channel by identifying the parameters of the corruptive elements that distort the information-carrying signal as it penetrates the propagation medium. The corruptive elements are in the form of multipath delay spread, Doppler spread due to motion, and signal fading of the frequency-selective and non-frequency-selective variety. The propagation channel is normally time-variant. In this chapter, we first model the channel as linear time-variant, and then characterize Doppler spread, multipath delay spread, and fading in terms of second order statistics.

2.1 MULTIPATH PROPAGATION ENVIRONMENT

The wireless propagation channel contains objects (particles) which randomly scatter the energy of the transmitted signal. The scattered signals arrive at the destination receiver out of step. These objects (particles) are referred to as scatterers. Scatterers introduce a variety of channel impairments including fading, multipath delay spread, Doppler spread, attenuation, etc., and the inherent background noise. Background noise can be approximated as thermal noise and treated as additive white Gaussian noise (AWGN). Digital transmission over practical wireless channels is mainly limited by interference and distortion other than AWGN.

Scattering by randomly located scatterers gives rise to different paths with different path lengths, resulting in multipath delay spread. Consider a point source (a single tone sinusoid) as a test signal. If the propagation channel does not exhibit multipath delay spread, the point source would appear at the front end of the receiver as another point source. A multipath situation arises when a transmitted point source is received as a multipoint source, with each of the individually received points experiencing a different transmission delay. A pictorial view of the scattering phenomenon is depicted in Figure 2.1. The effect of multipath propagation on digital transmission can be characterized by time dispersion and fading.

Time Dispersion. Because multiple propagation paths have different propagation delays, the transmitted point source will be received as a smeared waveform. Nonoverlapping scatterers give rise to distinct multiple paths, which are characterized by their locations in the scattering medium. As depicted in Figure 2.2, all scatterers are located on ellipses with the transmitter (Tx) and receiver (Rx) as the foci. One ellipse is associated with one path length. Therefore, signals reflected by scatterers located on the same ellipse will experience the same propagation delay. The signal components from these multiple paths are indistinguishable at the receiver. Signals that are reflected by scatterers located on different ellipses will arrive at the receiver with differential

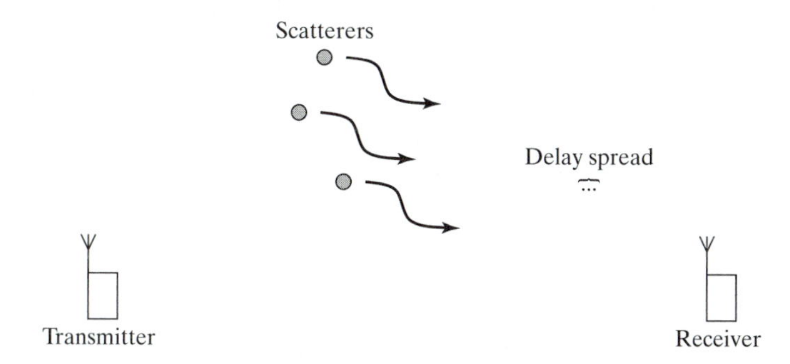

Figure 2.1 Multipath spread due to channel scattering.

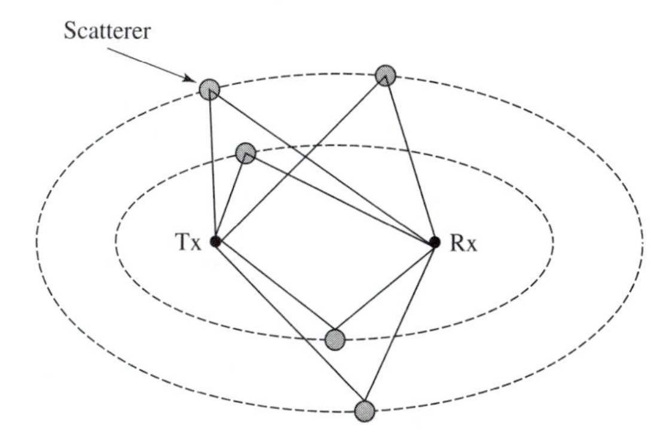

Figure 2.2 Ellipsoidal portrayal of scatterer location.

delays. If the maximum differential delay spread is small compared with the symbol duration of the transmitted signal, the channel is said to exhibit flat fading. If the differential delay spread is large compared with the symbol interval, the channel exhibits frequency-selective fading. In the time domain, the received signals corresponding to successive transmitted symbols will overlap, giving rise to a phenomenon known as intersymbol interference (ISI). ISI is a signal-dependent distortion. The severity of ISI increases with the width of the delay spread. The ISI distortion in the time domain can also be examined in the frequency domain. ISI degrades transmission performance. Channel equalization techniques can be used to combat ISI, as discussed in Chapter 4.

Fading. When the delay differences among various distinct propagation paths are very small compared with the symbol interval in digital transmission, the multipath components are almost indistinguishable at the receiver. These multipath components can add constructively or destructively, depending on the carrier frequency and delay differences. In addition, as the mobile station moves, the position of each scatterer with respect to the transmitter and receiver may change. The overall effect is that the received signal level fluctuates with time, a phenomenon called fading. As an example, consider the transmission of a single-tone sinusoidal signal with frequency f_c over a channel with two distinct paths, as shown in Figure 2.3. For simplicity, the delay of the line-of-sight (LOS) or direct path is assumed to be zero, and the delay of the non-line-of-sight (NLOS) or reflected path is τ. The received signal, in the absence of noise, can be represented as

$$r(t) = \alpha_1 \cos(2\pi f_c t) + \alpha_2 \cos(2\pi f_c (t - \tau)), \tag{2.1.1}$$

where α_1 and α_2 are the amplitudes of the signal components from the two paths respectively. The received signal can also be represented as

$$r(t) = \alpha \cos(2\pi f_c t + \phi), \tag{2.1.2}$$

where

$$\alpha = \sqrt{\alpha_1^2 + \alpha_2^2 + 2\alpha_1\alpha_2 \cos(2\pi f_c \tau)}$$

and

$$\phi = -\tan^{-1}\left[\frac{\alpha_2 \sin(2\pi f_c \tau)}{\alpha_1 + \alpha_2 \cos(2\pi f_c \tau)}\right]$$

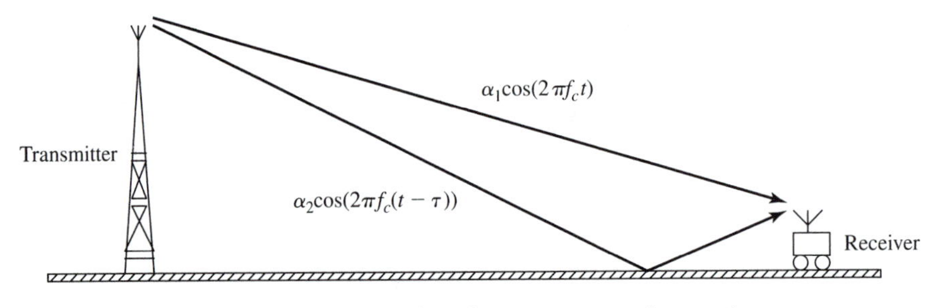

Figure 2.3 A channel with two propagation paths.

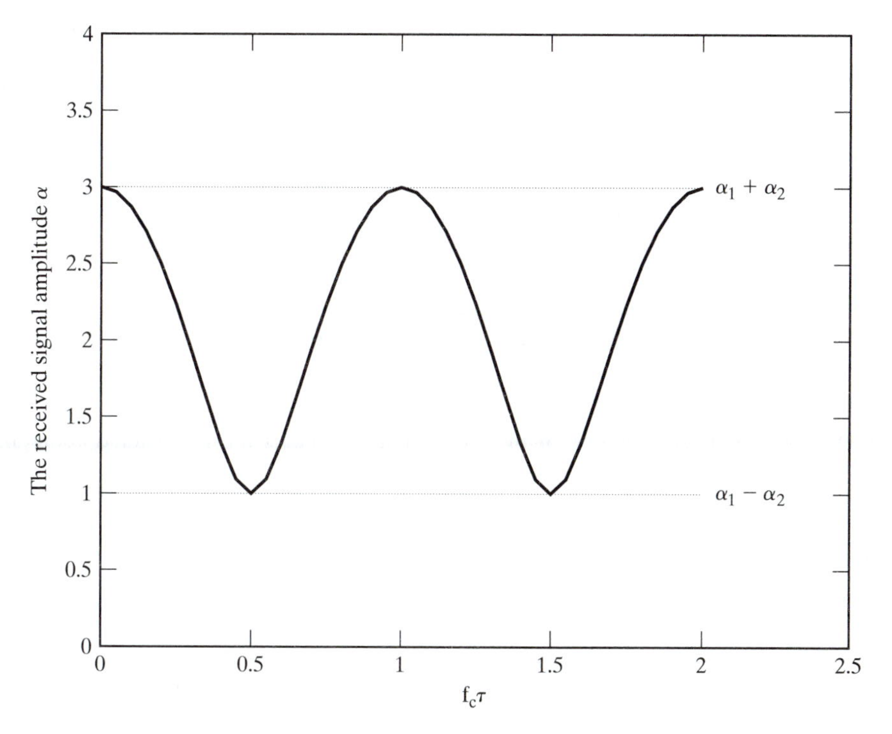

Figure 2.4 The amplitude fluctuation of the two-path channel with $\alpha_1 = 2$ and $\alpha_2 = 1$.

are the amplitude and phase of the received signal. Both α and ϕ are functions of α_1, α_2, and τ. Figure 2.4 plots the amplitude α as a function of $f_c\tau$, where $\alpha_1 = 2$ and $\alpha_2 = 1$. It can be observed that the two received signal components add constructively when $f_c\tau = 0, 1, 2, \ldots$ and destructively when $f_c\tau = 0.5, 1.5, 2.5, \ldots$. In general, as the mobile station moves, α_1, α_2, and τ change with time. The received signal amplitude α and phase ϕ also change with time. When the signal components from the two paths add destructively, the transmitted signal experiences deep fading with a small value of the amplitude α. During each deep fade, the instantaneously received signal power is very low, resulting in poor transmission quality (i.e., high transmission error rate). Diversity and error-correction coding (with interleaving) are effective to combat channel fading for better transmission accuracy, as discussed in Chapter 4.

The channel fading can be classified as long-term fading or short-term fading. The former is a large-scale path loss, characterizing the local average of the path loss, as discussed in Section 2.4. The latter describes the instantaneously received signal level variations with respect to the local average, as discussed in Section 2.5.

In summary, multipath propagation in the wireless mobile environment results in a fading dispersive channel. The signal propagation environment changes as the mobile station moves and/or as any surrounding scatterers move. Therefore, the channel is time varying and can be modeled as a linear time-variant (LTV) system. In the following sections, we first study how to describe a wireless channel using the LTV model in Section 2.2, and then focus on the correlation functions of the LTV channel in Section 2.3. The long-term and short-term fading models are then studied in Sections 2.4 and 2.5 respectively.

2.2 LINEAR TIME-VARIANT CHANNEL MODEL

Consider a multipath propagation environment with N distinct scatterers. The path associated with the nth distinct scatterer is characterized by the 2-tuple $(\alpha_n(t), \tau_n(t))$, where $\alpha_n(t)$ represents the amplitude fluctuation introduced to the transmitted signal by the scatterer at time t, $\tau_n(t)$ is the associated propagation delay, and $n = 1, 2, \ldots, N$. Consider a narrowband signal $\tilde{x}(t)$ transmitted over the wireless channel at a carrier frequency f_c, such that

$$\tilde{x}(t) = \Re\{x(t)e^{j2\pi f_c t}\}, \tag{2.2.1}$$

where $x(t)$ is the complex envelope of the signal and $\Re$ denotes the real-valued component. In the absence of background noise, the received signal at the channel output is

$$\begin{aligned}\tilde{r}(t) &= \Re\left\{\sum_{n=1}^{N} \alpha_n(t)x(t-\tau_n(t))e^{j2\pi f_c(t-\tau_n(t))}\right\}\\ &= \Re\{r(t)e^{j2\pi f_c t}\},\end{aligned}$$

where $r(t)$ is the complex envelope of the received signal and can be represented as

$$r(t) = \sum_{n=1}^{N} \alpha_n(t)e^{-j2\pi f_c \tau_n(t)}x(t-\tau_n(t)). \tag{2.2.2}$$

Note that the complex envelopes, $x(t)$ and $r(t)$, are respectively the equivalent representations of the transmitted and received narrowband signals at baseband. The channel can be characterized equivalently by its impulse response at baseband. Since the channel is time varying, the impulse response depends on the instant that the impulse is applied to the channel.

2.2.1 Channel Impulse Response

Let us first review the impulse response of a linear time-invariant (LTI) channel, as shown in Figure 2.5(a). Let $h(t)$ denote the impulse response (i.e., the channel output when the channel input is an impulse applied at $t = 0$, $\delta(t)$). Here $\delta(t)$ is the Dirac delta function and is defined by

$$\delta(t) = \begin{cases} 0, & t \neq 0 \\ \infty, & t = 0 \end{cases} \tag{2.2.3}$$

and

$$\int_{-\infty}^{\infty} \delta(t) = 1. \tag{2.2.4}$$

Because the channel is time invariant, if the input is delayed by t_1, the output is also delayed by t_1 correspondingly. That is, the channel output in response to an input applied at time t_1, $\delta(t - t_1)$,

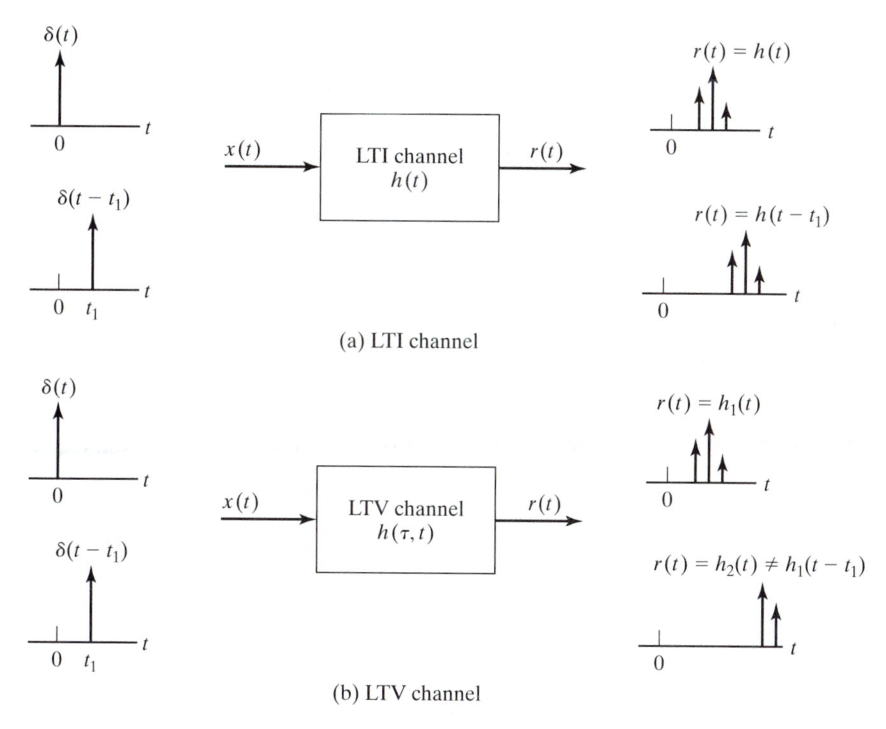

(a) LTI channel

(b) LTV channel

Figure 2.5 The linear channel model.

is $h(t - t_1)$. As a result, we can use the impulse response $h(t)$ to completely characterize the channel. The time variable t in the impulse response actually represents the propagation delay of the channel. The channel output $r(t)$ is given by

$$r(t) = h(t) * x(t) = \int_{-\infty}^{\infty} x(t - \tau)h(\tau)\, d\tau, \tag{2.2.5}$$

where $\star$ denotes the convolution operation.

Next consider an LTV channel, as shown in Figure 2.5(b), where $h_1(t)$ and $h_2(t)$ denote the channel responses to the inputs $\delta(t)$ and $\delta(t - t_1)$ respectively. As the channel propagation environment changes over the time duration $[0,\ t_1]$, the channel output $h_2(t)$ is not simply $h_1(t)$ delayed by t_1 (i.e., $h_2(t) \neq h_1(t - t_1)$). As a result, in order to characterize an LTV channel, we should define a channel impulse response as a function of two time variables, one describing the instant when the impulse is applied to the channel (initial time), and the other describing the instant of observing the channel output (final time).

Definition 2.1 The impulse response of an LTV channel, $h(\tau, t)$, is the channel output at t in response to an impulse applied to the channel at $t - \tau$.

In Definition 2.1, the variable τ represents the propagation delay. From the definition and Eq. (2.2.5), the channel output can be represented in terms of the impulse response and the channel input by

$$r(t) = \int_{-\infty}^{\infty} h(\tau, t)x(t - \tau)\, d\tau. \tag{2.2.6}$$

The channel impulse response for the channel with N distinct scatterers is then

$$h(\tau, t) = \sum_{n=1}^{N} \alpha_n(t)e^{-j\theta_n(t)}\delta(\tau - \tau_n(t)), \tag{2.2.7}$$

where $\theta_n(t) = 2\pi f_c \tau_n(t)$ represents the carrier phase distortion introduced by the nth scatterer. Thus, $\theta_n(t)$ changes by 2π radians whenever $\tau_n(t)$ changes by $1/f_c$, which is usually very small. This means that changes in $\theta_n(t)$ have a far greater effect on the transmitted signal than changes in $\alpha_n(t)$, as a small change (such as motion) in the scatterer can cause a significant change in the phase $\theta_n(t)$, but may not cause significant changes in the amplitude $\alpha_n(t)$. Substituting Eq. (2.2.7) into Eq. (2.2.6), we obtain Eq. (2.2.2).

2.2.2 Time-Variant Transfer Function

With the multipath channel characterized as a linear system, the channel behavior can also be examined in the frequency domain via a Fourier transformation. Time and frequency have an inverse relationship.

Definition 2.2 The time-variant transfer function of an LTV channel is the Fourier transform of the impulse response, $h(\tau, t)$, with respect to the delay variable τ.

Let $H(f, t)$ denote the channel transfer function, as shown in Figure 2.6. We have the Fourier transform pair

$$\begin{cases} H(f, t) = \mathcal{F}_\tau[h(\tau, t)] = \int_{-\infty}^{\infty} h(\tau, t)e^{-j2\pi f\tau} d\tau \\ h(\tau, t) = \mathcal{F}_f^{-1}[H(f, t)] = \int_{-\infty}^{\infty} H(f, t)e^{+j2\pi f\tau} df \end{cases},$$

where the time variable t can be viewed as a parameter. The received signal can be represented in terms of the transmitted signal and the transfer function as

$$r(t) = \int_{-\infty}^{\infty} R(f, t)e^{j2\pi ft} df, \tag{2.2.8}$$

where

$$R(f, t) = H(f, t)X(f)$$

and

$$X(f) = \mathcal{F}[x(t)].$$

At any instant, say $t = t_0$, the transfer function $H(f, t_0)$ characterizes the channel in the frequency domain. As the channel changes with t, the frequency domain representation also

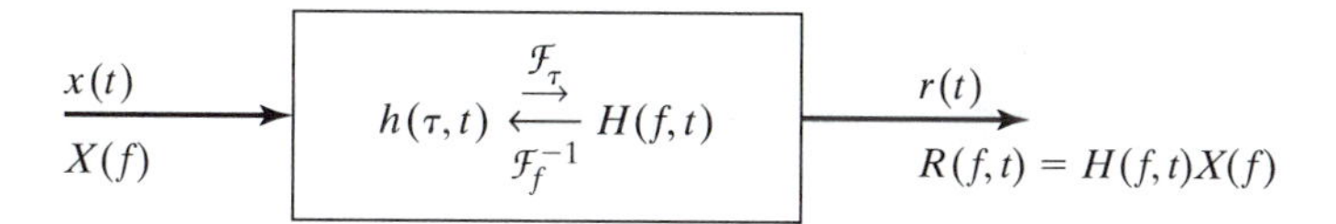

Figure 2.6 Frequency-time channel representation.

changes with t. Therefore, we have the channel time-varying transfer function. If the channel is time invariant, then the impulse response is a function of the delay variable τ and is independent of the time variable t; thus the transfer function varies only with the frequency variable f and is independent of t. This is consistent with the impulse response and transfer function of an LTI channel.

2.2.3 Doppler Spread Function and Delay-Doppler Spread Function

Doppler Shifts. In general, the output signal of an LTI system does not have frequency components different from those of the input signal. That is, an LTI system does not introduce frequency shifts to its input signal. On the other hand, both nonlinear and time-varying systems introduce new frequency components other than those existing in the input signal. For a wireless propagation environment, due to the mobility of mobile users and/or the surrounding scatterers, the channel is linear but time variant. As a result, a wireless channel introduces frequency shifts to the transmitted signal, a phenomenon called the Doppler effect and the introduced frequency shifts called the Doppler shifts.

Consider a scenario shown in Figure 2.7, where the base station (BS) transmits a single-tone pilot signal at frequency f_c, and the mobile station (MS) is moving along the x-axis with a constant velocity V. Let $\theta(t)$ denote the angle of the incoming pilot signal viewed from the

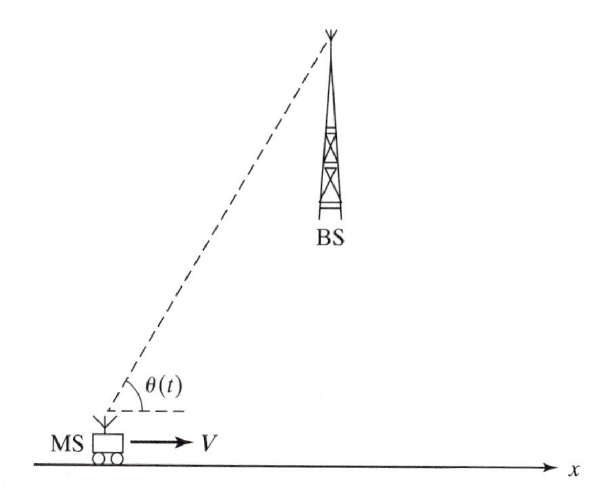

Figure 2.7 The Doppler effect.

mobile receiver at time t, with respect to the x-axis. The received signal at the mobile terminal at time t has a frequency of $f_c + \nu(t)$. $\nu(t)$ is the Doppler shift and is given by

$$\nu(t) = \frac{Vf_c}{c} \cos\theta(t), \tag{2.2.9}$$

where c is the velocity of light. From Eq. (2.2.9), it is observed that the Doppler shift increases with the signal frequency f_c and user velocity V.

As a wireless channel can be characterized equivalently in both time and frequency domains, a channel being time varying in the time domain means a channel introducing Doppler shifts in the frequency domain. In fact, as a wireless channel usually introduces continuous Doppler shifts in a certain range including the zero Doppler shift, the effect of the channel on the transmitted signal in the frequency domain is more spectral broadening than a simple spectral shift. This phenomenon can be illustrated by the following simple example.

Consider the wireless channel with N distinct scatterers as described by Eq. (2.2.7). Assume that the delay spread is negligible as compared with the symbol interval of the transmitted signal. As a result, the propagation delay can be approximated by its mean value $\bar{\tau}$. For simplicity, we further assume that this mean delay does not change with time. The time-variant impulse response of the channel can be approximately described in the form

$$h(\tau, t) \approx Z(t)\delta(\tau - \bar{\tau}), \tag{2.2.10}$$

where $Z(t) = \sum_{n=1}^{N} \alpha_n(t) \exp[-j2\pi f_c \tau_n(t)]$. Given that the transmitted signal is $x(t)$, the received signal in the absence of background noise is

$$\begin{aligned} r(t) &= \int_{-\infty}^{\infty} h(\tau, t)x(t - \tau)\, d\tau \\ &= \int_{-\infty}^{\infty} [Z(t)\delta(\tau - \bar{\tau})]x(t - \tau)\, d\tau \\ &= Z(t)x(t - \bar{\tau}). \end{aligned}$$

The overall effect of the channel is to provide a complex time-varying gain, $Z(t)$, and a transmission delay, $\bar{\tau}$, to the transmitted signal. In the frequency domain, the received signal is

$$\begin{aligned} R(f) &= \mathcal{F}[r(t)] \\ &= \mathcal{F}[Z(t)x(t - \bar{\tau})] \\ &= \mathcal{F}[Z(t)] \star \mathcal{F}[x(t - \bar{\tau})] \\ &= \mathcal{F}[Z(t)] \star [X(f)e^{-j2\pi f \bar{\tau}}], \end{aligned}$$

where $X(f) = \mathcal{F}[x(t)]$. Given that the channel gain $Z(t)$ changes with time, its Fourier transform $\mathcal{F}[Z(t)]$ has a finite but nonzero pulse width in the frequency domain. As a result, the pulse width of $R(f)$ is larger than the pulse width of $X(f)$ due to the convolution operation. This means that the channel indeed broadens the transmitted signal spectrum by introducing new frequency components, a phenomenon referred to as frequency dispersion.

Doppler Spread Function $H(f, \nu)$. An LTV wireless channel can be characterized by the Doppler spread function $H(f, \nu)$ which is defined by the relation between the channel input signal and output signal in the frequency domain. Let $X(f)$ and $R(f)$ denote the Fourier transforms of the transmitted signal $x(t)$ and received signal $r(t)$, respectively. Then $H(f, \nu)$ is defined by the following equation

$$R(f) = \int_{-\infty}^{\infty} X(f - \nu)H(f - \nu, \nu)\, d\nu, \tag{2.2.11}$$

where ν is a variable describing the Doppler shift introduced by the channel. From Eq. (2.2.11), $H(f, \nu)$ is the channel gain associated with Doppler shift ν to the input signal component at frequency f. Since both the time-variant transfer function $H(f, t)$ and the Doppler spread function $H(f, \nu)$ can be used to describe the same channel, there exists a relation between the two channel functions. It can be shown that

$$\begin{cases} H(f, \nu) = \mathcal{F}_t[H(f, t)] = \int_{-\infty}^{\infty} H(f, t)e^{-j2\pi\nu t} dt \\ H(f, t) = \mathcal{F}_\nu^{-1}[H(f, \nu)] = \int_{-\infty}^{\infty} H(f, \nu)e^{+j2\pi\nu t} d\nu \end{cases},$$

where the frequency variable f can be viewed as a parameter. The preceding Fourier transform relation verifies that *being time-variant in the time domain* can be equivalently described by *having Doppler shifts in the frequency domain.*

Delay-Doppler Spread Function. A wireless channel can be characterized by the delay-Doppler spread function $H(\tau, \nu)$ defined as the Fourier transform of the channel impulse response $h(\tau, t)$ with respect to t, as follows:

$$H(\tau, \nu) = \mathcal{F}_t[h(\tau, t)] = \int_{-\infty}^{\infty} h(\tau, t)e^{-j2\pi\nu t} dt. \tag{2.2.12}$$

Given the channel input signal $x(t)$, it can be shown that the channel output signal is

$$r(t) = \int_{-\infty}^{\infty} \int_{-\infty}^{\infty} x(t - \tau)H(\tau, \nu)e^{j2\pi\nu t} d\nu\, d\tau. \tag{2.2.13}$$

As both the transmitted signal and received signal can be represented either in the time domain or in the frequency domain, we have four channel functions, $h(\tau, t)$, $H(f, t)$, $H(\tau, \nu)$, and $H(f, \nu)$ to characterize the relation between the transmitted signal and the received signal. These four functions are equivalent in describing the LTV channel. Preference in selecting any one of the functions depends on whether the transmitted and received signals are represented

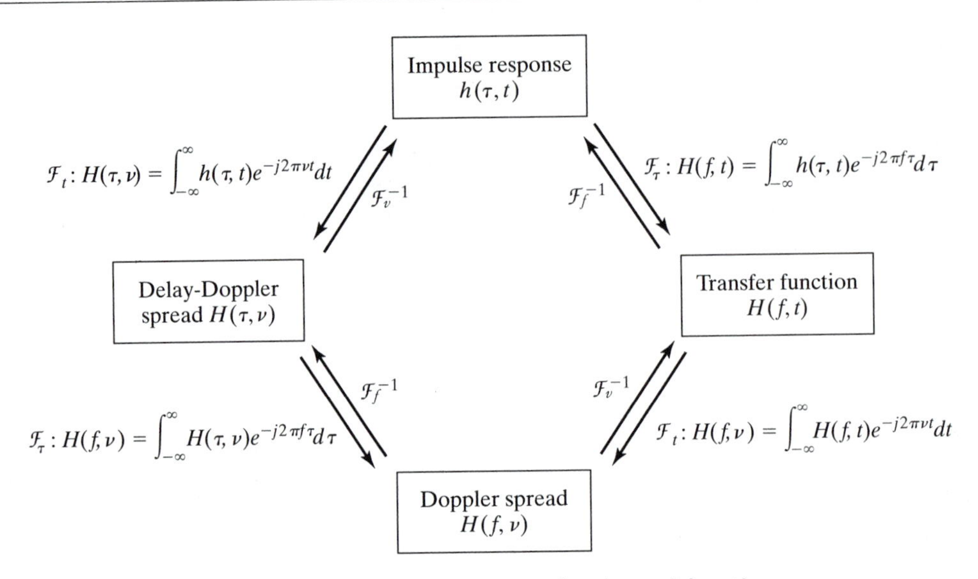

Figure 2.8 Relationships among the channel functions.

in the time or frequency domain. Figure 2.8 summarizes the relationships among the channel functions.

2.2.4 Example on the Channel Functions

Example 2.1 LTV Channel Model

Consider an LTV channel with impulse response given by

$$h(\tau,t) = 4\exp(-\tau/T)\cos(\Omega t), \quad \tau \geq 0, \tag{2.2.14}$$

where $T = 0.1$ ms and $\Omega = 10\pi$.

a. Find the channel time-variant transfer function $H(f,t)$ and the Doppler spread function $H(f,\nu)$.
b. Given that the transmitted signal is

$$x_1(t) = \begin{cases} 1, & |t| \leq T_0 \\ 0, & |t| > T_0 \end{cases}, \tag{2.2.15}$$

where $T_0 = 0.025$ ms, find the received signal in the absence of background noise.

c. Repeat part (*b*) if the transmitted signal is

$$x_2(t) = x_1(t - T_1), \tag{2.2.16}$$

where $T_1 = 0.05$ ms.

d. What do you observe from the results of parts (*b*) and (*c*)?

Solution

a. The time-variant transfer function is

$$
\begin{aligned}
H(f,t) &= \mathcal{F}_\tau[h(\tau,t)] \\
&= \mathcal{F}_\tau[4\exp(-\tau/T)\cos(\Omega t)], \quad \tau \ge 0 \\
&= 4\cos(\Omega t)\mathcal{F}_\tau[\exp(-\tau/T)], \quad \tau \ge 0 \\
&= \frac{4T\cos(\Omega t)}{1+j2\pi fT}.
\end{aligned}
$$

The Doppler spread function is

$$
\begin{aligned}
H(f,\nu) &= \mathcal{F}_t[H(f,t)] \\
&= \mathcal{F}_t\left[\frac{4T\cos(\Omega t)}{1+j2\pi fT}\right] \\
&= \frac{4T}{1+j2\pi fT}\mathcal{F}_t[\cos(\Omega t)] \\
&= \frac{2T}{1+j2\pi fT}[\delta(2\pi\nu+\Omega)+\delta(2\pi\nu-\Omega)].
\end{aligned}
$$

b. The received signal is calculated as follows:

$$
\begin{aligned}
r_1(t) &= \int_{-\infty}^{\infty} h(\tau,t)x_1(t-\tau)\,d\tau \\
&= \int_0^{\infty} 4\exp(-\tau/T)\cos(\Omega t)x_1(t-\tau)\,d\tau \\
&= \begin{cases} 0, & t \le -T_0 \\ 4T\cos(\Omega t)\left[1-\exp\left(-\dfrac{t+T_0}{T}\right)\right], & -T_0 < t < T_0 \\ 4T\cos(\Omega t)\left[\exp\left(-\dfrac{t-T_0}{T}\right)-\exp\left(-\dfrac{t+T_0}{T}\right)\right], & t \ge T_0 \end{cases}
\end{aligned}
$$

The transmitted and received signals are plotted in Figure 2.9(a).

c. Similar to part (*b*), the received signal is calculated as follows:

$$
\begin{aligned}
r_2(t) &= \int_{-\infty}^{\infty} h(\tau,t)x_2(t-\tau)\,d\tau \\
&= \int_0^{\infty} 4\exp(-\tau/T)\cos(\Omega t)x_1(t-T_1-\tau)\,d\tau
\end{aligned}
$$

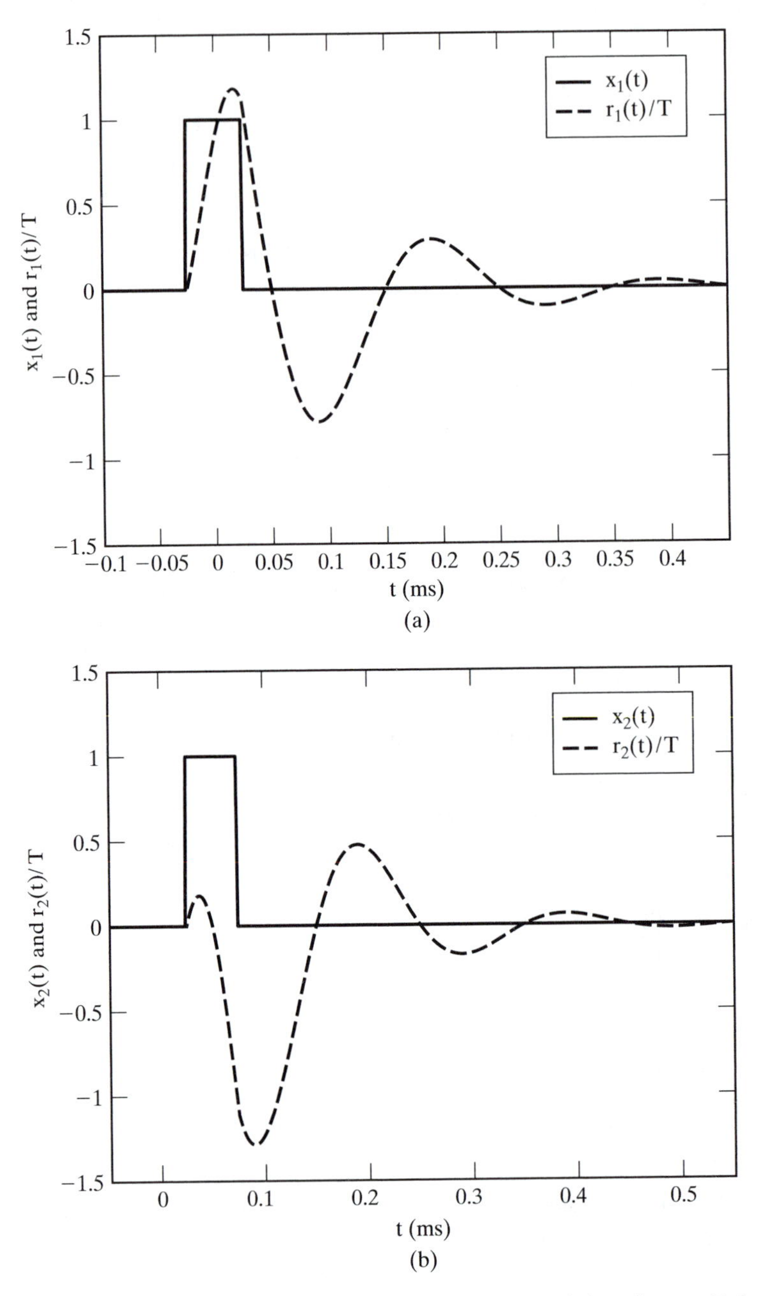

Figure 2.9 The transmitted signals $x_i(t)$ and normalized received signals $r_i(t)/T$ ($i = 1$ and 2) in Example 2.1.

$$= \begin{cases} 0, & t \leq T_1 - T_0 \\ 4T\cos(\Omega t)\left[1 - \exp\left(-\frac{t - T_1 + T_0}{T}\right)\right], & T_1 - T_0 < t < T_1 + T_0 \\ 4T\cos(\Omega t)\left[\exp\left(-\frac{t - T_1 - T_0}{T}\right) - \exp\left(-\frac{t - T_1 + T_0}{T}\right)\right], & t \geq T_1 + T_0 \end{cases}$$

The transmitted and received signals are plotted in Figure 2.9(b).

d. From Figure 2.9, it is observed that: (1) the received signals have a larger pulse width than the corresponding transmitted signals because the channel is time dispersive; and (2) even though the transmitted signal $x_2(t)$ is $x_1(t)$ delayed by T_1, the received signal $r_2(t)$ is not $r_1(t)$ delayed by T_1 because the channel is time varying.

2.3 CHANNEL CORRELATION FUNCTIONS

When the channel changes with time randomly, the channel impulse response $h(\tau, t)$, time-variant transfer function $H(f, t)$, Doppler spread function $H(f, \nu)$, and delay-Doppler spread function $H(\tau, \nu)$ are random processes and are difficult to characterize. Under the assumption that the random processes have zero mean, we are interested in the correlation functions of the random processes. For simplicity of analysis, we assume that

a. the channel impulse response $h(\tau, t)$ is a wide-sense stationary (WSS) process;
b. the channel impulse responses at τ_1 and τ_2, $h(\tau_1, t)$ and $h(\tau_2, t)$, are uncorrelated if $\tau_1 \neq \tau_2$ for any t.

A channel under assumptions (a) and (b) is said to be a wide-sense stationary uncorrelated scattering (WSSUS) channel.

2.3.1 Delay Power Spectral Density

Under assumption (a), the autocorrelation function of $h(\tau, t)$, defined as

$$\frac{1}{2}E[h^*(\tau_1, t)h(\tau_2, t + \Delta t)],$$

is a function of τ_1, τ_2, and Δt, and does not depend on t. The superscript (*) denotes complex conjugation. The correlation function can be represented as

$$\phi_h(\tau_1, \tau_2, \Delta t) \triangleq \frac{1}{2}E[h^*(\tau_1, t)h(\tau_2, t + \Delta t)]. \tag{2.3.1}$$

Furthermore, under assumption (b), the autocorrelation function can be represented in the form [12]

$$\phi_h(\tau_1, \tau_2, \Delta t) = \phi_h(\tau_1, \Delta t)\delta(\tau_1 - \tau_2), \tag{2.3.2}$$

or equivalently,

$$\phi_h(\tau, \tau + \Delta\tau, \Delta t) = \phi_h(\tau, \Delta t)\delta(\Delta\tau), \tag{2.3.3}$$

where

$$\phi_h(\tau, \Delta t) = \int \phi_h(\tau, \tau + \Delta\tau, \Delta t) d\Delta\tau.$$

At $\Delta t = 0$, we define

$$\phi_h(\tau) \stackrel{\Delta}{=} \phi_h(\tau, 0). \tag{2.3.4}$$

From Eqs. (2.3.1)–(2.3.4), we have

$$\begin{aligned} \phi_h(\tau) &= \mathcal{F}_{\Delta\tau}[\phi_h(\tau, \tau + \Delta\tau, \Delta t)]|_{\Delta t=0} \\ &= \mathcal{F}_{\Delta\tau}\left\{\frac{1}{2}E[h^*(\tau, t)h(\tau + \Delta\tau, t)]\right\}. \end{aligned} \tag{2.3.5}$$

From Eq. (2.3.5), we observe that $\phi_h(\tau)$ is the Fourier transform of the correlation function. According to the Wiener-Khintchine relations[1], the function $\phi_h(\tau)$ represents power spectral density (psd). It measures the average psd at the channel output as a function of the propagation delay, τ, and is therefore called the delay psd of the channel, also known as the multipath intensity profile. Figure 2.10 illustrates an example of the psd, where the minimum delay is assumed to be zero. The nominal width of the delay psd pulse is called the multipath delay spread, denoted by T_m.

From the delay psd, we can calculate statistics describing the time dispersive characteristics of the channel. The nth moment of the delay, denoted by $\bar{\tau^n}$, is given by

$$\bar{\tau^n} = \frac{\int \tau^n \phi_h(\tau)\, d\tau}{\int \phi_h(\tau)\, d\tau}. \tag{2.3.6}$$

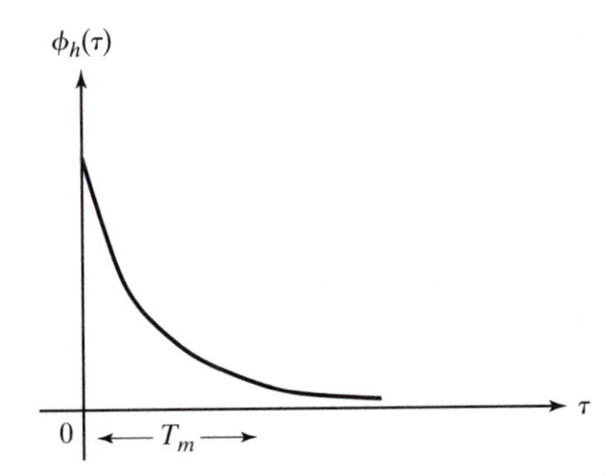

Figure 2.10 Delay power spectral density.

[1]For a WSS random process, its power spectral density in the frequency domain is the Fourier transform of its autocorrelation function in the time domain. The relations between the autocorrelation function and the power spectral density are called the Wiener-Khintchine relations.

The mean propagation delay, or first moment, denoted by $\bar{\tau}$, is

$$\bar{\tau} = \frac{\int \tau \phi_h(\tau)\,d\tau}{\int \phi_h(\tau)\,d\tau}, \tag{2.3.7}$$

and the rms (root-mean-square) delay spread, denoted by σ_τ, is

$$\sigma_\tau = \left[\frac{\int (\tau - \bar{\tau})^2 \phi_h(\tau)\,d\tau}{\int \phi_h(\tau)\,d\tau} \right]^{1/2}. \tag{2.3.8}$$

In calculating a value for the multipath delay spread, it is usually assumed that

$$T_m \approx \sigma_\tau.$$

In the preceding calculations, the contribution of each delay τ to the statistics is weighted by the psd level associated with the delay, and is normalized by the total power $\int \phi_h(\tau)\,d\tau$.

2.3.2 Frequency and Time Correlation Functions

From the relationship between $h(\tau, t)$ and $H(f, t)$, we have the following:

a. If $h(\tau, t)$ is WSS, then $H(f, t)$ is also WSS with respect to t. As a result, we can define the autocorrelation function of the time-variant transfer function $H(f, t)$ as

$$\phi_H(f_1, f_2, \Delta t) \triangleq \frac{1}{2} E[H^*(f_1, t) H(f_2, t + \Delta t)]. \tag{2.3.9}$$

b. The correlation function $\phi_H(f_1, f_2, \Delta t)$ can be represented in terms of the correlation function $\phi_h(\tau, \Delta t)$.

$$\begin{aligned} \phi_H(f_1, f_2, \Delta t) &= \int_{-\infty}^{\infty} \phi_h(\tau, \Delta t) e^{-j2\pi(f_2 - f_1)\tau}\,d\tau \\ &= \int_{-\infty}^{\infty} \phi_h(\tau, \Delta t) e^{-j2\pi(\Delta f)\tau}\,d\tau \\ &\triangleq \phi_H(\Delta f, \Delta t), \end{aligned}$$

where $\Delta f = f_2 - f_1$.

The function $\phi_H(\Delta f, \Delta t)$ is a time-frequency correlation function. Letting $\Delta t = 0$, we can write the frequency domain representation as a Fourier transform of the delay psd, i.e.,

$$\begin{aligned} \phi_H(\Delta f) &\triangleq \frac{1}{2} E[H^*(f, t) H(f + \Delta f, t)] \\ &= \int_{-\infty}^{\infty} \phi_h(\tau) e^{-j2\pi(\Delta f)\tau}\,d\tau. \end{aligned} \tag{2.3.10}$$

$\phi_H(\Delta f)$ is called the frequency correlation function of the wireless channel. The delay psd, $\phi_h(\tau)$, portrays the time domain behavior of the fading channel, whereas the frequency correlation function, $\phi_H(\Delta f)$, portrays the frequency domain behavior. The nominal width of $\phi_H(\Delta f)$, denoted by $(\Delta f)_c$, is called the channel coherence bandwidth, as shown in Figure 2.11. Since time and frequency have an inverse relationship, we have

$$(\Delta f)_c \approx \frac{1}{T_m} \tag{2.3.11}$$

(i.e., a large multipath delay spread also means a small channel coherence bandwidth).

Over the channel coherence bandwidth, all signal frequency components experience the correlated perturbation. When a signal propagates through the channel, the frequency components of the signal separated by a frequency width greater than the channel coherence bandwidth are distorted by the channel in an uncorrelated manner. The degree of fading experienced by the transmitted signal depends on the relationship between the channel coherence bandwidth and the transmitted signal bandwidth. Let W_s denote the bandwidth of the transmitted signal. The fading channel can be grossly categorized as follows:

If $(\Delta f)_c < W_s$, the channel is said to exhibit frequency selective fading which introduces severe ISI to the received signal;

If $(\Delta f)_c \gg W_s$, the channel is said to exhibit frequency nonselective fading or flat fading which introduces negligible ISI.

In the time-frequency correlation function $\phi_H(\Delta f, \Delta t)$, letting $\Delta f = 0$, we have

$$\phi_H(\Delta t) \triangleq \phi_H(0, \Delta t) = \frac{1}{2} E[H^*(f,t) H(f, t + \Delta t)]. \tag{2.3.12}$$

$\phi_H(\Delta t)$ is called the time correlation function of the channel. It characterizes, on average, how fast the channel transfer function changes with time at each frequency. The nominal width of $\phi_H(\Delta t)$ is called the coherence time of the fading channel, denoted by $(\Delta t)_c$, as shown in Figure 2.12.

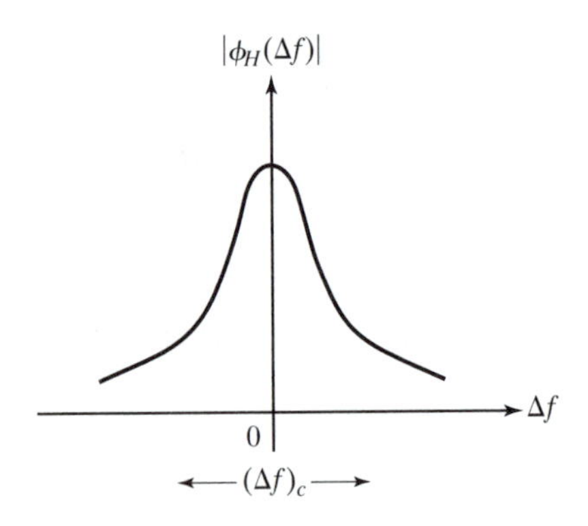

Figure 2.11 Frequency-correlation function and channel coherence bandwidth.

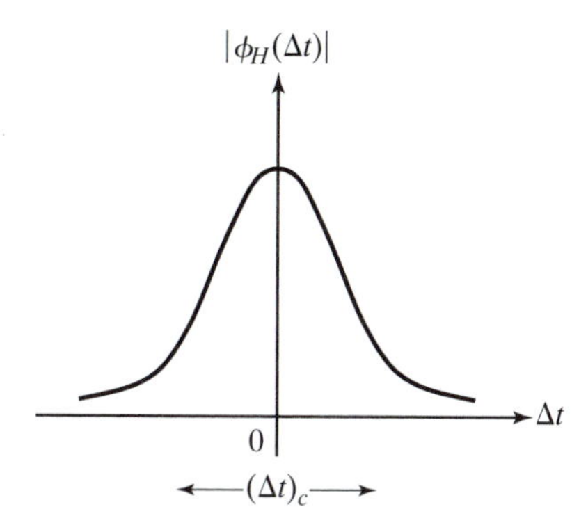

Figure 2.12 Time-correlation function and channel coherence time.

The channel fading at time t will be very different from that at time $t - (\Delta t)_c$ or earlier. If the channel coherence time is much larger than the symbol interval of the transmitted signal, the channel exhibits slow fading to the signal because the channel changing rate is much smaller than the transmission symbol rate. The time correlation function is independent of the frequency variable f due to the assumption that the channel has uncorrelated scattering. In other words, a channel exhibiting uncorrelated scattering in the time domain can be described equivalently as the condition that the channel transfer function is WSS in the frequency domain (with respect to f).

2.3.3 Doppler Power Spectral Density

The correlation function of the Doppler spread function $H(f, \nu)$ is defined as

$$\frac{1}{2}E[H^*(f_1, \nu_1)H(f_2, \nu_2)].$$

For a WSSUS channel, the correlation function can be represented in the form [12]

$$\Phi_H(\Delta f, \nu_1)\delta(\nu_1 - \nu_2),$$

where $\Delta f = f_2 - f_1$ and $\Phi_H(\Delta f, \nu)$ can be obtained by Fourier transformation of the frequency-time correlation function $\phi_H(\Delta f, \Delta t)$ with respect to Δt. Therefore,

$$\Phi_H(\Delta f, \nu) = \int_{-\infty}^{\infty} \phi_H(\Delta f, \Delta t)e^{-j2\pi\nu\Delta t}d\Delta t. \tag{2.3.13}$$

At $\Delta f = 0$, we have

$$\Phi_H(\nu) \triangleq \Phi_H(0, \nu) = \int_{-\infty}^{\infty} \phi_H(\Delta t)e^{-j2\pi\nu\Delta t}d\Delta t. \tag{2.3.14}$$

Equation (2.3.14) shows that the correlation function $\Phi_H(\nu)$ is the Fourier transform of the channel correlation function $\phi_H(\Delta t)$. Based on the Wiener-Khintchine relations, $\Phi_H(\nu)$ is a psd in terms

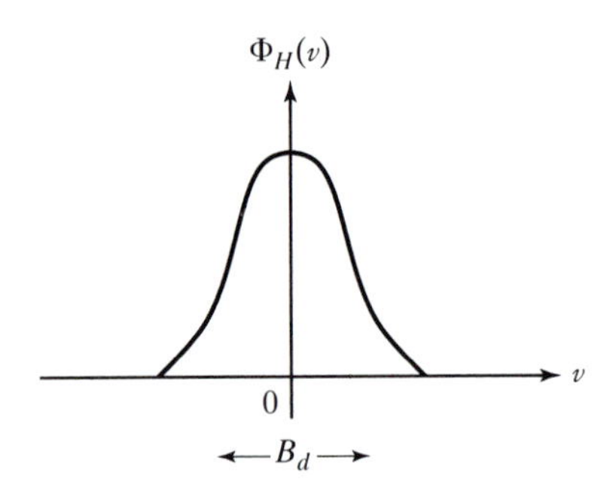

Figure 2.13 Doppler power spectral density and Doppler spread.

of the Doppler shift ν. As a result, the function $\Phi_H(\nu)$ is called the Doppler psd. The nominal width of the Doppler psd, denoted by B_d, is called the Doppler spread, as shown in Figure 2.13. From the relation that $\phi_H(\Delta t)$ and $\Phi_H(\nu)$ are a Fourier transform pair, we see that the coherence time is inversely proportional to the Doppler spread,

$$(\Delta t)_c \approx \frac{1}{B_d}. \tag{2.3.15}$$

With the Doppler psd, the nth moment of Doppler shift, denoted by $\bar{\nu^n}$, can be calculated by

$$\bar{\nu^n} = \frac{\int \nu^n \Phi_H(\nu)\, d\nu}{\int \Phi_H(\nu)\, d\nu}. \tag{2.3.16}$$

The mean Doppler shift, or the first moment, denoted by $\bar{\nu}$, is

$$\bar{\nu} = \frac{\int \nu \Phi_H(\nu) d\nu}{\int \Phi_H(\nu) d\nu}, \tag{2.3.17}$$

and the rms Doppler spread, denoted by σ_ν, is

$$\sigma_\nu = \left[\frac{\int (\nu - \bar{\nu})^2 \Phi_H(\nu) d\nu}{\int \Phi_H(\nu) d\nu} \right]^{1/2}. \tag{2.3.18}$$

As an approximation, it is usually assumed that

$$B_d \approx \sigma_\nu.$$

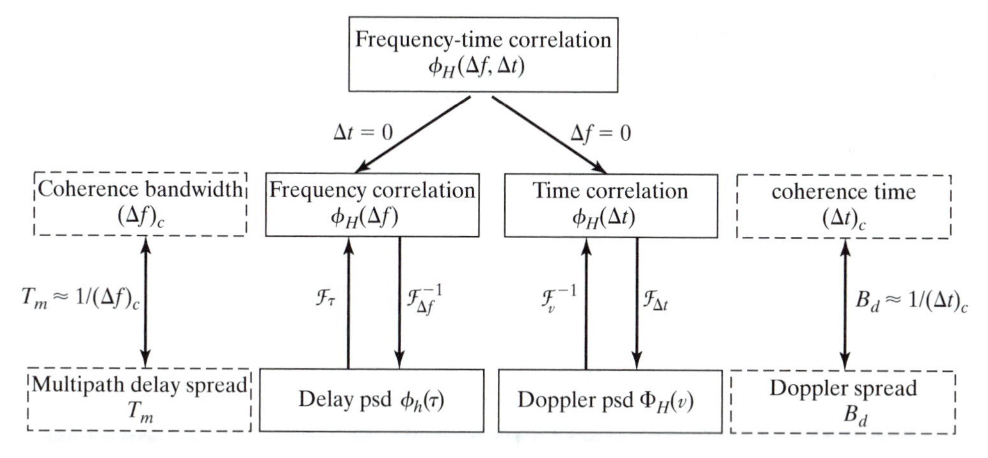

Figure 2.14 Relationships between the channel correlation functions and between the channel parameters.

Figure 2.14 summarizes the relationships between the correlation functions and power spectral densities and between the channel parameters.

2.3.4 Examples on the Channel Correlation Functions

Example 2.2 Calculation of the Channel Parameters

Consider a fading channel which exhibits a Doppler frequency shift uniformly distributed between -10 Hz and 10 Hz. Determine

a. the mean Doppler shift,
b. the rms Doppler spread, and
c. the coherence time.

Solution The Doppler psd normalized with respect to the total power is

$$\Phi_H(\nu) = \begin{cases} \dfrac{1}{20}, & -10 < \nu < 10 \\ 0, & \text{elsewhere.} \end{cases}$$

a. The mean Doppler shift is

$$\bar{\nu} = \frac{\displaystyle\int_{-10}^{10} \frac{1}{20} \nu \, d\nu}{\displaystyle\int_{-10}^{10} \frac{1}{20} \, d\nu} = \frac{\left.\dfrac{\nu^2}{40}\right|_{-10}^{10}}{1} = 0.$$

b. The rms Doppler spread is

$$\sigma_\nu = \left[\frac{\int (\nu - \bar{\nu})^2 \Phi_H(\nu) d\nu}{\int \Phi_H(\nu) d\nu} \right]^{1/2}$$

$$= \left[\int_{-10}^{10} \frac{1}{20} \nu^2 d\nu \right]^{1/2}$$

$$= \left[\frac{\nu^3}{60} \Big|_{-10}^{10} \right]^{1/2}$$

$$= 5.77 \text{ Hz}.$$

c. The channel coherence time is

$$(\Delta t)_c \simeq \frac{1}{B_d} \approx \frac{1}{\sigma_\nu} = \frac{1}{5.77} = 0.1733 \text{ s}.$$

Example 2.3 Delay PSD and Frequency-Correlation Function

Consider a WSSUS channel whose time-variant impulse response is given by

$$h(\tau, t) = \exp(-\tau/T) n(\tau) \cos(\Omega t + \Theta), \quad \tau \geq 0,$$

where T and Ω are constants, Θ is a random variable uniformly distributed in $[-\pi, +\pi]$, and $n(\tau)$ is a real-valued random process independent of Θ, with $E[n(\tau)] = \mu_n$ and $E[n(\tau_1)n(\tau_2)] = \delta(\tau_1 - \tau_2)$.

a. Calculate the delay psd and the multipath delay spread.
b. Calculate the frequency correlation function and the channel coherence bandwidth.
c. Determine whether the channel exhibits frequency-selective fading for GSM systems with $T = 0.1$ ms.

Solution

a. From Eq. (2.3.5), we have

$$\phi_h(\tau) = \mathcal{F}_{\Delta\tau} \left\{ \frac{1}{2} E[h^*(\tau, t) h(\tau + \Delta\tau, t] \right\}$$

$$= \mathcal{F}_{\Delta\tau} \left\{ \frac{1}{2} E[n(\tau) n(\tau + \Delta\tau)] E[e^{-(2\tau + \Delta\tau)/T} \cos^2(\Omega t + \Theta)] \right\}, \quad \tau \geq 0$$

$$= \mathcal{F}_{\Delta\tau} \left\{ \frac{1}{4} e^{-(2\tau + \Delta\tau)/T} \delta(\Delta\tau) E[1 + \cos(2\Omega t + 2\Theta)] \right\}, \quad \tau \geq 0$$

$$= \frac{1}{4} e^{-2\tau/T}, \quad \tau \geq 0,$$

where

$$E[\cos(2\Omega t + 2\Theta)] = \int_{-\pi}^{+\pi} \cos(2\Omega t + 2\theta)\frac{1}{2\pi}d\theta = 0.$$

For $\tau < 0$, $\phi_h(\tau) = 0$.
The mean propagation delay is

$$\bar{\tau} = \frac{\int_0^\infty \tau\phi_h(\tau)d\tau}{\int_0^\infty \phi_h(\tau)d\tau} = \frac{\int_0^\infty \tau\frac{1}{4}e^{-2\tau/T}d\tau}{\int_0^\infty \frac{1}{4}e^{-2\tau/T}d\tau} = \frac{T}{2}$$

and the multipath delay spread is

$$T_m \approx \sigma_\tau = \left[\frac{\int(\tau-\bar{\tau})^2\phi_h(\tau)d\tau}{\int\phi_h(\tau)d\tau}\right]^{1/2} = \left[\frac{\int\tau^2\phi_h(\tau)d\tau}{\int\phi_h(\tau)d\tau} - \bar{\tau}^2\right]^{1/2} = \frac{T}{2}.$$

b. The frequency correlation function is

$$\begin{aligned}\phi_H(\Delta f) &= \mathcal{F}[\phi_h(\tau)] \\ &= \int_0^\infty \frac{1}{4}e^{-2\tau/T}e^{-j2\pi(\Delta f)\tau}d\tau \\ &= \frac{T}{8 + j8\pi T(\Delta f)}.\end{aligned}$$

The coherence bandwidth is

$$(\Delta f)_c \approx 1/T_m = \frac{2}{T}.$$

c. With $T = 0.1$ ms, we have $(\Delta f)_c = 20$ kHz. The GSM channels have a bandwidth of 200 kHz. Since $(\Delta f)_c \ll 200$ kHz, the channel fading is frequency selective.

Example 2.4 Doppler PSD

For the channel specified in Example 2.3 with $\Omega = 10\pi$, find

a. the Doppler psd,
b. the mean Doppler shift and the rms Doppler spread,
c. the channel coherence time, and
d. whether the channel exhibits slow fading for GSM systems.

Solution

a. The Doppler psd can be calculated by taking the Fourier transform of the time correlation function $\phi_H(\Delta t)$. In this way, we need to calculate the correlation function $\phi_h(\tau, \Delta t)$ first.

For the WSS channel, from Eqs. (2.3.1)–(2.3.5), we have for $\tau \geq 0$

$$
\begin{aligned}
\phi_h(\tau, \Delta t) &= \mathcal{F}_{\Delta\tau}\left\{\frac{1}{2}E[h^*(\tau, t)h(\tau+\Delta\tau, t+\Delta t)]\right\} \\
&= \mathcal{F}_{\Delta\tau}\left\{\frac{1}{2}E[e^{-\tau/T}n(\tau)\cos(\Omega t+\Theta)\right. \\
&\qquad \left.\cdot\, e^{-(\tau+\Delta\tau)/T}n(\tau+\Delta\tau)\cos(\Omega t+\Omega\Delta t+\Theta)]\right\} \\
&= \mathcal{F}_{\Delta\tau}\left\{\frac{1}{4}e^{-(2\tau+\Delta\tau)/T}E[n(\tau)n(\tau+\Delta\tau)]E[\cos(\Omega\Delta t)\right. \\
&\qquad \left.+\cos(2\Omega t+\Omega\Delta t+2\Theta)]\right\} \\
&= \mathcal{F}_{\Delta\tau}\left\{\frac{1}{4}e^{-(2\tau+\Delta\tau)/T}\delta(\Delta\tau)\cos(\Omega\Delta t)\right\} \\
&= \frac{1}{4}e^{-2\tau/T}\cos(\Omega\Delta t), \qquad \tau \geq 0.
\end{aligned}
$$

The time correlation function is then

$$
\begin{aligned}
\phi_H(\Delta t) &= \phi_H(\Delta f, \Delta t)|_{\Delta f=0} \\
&= \int_{-\infty}^{\infty}\phi_h(\tau, \Delta t)\,d\tau \\
&= \frac{1}{4}\cos(\Omega\Delta t)\int_0^{\infty}e^{-2\tau/T}d\tau \\
&= \frac{T}{8}\cos(\Omega\Delta t).
\end{aligned}
$$

The Doppler psd is

$$
\begin{aligned}
\Phi_H(\nu) &= \mathcal{F}[\phi_H(\Delta t)] \\
&= \mathcal{F}\left[\frac{T}{8}\cos(\Omega\Delta t)\right] \\
&= \frac{T}{16}[\delta(2\pi\nu-\Omega)+\delta(2\pi\nu+\Omega)].
\end{aligned}
$$

That is, the channel introduces two Doppler shifts, $\pm\Omega/2\pi = \pm 5$ Hz, with equal psd.

b. The mean Doppler shift is zero as the two Doppler shifts are negative of each other and have the same psd. This can also be verified by using Eq. (2.3.17) and the psd computed in (a). From Eq. (2.3.18), the rms Doppler spread is

$$\sigma_\nu = \left\{ \frac{\int_{-\infty}^{\infty} \nu^2 \cdot \frac{T}{16}[\delta(2\pi\nu - \Omega) + \delta(2\pi\nu + \Omega)]d\nu}{\int_{-\infty}^{\infty} \frac{T}{16}[\delta(2\pi\nu - \Omega) + \delta(2\pi\nu + \Omega)]d\nu} \right\}^{1/2}$$

$$= \left\{ \frac{\frac{T}{16}\left[\left(\frac{\Omega}{2\pi}\right)^2 + \left(-\frac{\Omega}{2\pi}\right)^2\right]}{\frac{T}{16}[1+1]} \right\}^{1/2}$$

$$= \frac{\Omega}{2\pi},$$

which is 5 Hz.

c. The coherence time is

$$(\Delta t)_c \approx \frac{1}{\sigma_\nu} = 0.2 \text{ s}.$$

d. In GSM systems, the data rate $R_s = 270.833$ kbps, which corresponds to a symbol interval

$$T_s = \frac{1}{R_s} \approx 3.7 \times 10^{-6} \text{ s}.$$

Since $T_s \ll (\Delta t)_c$, the channel exhibits slow fading.

2.4 LARGE-SCALE PATH LOSS AND SHADOWING

The channel functions in a wireless environment are random processes and are very difficult to characterize. As a result, in Section 2.3, we focus only on the channel correlation functions. In the following, we will study the channel from a different point of view: At any instant (or distance), the channel impulse response is a random variable. We are interested in describing the channel at any time t (or distance d) using a probability density function (pdf) under some assumptions.

Consider a flat fading channel where the multipath delay spread is very small compared with the symbol interval of the transmitted signal. The channel impulse response can be approximated by

$$h(\tau, t) \approx h(\bar{\tau}, t) \triangleq g(t)\delta(\tau - \bar{\tau}), \quad (2.4.1)$$

where $\bar{\tau}$ is the mean delay and is assumed to be time invariant. Given that the transmitted signal is $x(t)$, from Eq. (2.2.6), the received signal is

$$r(t) = \int_{-\infty}^{\infty} h(\tau, t)x(t - \tau)\, d\tau \approx g(t)x(t - \bar{\tau}). \quad (2.4.2)$$

In addition to the delay $\bar{\tau}$, the channel provides a time-varying gain $g(t)$ to the transmitted signal. In general, the channel gain can be decomposed into a small-scale (or short-term) fading component, $Z(t)$, and a path loss with large-scale (or long-term) shadowing component (representing the local mean), as depicted in Figure 2.15. The short-term fading is due to multipath propagation and is independent of the distance between the transmitter and receiver. It can be characterized by a Rayleigh, Rician or Nakagami distribution [103]. The path loss represents the local mean of the channel gain and is therefore dependent on the distance between the transmitter and receiver. It also depends on the propagation environment. A path loss model is important for determination of the base station (cell) coverage area. Various path loss models have been proposed, mainly based on field measurements [106, 60, 81]. In this section, we study a few popular path loss models. To illustrate the distance-dependence of the propagation loss, we start with the propagation in free space.

2.4.1 Free Space Propagation Model

When the distance between the transmitting antenna and receiving antenna is much larger than the wavelength of the transmitted wave and the largest physical linear dimension of the antennas, the power P_r at the output of the receiving antenna is given by [122]

$$P_r = P_t G_t G_r \left(\frac{\lambda}{4\pi d} \right)^2,$$

where

$$\begin{aligned} P_t &= \text{total power radiated by an isotropic source,} \\ G_t &= \text{transmitting antenna gain,} \\ G_r &= \text{receiving antenna gain,} \\ d &= \text{distance between transmitting and receiving antennas,} \\ \lambda &= \text{wavelength of the carrier signal} = c/f_c, \\ c &= 3 \times 10^8 \text{ m/s (velocity of light),} \\ f_c &= \text{carrier frequency, and} \\ P_t G_t &\triangleq \text{effective isotropically radiated power (EIRP).} \end{aligned}$$

The term $(4\pi d/\lambda)^2$ is known as the *free-space path loss* denoted by $L_p(d)$, which is

$$\begin{aligned} L_p(d) &= \frac{\text{EIRP} \times \text{Receiving antenna gain}}{\text{Received power}} \\ &= -10 \log_{10} \left[\frac{\lambda^2}{(4\pi d)^2} \right] \quad \text{(dB)} \\ &= -20 \log_{10} \left(\frac{c/f_c}{4\pi d} \right) \quad \text{(dB).} \end{aligned}$$

In other words, the path loss is

$$L_p(d) = 20 \log_{10} f_c + 20 \log_{10} d - 147.56 \quad \text{(dB).} \tag{2.4.3}$$

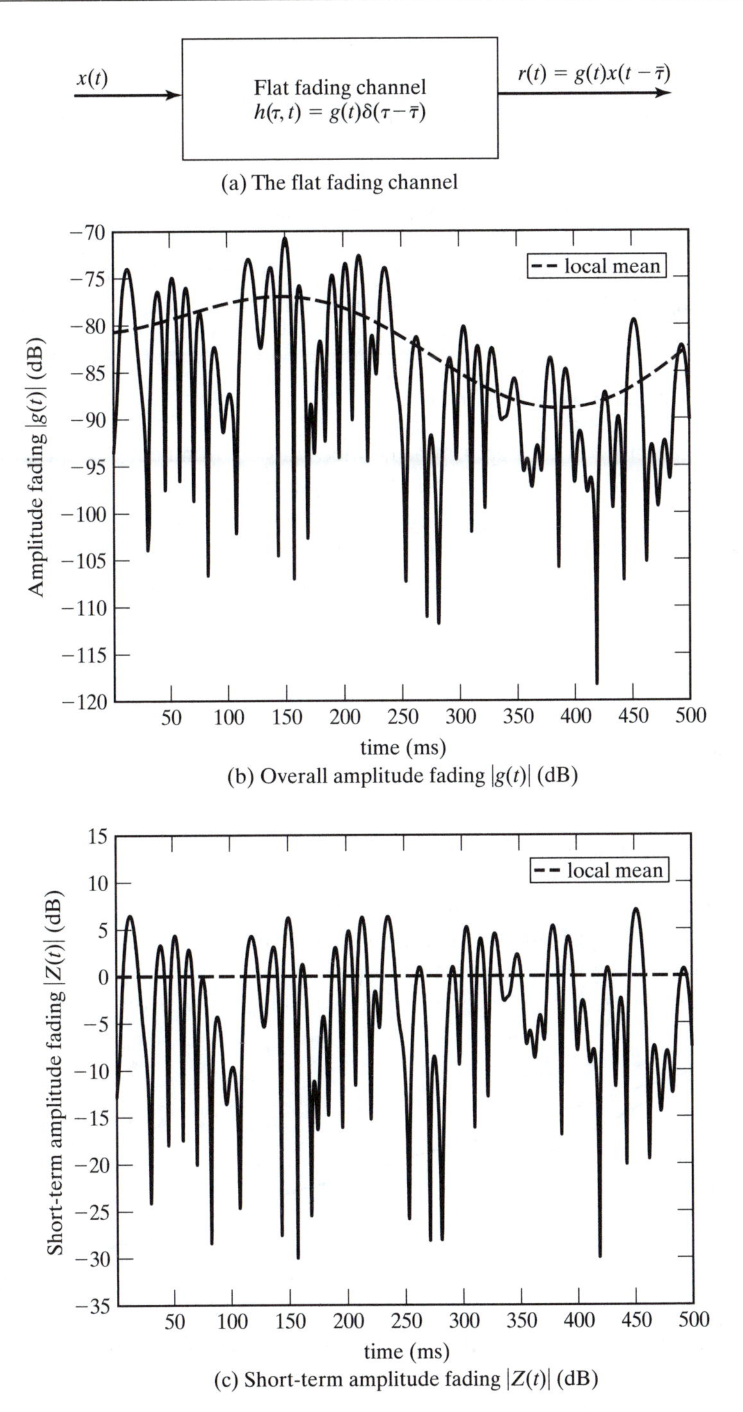

Figure 2.15 Representation of long-term and short-term fading components.

Note that the free-space path loss increases by 6 dB for every doubling of the distance and also for every doubling of the radio frequency.

With the path loss $L_p(d)$ in dB, if the transmitted power P_t is in watts, then the received signal power in dBW is

$$P_r(d)_{\text{(dBW)}} = 10\log_{10}(P_t G_t G_r)_{\text{(dBW)}} - L_p(d)_{\text{(dB)}}. \tag{2.4.4}$$

The term dBW denotes dB greater or less than 1 watt (0 dBW). The units dBW and dBm (dB greater or less than 1 mW) are widely used in communication engineering. Since 1 W is equal to 1000 mW, x dBW is equal to $x + 10\log_{10} 1000$ $(= x + 30)$ dBm.

2.4.2 Propagation Over Smooth Plane

Free space propagation does not apply in a mobile radio environment and the propagation path loss depends not only on the distance and wavelength, but also on the antenna heights of the mobile station and the base station, and the local terrain characteristics such as buildings and hills. A simple two-path model can be used to illustrate the effect of transmitting and receiving antenna heights. Consider the signal transmission over a smooth, reflecting, and flat plane (such as earth or water surface), as shown in Figure 2.16. It can easily be derived that the propagation distance of the direct path is

$$d_1 = \sqrt{d^2 + (h_t - h_r)^2} \tag{2.4.5}$$

and the propagation distance of the reflected path is

$$d_2 = \sqrt{d^2 + (h_t + h_r)^2}. \tag{2.4.6}$$

In the free space, the received signal power is inversely proportional to the square of the propagation distance. Given that $d \gg h_t h_r$, we have $d_1 \approx d$ and $d_2 \approx d$. However, since the carrier wavelength λ is very small as compared with d, a slight change in the propagation distance can cause a significant change in the received signal carrier phase. Depending on the difference between the carrier phases of the signals from the two paths, the received signal components from

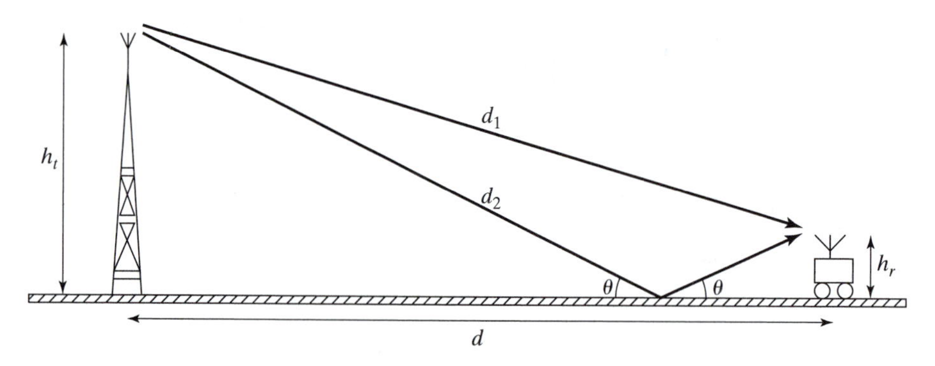

Figure 2.16 Two-path propagation over a flat plane.

the two paths may enhance each other or cancel each other. The carrier phase difference is related to the difference between the two propagation distances d_2 and d_1 by

$$\phi_2 - \phi_1 = \frac{2\pi}{\lambda}(d_2 - d_1). \tag{2.4.7}$$

Taking into account the phase difference, the received signal power is

$$P_r(d) = P_t G_t G_r \left(\frac{\lambda}{4\pi d}\right)^2 |1 + \alpha_f e^{-j\beta_f} \exp[j(\phi_2 - \phi_1)]|^2, \tag{2.4.8}$$

where α_f and β_f are the amplitude attenuation and carrier phase shift, respectively, introduced by the reflection. If $\alpha_f \approx 1$ (i.e., the reflection loss is negligible) and $\beta_f = \pi$ for $\theta \ll 1$, then

$$\begin{aligned} P_r(d) &= P_t G_t G_r \left(\frac{\lambda}{4\pi d}\right)^2 \left|1 - \cos\left(\frac{2\pi \Delta d}{\lambda}\right) - j\sin\left(\frac{2\pi \Delta d}{\lambda}\right)\right|^2 \\ &= P_t G_t G_r \left(\frac{\lambda}{4\pi d}\right)^2 \left[2 - 2\cos\left(\frac{2\pi \Delta d}{\lambda}\right)\right] \\ &= 4P_t G_t G_r \left(\frac{\lambda}{4\pi d}\right)^2 \sin^2\left(\frac{\pi \Delta d}{\lambda}\right), \end{aligned} \tag{2.4.9}$$

where $\Delta d = d_2 - d_1$. Given that $d \gg h_t$ and $d \gg h_r$, from Eqs. (2.4.5)–(2.4.6), it can be derived that

$$\Delta d \approx \frac{2h_t h_r}{d}. \tag{2.4.10}$$

As a result, the received signal power is

$$P_r(d) \approx 4P_t G_t G_r \left(\frac{\lambda}{4\pi d}\right)^2 \sin^2\left(\frac{2\pi h_t h_r}{\lambda d}\right). \tag{2.4.11}$$

The corresponding path loss is

$$\begin{aligned} L_p(d) &= \left[4\left(\frac{\lambda}{4\pi d}\right)^2 \sin^2\left(\frac{2\pi h_t h_r}{\lambda d}\right)\right]^{-1} \\ &= -10\log_{10}\left[4\left(\frac{\lambda}{4\pi d}\right)^2 \sin^2\left(\frac{2\pi h_t h_r}{\lambda d}\right)\right] \quad \text{(dB)}. \end{aligned} \tag{2.4.12}$$

Figure 2.17 illustrates the path loss $L_p(d)$ in dB as a function of the distance d, where $f_c =$ 900 MHz, $h_t = 35$ m, and $h_r = 3$ m. It is observed that (1) the path loss has alternate minima and maxima as the distance between the transmitter and receiver increases, and (2) in general, the path loss increases with the distance d.

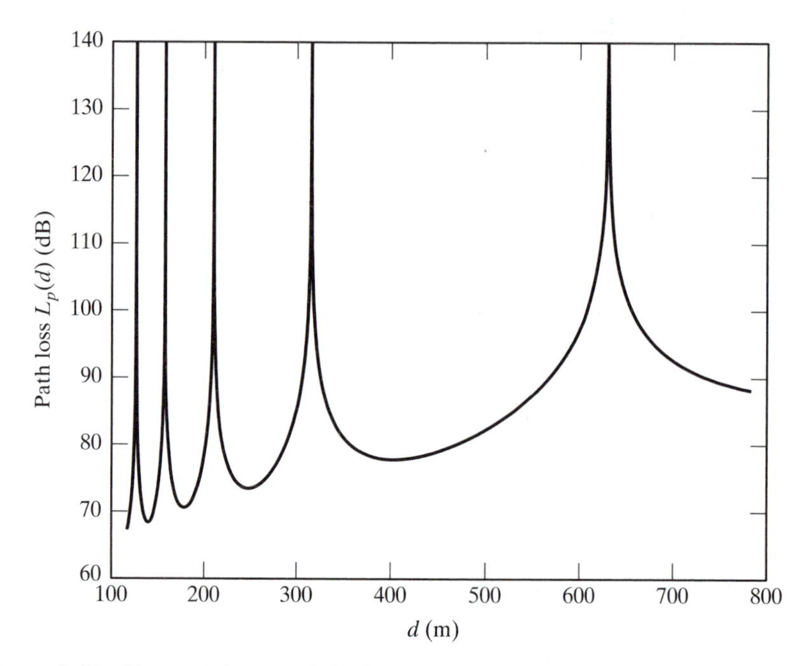

Figure 2.17 The path loss $L_p(d)$ in dB versus distance d in the two-path model.

2.4.3 Log-Distance Path Loss with Shadowing

As a mobile user moves away from its base station, the received signal becomes weaker because of the growing propagation attenuation with the distance. Let $\bar{L}_p(d)$ denote the log-distance path loss, which is a function of the distance d separating the transmitter and the receiver. Then,

$$\bar{L}_p(d) \propto \left(\frac{d}{d_0}\right)^{\kappa}, \quad d \geq d_0$$

or equivalently,

$$\bar{L}_p(d) = \bar{L}_p(d_0) + 10\kappa \log_{10}\left(\frac{d}{d_0}\right) \text{ dB}, \quad d \geq d_0 \tag{2.4.13}$$

where κ is the path loss exponent and d_0 is the close-in reference distance. Typically, d_0 is 1 km for macrocells, 100 m for outdoor microcells, and 1 m for indoor picocells. Given d_0, the value $\bar{L}_p(d_0)$ depends on the carrier frequency, antenna heights and gains, and other factors. Table 2.1 gives typical values of the path loss exponent in different propagation environments.

Furthermore, as the mobile moves in uneven terrain, it often travels into a propagation shadow behind a building or a hill or other obstacle much larger than the wavelength of the transmitted signal, and the associated received signal level is attenuated significantly. This phenomenon is called shadowing. A log-normal distribution is a popular model for characterizing the shadowing

Table 2.1 Path Loss Exponents for Different Environments

Environment	Path Loss Exponent, κ
free space	2
urban cellular radio	2.7 to 3.5
shadowed urban cellular radio	3 to 5
in building with line of sight	1.6 to 1.8
obstructed in building	4 to 6

process. As a result, long-term fading is a combination of log-distance path loss and log-normal shadowing. Let $\epsilon_{(\mathrm{dB})}$ be a zero-mean Gaussian distributed random variable (in dB) with standard deviation σ_ϵ (in dB). The pdf of $\epsilon_{(\mathrm{dB})}$ is given by

$$f_{\epsilon(\mathrm{dB})}(x) = \frac{1}{\sqrt{2\pi}\sigma_\epsilon} \exp\left(-\frac{x^2}{2\sigma_\epsilon^2}\right). \tag{2.4.14}$$

Let $L_p(d)$ denote the overall path loss with shadowing (long-term fading) in dB. Then,

$$\begin{aligned} L_p(d) &= \bar{L}_p(d) + \epsilon_{(\mathrm{dB})} \\ &= \bar{L}_p(d_0) + 10\kappa \log_{10}\left(\frac{d}{d_0}\right) + \epsilon_{(\mathrm{dB})} \ (\mathrm{dB}), \quad d \geq d_0. \end{aligned} \tag{2.4.15}$$

Since $\epsilon_{(\mathrm{dB})}$ follows the Gaussian (normal) distribution with pdf given by Eq. (2.4.14), ϵ in linear scale is said to follow a log-normal distribution with pdf given by

$$f_\epsilon(y) = \frac{20/\ln 10}{\sqrt{2\pi} y \sigma_\epsilon} \exp\left[-\frac{(20 \log_{10} y)^2}{2\sigma_\epsilon^2}\right]. \tag{2.4.16}$$

The first-order statistics of log-normal shadowing are characterized by the standard deviation σ_ϵ, which can be obtained from measurements. For example, 8 dB is a typical value for σ_ϵ in an outdoor cellular system and 5 dB is a value for an indoor environment.

2.4.4 Okumura–Hara Path Loss Model

The Okumura–Hara path loss model [106, 60] was developed by curve fitting the measurement data collected in Tokyo, Japan. It is suitable for outdoor macrocells. The path loss is represented as a function of (a) the carrier frequency $f_c \in [150,\ 1000]$ MHz, (b) antenna heights of base station, $h_b \in [30,\ 200]$ m, and mobile station, $h_m \in [1,\ 10]$ m, and (c) the distance between the base station and mobile station $d \in [1,\ 20]$ km. The path loss in dB is given by

$$L_p(d) = \begin{cases} A + B\log_{10}(d) & \text{for urban area} \\ A + B\log_{10}(d) - C & \text{for suburban area ,} \\ A + B\log_{10}(d) - D & \text{for open area} \end{cases} \tag{2.4.17}$$

where

$$A = 69.55 + 26.16\log_{10}(f_c) - 13.82\log_{10}(h_b) - a(h_m),$$
$$B = 44.9 - 6.55\log_{10}(h_b),$$
$$C = 5.4 + 2[\log_{10}(f_c/28)]^2,$$
$$D = 40.94 + 4.78[\log_{10}(f_c)]^2 - 18.33\log_{10}(f_c),$$

and $a(h_m)$ is the correction factor for mobile antenna height, and is given by

$$a(h_m) = [1.1\log_{10}(f_c) - 0.7]h_m - [1.56\log_{10}(f_c) - 0.8]$$

for a small to medium city and

$$a(h_m) = \begin{cases} 8.29[\log_{10}(1.54h_m)]^2 - 1.1 & \text{for } f_c \leq 200 \text{ MHz} \\ 3.2[\log_{10}(11.75h_m)]^2 - 4.97 & \text{for } f_c \geq 400 \text{ MHz} \end{cases}$$

for a large city.

Figure 2.18 shows the Okumura–Hara path loss in large city, suburban and open areas respectively as a function of d in km, with parameters $f_c = 900$ MHz, $h_b = 50$ m, and $h_m = 3$ m.

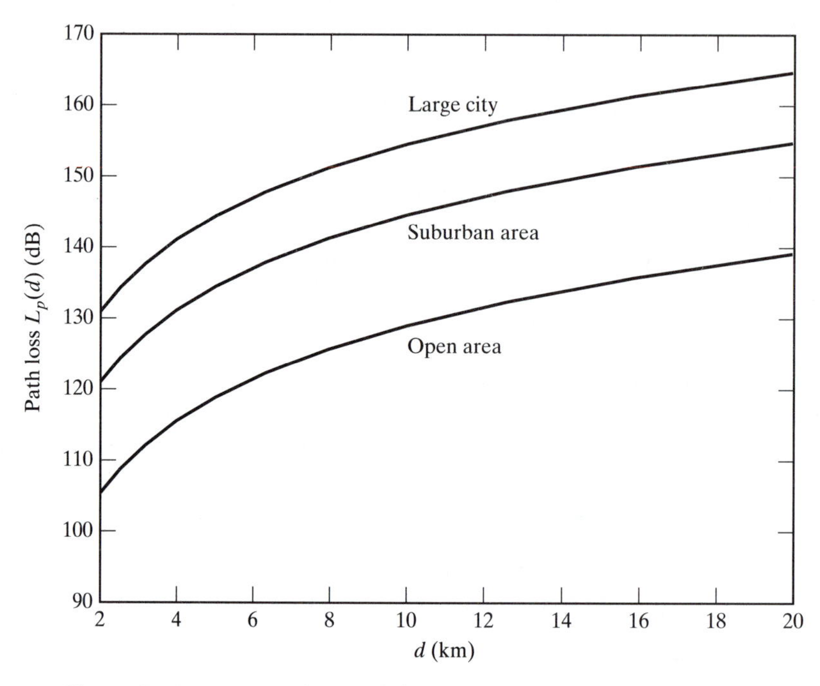

Figure 2.18 Propagation path loss using Okumura–Hara model.

2.4.5 Lee's Path Loss Model

Lee's model [81] can be used to predict area-to-area path loss. The model consists of two parts: (*a*) path loss prediction for a specified set of conditions, and (*b*) adjustment factors for a set of conditions different from the specified one. The model requires two parameters: (*a*) the power at a 1.6 km (1 mile) of interception P_0 in dBm, and (*b*) the path-loss exponent κ. The specified set of conditions is as follows:

Carrier frequency $f_c = 900$ MHz

Base station antenna height = 30.48 m (100 ft)

Base station power at the antenna = 10 W

Base station antenna gain = 6 dB above dipole gain

Mobile station antenna height = 3 m (10 ft)

Mobile station antenna gain = 0 dB above dipole gain

The received signal power in dBm is represented by

$$P_r(d) = P_{0(\text{dBm})} - 10\kappa \log_{10}(d/d_0) - 10n \log_{10}(f/f_c) + a_{0(\text{dB})}, \tag{2.4.18}$$

where $d_0 = 1.6$ km, d ($\geq d_0$) is the distance between the mobile station and the base station in km, and n is a constant between 2 and 3 dependent on the geographical locations and the operating frequency ranges. $n = 2$ is recommended for a suburban or open area with $f < 450$ MHz and $n = 3$ for an urban area with $f > 450$ MHz. The parameter $\alpha_{0(\text{dB})}$ is an adjustment factor for a different set of conditions

$$a_{0(\text{dB})} = 10 \log_{10}(a_1 a_2 a_3 a_4 a_5), \tag{2.4.19}$$

where

$$a_1 = \left[\frac{\text{new base station antenna height (m)}}{30.48 \text{ (m)}}\right]^2$$

$$a_2 = \left[\frac{\text{new mobile station antenna height (m)}}{3 \text{ (m)}}\right]^\nu$$

$$a_3 = \frac{\text{new transmitter power (W)}}{10 \text{ (W)}}$$

$$a_4 = \frac{\text{new base station antenna gain with respect to } \lambda/2 \text{ dipole}}{4}$$

$$a_5 = \text{different antenna gain correction factor at the mobile station}$$

The value ν in a_2 is obtained from empirical data and is given by

$$\nu = \begin{cases} 2 & \text{for new mobile station antenna height} > 10 \text{ m} \\ 1 & \text{for new mobile station antenna height} < 3 \text{ m} \end{cases}.$$

Table 2.2 The Parameters in Lee's Path Loss Model in Various Propagation Environments

terrain	$P_{0(\mathrm{dBm})}$	κ
free space	−45	2.00
open area	−49	4.35
suburban areas	−61.7	3.84
urban area (Philadelphia)	−70	3.68
urban area (Newark)	−64	4.31
urban area (Tokyo)	−84	3.05

A 2 dB signal gain is provided by an actual 4 dB gain antenna at the mobile unit in a suburban area, and less than 1 dB gain received from the same antenna in an urban area for adjusting a_5. The two required parameters $P_{0(\mathrm{dBm})}$ and κ have been determined for various propagation environments based on empirical data and are given in Table 2.2.

The propagation path loss using Lee's model for some environments given in Table 2.2 is plotted in Figure 2.19, where the base station antenna height is 50 m with a gain of 6 dB with respect to a $\lambda/2$ dipole, the transmitter power is 10 W, the mobile station antenna height is 3 m, and the carrier frequency f_c is 900 MHz.

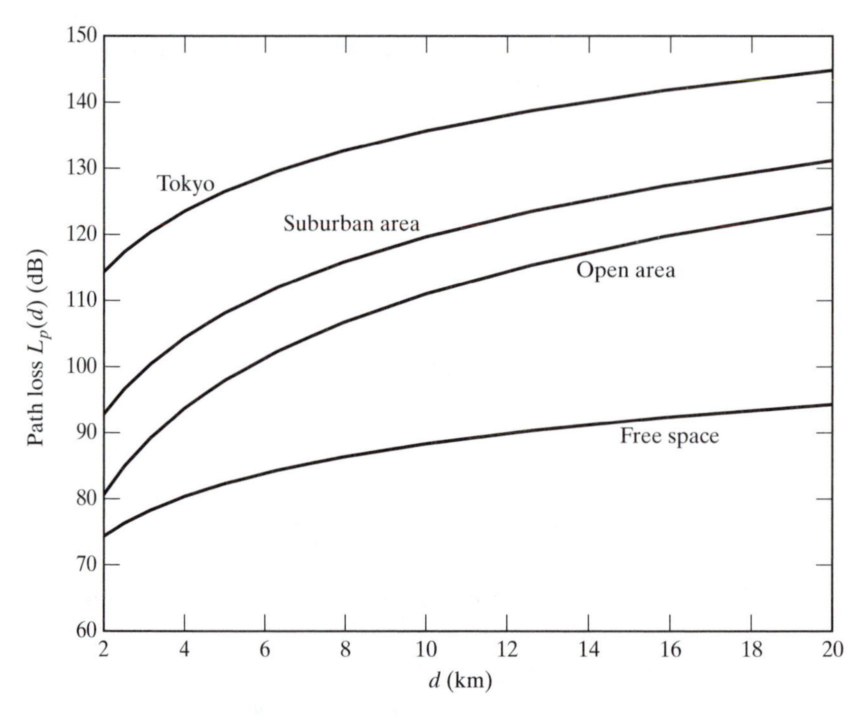

Figure 2.19 Propagation path loss using Lee's model.

2.4.6 Radio Cell Coverage

Radio cell coverage is the service area supported by each base station. The coverage depends on (*a*) service quality requirements, such as the required ratio of the signal power to interference-plus-noise power (see Chapter 5 for more details), or the required minimum received signal power level given the transmitted signal power, and (*b*) the propagation environment. For example, for free space transmission with an omnidirectional antenna, the cell coverage is a circle centered at the base station (transmitter) with a radius depending on the propagation loss. Given the transmitted signal power level, the minimum required received power level can be mapped to the maximum allowed path loss. Then from Eq. (2.4.3), the cell radius (d) can be determined. In practice, the path loss depends on the propagation environment, including the transmitter and receiver antenna heights, and may differ from angle to angle as seen from the base station transmitter antenna. As a result, the cell coverage will generally have an irregular shape (not a circle). Furthermore, because of the random nature in the path loss due to factors such as shadowing, the cell coverage is not deterministic but should be specified based on statistic parameters. The following example illustrates how to determine the cell coverage for a given propagation model, where the service quality criterion is specified in terms of the propagation path loss.

Example 2.5 Radio Cell Coverage for the Log-Distance Path Loss

Consider the log-distance path loss model. Determine the cell coverage for the following two situations:

a. Without shadowing, the path loss is given by Eq. (2.4.13). It is required that, at the cell border, the path loss cannot be γ dB larger than that at the reference distance d_0.
b. With shadowing, the path loss is given by Eq. (2.4.15). Due to the random nature of the propagation path loss, the cell coverage is defined as the service area of the base station over which the path loss (over that at the reference distance d_0) is limited to γ dB with a pre-defined probability. Assume the reference distance d_0 is very small compared with the cell radius.

Solution

a. The path loss in dB at a distance d ($> d_0$) from the base station is

$$\bar{L}_p(d) = \bar{L}_p(d_0) + 10\kappa \log_{10}\left(\frac{d}{d_0}\right).$$

As the path loss monotonically increases with the distance d and is independent of the propagation direction as observed at the base station, the cell coverage is a circle centered at the base station with radius R specified by

$$\Delta\bar{L}_p(d)|_{d=R} \stackrel{\triangle}{=} \bar{L}_p(d)|_{d=R} - \bar{L}_p(d_0) = 10\kappa \log_{10}\left(\frac{d}{d_0}\right)\bigg|_{d=R} \leq \gamma.$$

Solving the above equation for the radius R, we obtain

$$R \leq d_0 \times 10^{0.1(\gamma/\kappa)}.$$

b. With shadowing, the path loss in dB at a distance d ($> d_0$) from the base station is

$$\bar{L}_p(d) = \bar{L}_p(d_0) + 10\kappa \log_{10}\left(\frac{d}{d_0}\right) + \epsilon_{(\mathrm{dB})},$$

where the random variable $\epsilon_{(\mathrm{dB})}$ characterizes the effect of shadowing and is modeled by a Gaussian random variable with zero mean and standard deviation σ_ϵ (in dB). The relative path loss at d ($> d_0$) with respect to the loss at d_0 is given by

$$\Delta\bar{L}_p(d) \triangleq \bar{L}_p(d) - \bar{L}_p(d_0) = 10\kappa \log_{10}\left(\frac{d}{d_0}\right) + \epsilon_{(\mathrm{dB})}.$$

The problem of estimating the cell coverage can be approached in two steps:

Step 1: Determine the probability a_1 that the path loss at location r ($> d_0$) is below the threshold γ, where the probability is averaged over the circumference;

Step 2: Calculate the probability a_2 of the circular area (defined by R) over which the path loss is below the threshold based on a_1, where the probability is averaged over the circular area (cell area).

At the distance $d = r$ ($> d_0$), the probability that the relative path loss is limited to γ dB is

$$\begin{aligned} a_1(r) &= P(\Delta\bar{L}_p(r) \le \gamma) \\ &= P(\epsilon_{(\mathrm{dB})} \le \gamma - 10\kappa \log_{10}(r/d_0)) \\ &= Q\left(\frac{\gamma - 10\kappa \log_{10}(r/d_0)}{\sigma_\epsilon}\right), \end{aligned} \tag{2.4.20}$$

where $Q(x) \triangleq \frac{1}{\sqrt{2\pi}}\int_x^\infty \exp(-t^2/2)dt$. Next, we want to find the probability a_2 of the event "$\Delta\bar{L}_p(r) \le \gamma$" over a circular area A centered at the base station with radius R, as shown in Figure 2.20. The probability is given by

$$\begin{aligned} a_2 &= P(\Delta\bar{L}_p(r) \le \gamma \text{ over the area } A) \\ &= \frac{1}{A}\int_A P(\Delta\bar{L}_p(r) \le \gamma \text{ over the area } dA)\, dA \\ &= \frac{1}{A}\int_A a_1(r)\, dA, \end{aligned} \tag{2.4.21}$$

where $A = \pi(R^2 - d_0^2) \approx \pi R^2$ for $R \gg d_0$ and $dA = r d\theta dr$ as shown in Figure 2.20. Substituting Eq. (2.4.20) into Eq. (2.4.21), we have

$$\begin{aligned} a_2 &\approx \frac{1}{\pi R^2}\int_0^{2\pi} d\theta \int_{d_0}^R a_1(r) r dr \\ &= 2\int_{d_0}^R Q\left(\frac{\gamma - 10\kappa \log_{10}(r/d_0)}{\sigma_\epsilon}\right)\frac{r}{R}\frac{dr}{R} \\ &= 2\int_{d_0/R}^1 Q\left(\frac{\gamma - 10\kappa \log_{10}(xR/d_0)}{\sigma_\epsilon}\right) x dx, \end{aligned}$$

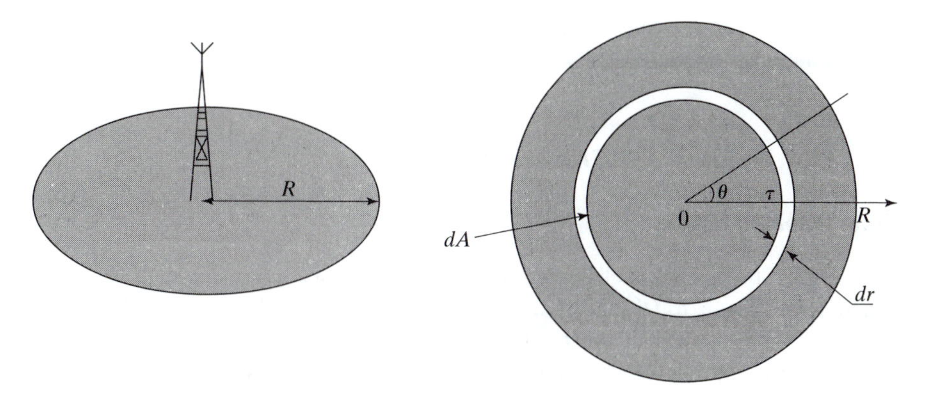

Figure 2.20 The cell coverage area.

where

$$\frac{\gamma - 10\kappa \log_{10}(xR/d_0)}{\sigma_\epsilon} = a - b\ln x,$$

with

$$a = \frac{\gamma - 10\kappa \log_{10}(R/d_0)}{\sigma_\epsilon}$$

and

$$b = \frac{10\kappa \log_{10} e}{\sigma_\epsilon}.$$

As a result,

$$\begin{aligned}
a_2 &= 2\int_{d_0/R}^{1}\left[\frac{1}{\sqrt{2\pi}}\int_{a-b\ln x}^{\infty} e^{-t^2/2}dt\right]xdx \\
&= 2\left\{\int_{a}^{a+b\ln(R/d_0)} \frac{1}{\sqrt{2\pi}}\left[\int_{\exp(\frac{a-t}{b})}^{1} xdx\right]e^{-t^2/2}dt\right. \\
&\quad \left. + \int_{a+b\ln(R/d_0)}^{\infty} \frac{1}{\sqrt{2\pi}}\left[\int_{d_0/R}^{1} xdx\right]e^{-t^2/2}dt\right\} \\
&= Q(a) - \left(\frac{d_0}{R}\right)^2 Q\left(a + b\ln\left(\frac{R}{d_0}\right)\right) \\
&\quad - \exp\left(\frac{2ab+2}{b^2}\right)\left[Q\left(a+\frac{2}{b}\right) - Q\left(a+\frac{2}{b}+b\ln\left(\frac{R}{d_0}\right)\right)\right].
\end{aligned}$$

If $d_0 \ll R$, we have

$$a_2 \approx Q(a) - \exp\left(\frac{2ab+2}{b^2}\right)\left[Q\left(a+\frac{2}{b}\right) - Q\left(a+\frac{2}{b}+b\ln\left(\frac{R}{d_0}\right)\right)\right].$$

Given the required a_2 value, the cell radius R can be determined numerically from the above equations.

2.5 SMALL-SCALE MULTIPATH FADING

2.5.1 First-Order Statistics

Suppose the multipath channel is characterized by N distinct scatterers in which the nth scatterer is associated with a gain $\alpha_n(t)$ and a delay $\tau_n(t)$, where $n = 1, 2, \ldots, N$. Consider digital transmission over the channel at the carrier frequency f_c with a symbol interval much larger than the channel multipath delay spread. From Eq. (2.2.2), the received signal at baseband, in the absence of background noise, is

$$r(t) = \sum_{n=1}^{N} \alpha_n(t) e^{-j2\pi f_c \tau_n(t)} x(t - \tau_n(t))$$

$$\approx \left[\sum_{n=1}^{N} \alpha_n(t) e^{-j2\pi f_c \tau_n(t)} \right] x(t - \bar{\tau}).$$

The approximation is reasonable as long as the delay spread is much smaller than the symbol interval. The complex gain of the channel is $Z(t) = \sum_{n=1}^{N} \alpha_n(t) e^{-j2\pi f_c \tau_n(t)}$. Let $Z_c(t)$ and $-Z_s(t)$ denote the real and imaginary components of the complex channel gain, respectively, so that $Z(t) = Z_c(t) - jZ_s(t)$. Then

$$Z_c(t) = \sum_{n=1}^{N} \alpha_n(t) \cos\theta_n(t)$$

$$Z_s(t) = \sum_{n=1}^{N} \alpha_n(t) \sin\theta_n(t),$$

where $\theta_n(t) = 2\pi f_c \tau_n(t)$.

Furthermore, let

$$\alpha(t) = \sqrt{Z_c^2(t) + Z_s^2(t)}, \quad \theta(t) = \tan^{-1}[Z_s(t)/Z_c(t)]$$

be the amplitude fading and carrier distortion introduced by the channel. The fading characteristics can be studied by examining the pdfs of the envelope $\alpha(t)$ and phase $\theta(t)$ at any time t. The fading characteristics depend on whether the transmitter and receiver are in line-of-sight or not in line-of-sight. The former case is called LOS scattering while the latter case is referred to as NLOS scattering. LOS scattering has a specular component (from the direct path), and can be modeled as a Rician distribution. NLOS scattering does not have a specular component, and can be modeled as a Rayleigh distribution. A pictorial view of LOS and NLOS scattering is depicted in Figure 2.21.

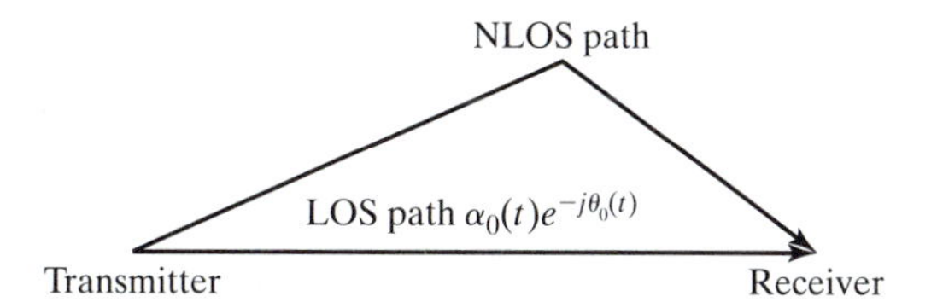

Figure 2.21 NLOS versus LOS scattering.

Rayleigh Fading (NLOS propagation). In this case,

$$E[Z_c(t)] = E[Z_s(t)] = 0. \tag{2.5.1}$$

Assume that, at any time t, for $n = 1, 2, \ldots, N$,

a. the values of $\theta_n(t)$ are statistically independent, each being uniformly distributed over $[0,\ 2\pi]$;
b. the values of $\alpha_n(t)$ are identically distributed random variables, independent of each other and of the $\theta_n(t)$'s.

According to the central limit theorem, $Z_c(t)$ and $Z_s(t)$ are approximately Gaussian random variables at any time t if N is sufficiently large. For simplicity of notation, let Z_c and Z_s denote $Z_c(t)$ and $Z_s(t)$ at any time t. It can be shown that Z_c and Z_s are independent Gaussian random variables with zero mean and equal variance $\sigma_z^2 = \frac{1}{2}\sum_{n=1}^{N} E[\alpha_n^2]$, where α_n denotes $\alpha_n(t)$ at any time t. As a result, the joint pdf of Z_c and Z_s is

$$f_{Z_cZ_s}(x, y) = \frac{1}{2\pi\sigma_z^2}\exp\left[-\frac{x^2+y^2}{2\sigma_z^2}\right], \quad -\infty < x < \infty, \ -\infty < y < \infty. \tag{2.5.2}$$

Let α and θ be the amplitude fading $\alpha(t)$ and carrier phase distortion $\theta(t)$ at any time t. Then it can easily be shown that

a. the amplitude fading, α, follows a Rayleigh distribution with parameter σ_z^2,

$$f_\alpha(x) = \begin{cases} \dfrac{x}{\sigma_z^2}\exp\left(-\dfrac{x^2}{2\sigma_z^2}\right), & x \geq 0 \\ 0, & x < 0 \end{cases}, \tag{2.5.3}$$

with $E[\alpha] = \sigma_z\sqrt{\pi/2}$ and $E(\alpha^2) = 2\sigma_z^2$;
b. the phase distortion follows a uniform distribution over $[0,\ 2\pi]$,

$$f_\theta(x) = \begin{cases} \dfrac{1}{2\pi}, & 0 \leq x \leq 2\pi \\ 0, & \text{elsewhere} \end{cases}; \tag{2.5.4}$$

c. the amplitude fading α and the phase distortion θ are independent.

The channel is called a Rayleigh fading channel.

Rician Fading (LOS propagation). If there exists an LOS path, the channel gain can be represented by

$$Z(t) = Z_c(t) - jZ_s(t) + \Gamma(t),$$

where $\Gamma(t) = \alpha_0(t)e^{-j\theta_0(t)}$ is the deterministic LOS component, and $Z_c(t) - jZ_s(t)$ represents all the NLOS components. With the LOS component, $E[Z(t)] = \Gamma(t) \neq 0$. The distribution of the envelope at any time t is given by the Rayleigh distribution modified by

a. a factor containing a non-centrality parameter, and
b. a zero-order modified Bessel function of the first kind.

The resultant pdf for the amplitude fading at any t, α, is known as the Rician distribution, given by (see Appendix D)

$$f_\alpha(x) = \underbrace{\frac{x}{\sigma_z^2}\exp\left(-\frac{x^2}{2\sigma_z^2}\right)}_{\text{Rayleigh}} \cdot \underbrace{\exp\left\{-\frac{\alpha_0^2}{2\sigma_z^2}\right\} \cdot I_0\left(\frac{\alpha_0 x}{\sigma_z^2}\right)}_{\text{modifier}}$$

$$= \frac{x}{\sigma_z^2}\exp\left(-\frac{x^2+\alpha_0^2}{2\sigma_z^2}\right) I_0\left(\frac{\alpha_0 x}{\sigma_z^2}\right), \quad x \geq 0, \tag{2.5.5}$$

where α_0 is $\alpha_0(t)$ at any t. α_0^2, is the power of the LOS component and is the non-centrality parameter, $I_0(\cdot)$ is the zero-order modified Bessel function of the first kind and is given by

$$I_0(x) = \frac{1}{2\pi}\int_0^{2\pi} \exp(x\cos\theta)d\theta. \tag{2.5.6}$$

The Rician fading channel has an important parameter called the K factor. It is defined as

$$K \triangleq \frac{\text{Power of the LOS component}}{\text{Total power of all other scattered components}} = \frac{\alpha_0^2}{2\sigma_z^2}.$$

As K approaches zero, the Rician distribution approaches the Rayleigh distribution. On the other hand, as K approaches infinity, only the dominant component matters and there is no fading. As a result, the wireless channel approaches an AWGN channel. Figure 2.22 shows the Rician distribution with $\sigma_z = 1$ and various K values. Assuming $\theta_0(t) = \pi/2$, it can be derived that, at a given t, the pdf of the carrier phase distortion $\theta(t)$ is given by

$$f_\theta(x) = \frac{1}{2\pi}\exp(-K) + \frac{1}{2}\sqrt{\frac{K}{\pi}}(\cos x)\exp(-K\sin^2 x)[1 + \text{erf}(\sqrt{K}\cos x)] \tag{2.5.7}$$

for $x \in [-\pi, +\pi]$, where $\text{erf}(x) = \frac{2}{\sqrt{\pi}}\int_0^x e^{-y^2}dy$ is the error function.

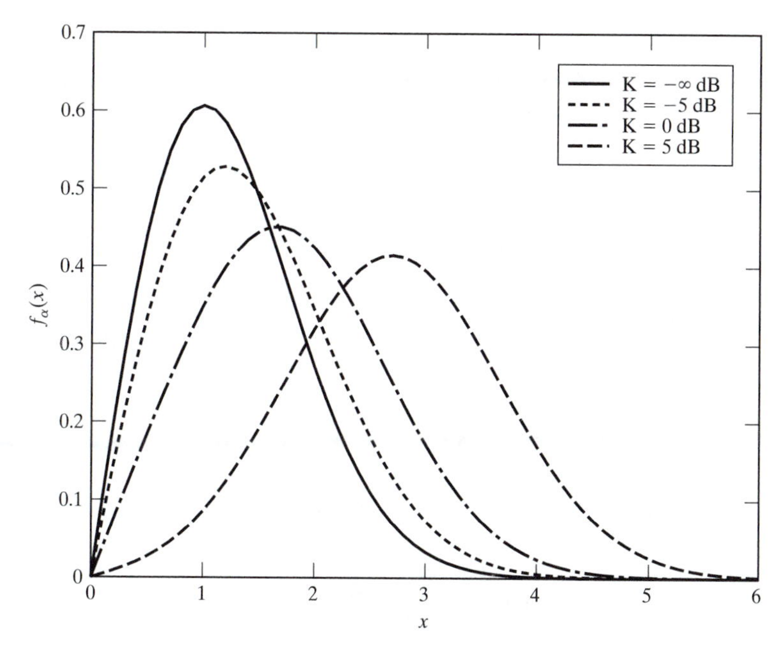

Figure 2.22 Rayleigh and Rician fading distributions with $\sigma_z = 1$.

2.5.2 Second-Order Statistics

The pdfs of the amplitude fluctuation and phase distortion of a fading channel in Section 2.4 tell us how the amplitude and phase will behave at each time instant, but do not tell us how they change with time. In designing efficient modulation and channel coding schemes to combat channel fading, just knowing the pdfs of the amplitude fading and carrier phase distortion introduced by the channel is not enough. How fast channel fading changes with time is also important. Level crossing rate (LCR) and average fade duration (AFD) are two statistics which describe the frequency of fading. They are closely related to the Doppler frequency shifts introduced by the channel. Although the second-order statistics provide the same information as the channel correlation functions (in particular, the channel coherence time and Doppler spread) given in Section 2.3, LCR and AFD are defined more specifically and, therefore, their mathematical expressions can be derived for a flat Rayleigh fading channel. In the following discussion, we are interested in LCR and AFD for the short-term amplitude fading and, in particular, for a WSS and ergodic[2] Rayleigh flat fading channel.

Level Crossing Rate

Definition 2.3 The crossing rate at level R of a flat fading channel is the expected number of times that the channel amplitude fading level, $\alpha(t)$, crosses the specified level R, with a positive slope, divided by the observation time interval.

[2]An ergodic process is one in which any state will occur with a non-zero probability; every sizable sample is equally representative of the whole.

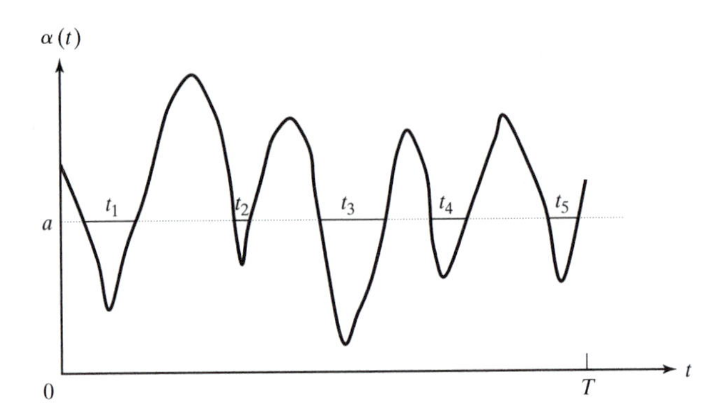

Figure 2.23 Level crossing rate and average duration of fades.

As an example, Figure 2.23 shows the channel amplitude fading level, where R is the chosen threshold. If the observation time interval is $[0, T]$, the total number of positive crossings is $M_T = 5$. The number of up crossings per second over the interval is then $M_T/T = 5/T$.

For an ergodic random process, statistical average is the same as time average when the time interval approaches infinity. Mathematically, the LCR at level R, denoted by N_R, is defined as the expectation of the positive time rate of the amplitude fading level change at the given threshold R. That is

$$N_R = \text{E[upward crossing rate at level } R]. \tag{2.5.8}$$

Let $\dot{\alpha}$ denote the amplitude fading rate, $d\alpha(t)/dt$, at any time t, and let $f_{\alpha\dot{\alpha}}(x, y)$ denote the joint pdf of the amplitude fading $\alpha(t)$ and its derivative $\dot{\alpha}(t)$ at any time t. Then $f_{\alpha\dot{\alpha}}(x, y)|_{x=R}$ gives the joint pdf at the amplitude level R. From the definition, LCR is the expectation of the positive rate (i.e., $\dot{\alpha} > 0$) and at the level R, which can be expressed by

$$N_R = \int_0^\infty y f_{\alpha\dot{\alpha}}(x, y)|_{x=R} dy. \tag{2.5.9}$$

For the Rayleigh fading environment studied in Subsection 2.5.1, it can be shown that [130]

$$f_{\alpha\dot{\alpha}}(x, y) = \frac{x}{\sqrt{2\pi\sigma_{\dot{\alpha}}^2\sigma_z^2}} \exp\left[-\frac{1}{2}\left(\frac{x^2}{\sigma_z^2} + \frac{y^2}{\sigma_{\dot{\alpha}}^2}\right)\right], \quad x \geq 0, -\infty < y < \infty, \tag{2.5.10}$$

where

$$\sigma_{\dot{\alpha}}^2 = \frac{1}{2}(2\pi\nu_m)^2\sigma_z^2$$

and ν_m is the maximum Doppler shift. Substituting Eq. (2.5.10) into Eq. (2.5.9), the LCR is

$$\begin{aligned} N_R &= \int_0^\infty y \cdot \frac{R}{\sqrt{2\pi\sigma_{\dot{\alpha}}^2\sigma_z^2}} \exp\left[-\frac{1}{2}\left(\frac{R^2}{\sigma_z^2} + \frac{y^2}{\sigma_{\dot{\alpha}}^2}\right)\right] dy \\ &= \sqrt{2\pi}\nu_m \left(\frac{R}{\sqrt{2}\sigma_z}\right) \exp\left(-\frac{R^2}{2\sigma_z^2}\right). \end{aligned}$$

Letting

$$\rho = \frac{R}{\sqrt{2}\sigma_z}$$

be the normalized threshold with respect to the rms value of α (i.e., $\sqrt{2}\sigma_z$), we have

$$N_R = \sqrt{2\pi}\,\nu_m \rho \exp(-\rho^2). \tag{2.5.11}$$

The LCR is a product of two terms. The first term, $\sqrt{2\pi}\,\nu_m$, is proportional to the maximum Doppler shift. Since $\nu_m = \frac{V f_c}{c}$, where V is the velocity of the mobile user, f_c is the carrier frequency, and c is the speed of light, LCR is proportional to the user speed and the carrier frequency.

The second term, $\rho \exp(-\rho^2)$, depends only on the normalized threshold. Figure 2.24 shows how the component changes with the normalized threshold ρ in dB. It is observed that the maximum value for LCR occurs at a value of ρ which is 3 dB below the rms value. This is always true and can be proved by differentiating Eq. (2.5.11) with respect to ρ and setting the result to zero. The sharp dropping of the LCR for values of ρ above the rms value, as shown in Figure 2.24, can be explained by Figure 2.25. Recall that the basic definition of LCR is the number of up crossings divided by the observation interval. In Figure 2.25, let the signal level c be the rms value. When the signal level moves from c to b, the number of crossings decreases rapidly. This is the reason why the LCR shown in Figure 2.24 drops so fast. Similarly, the LCR value decreases when the value of ρ is more than 3 dB smaller than the rms value, as can be

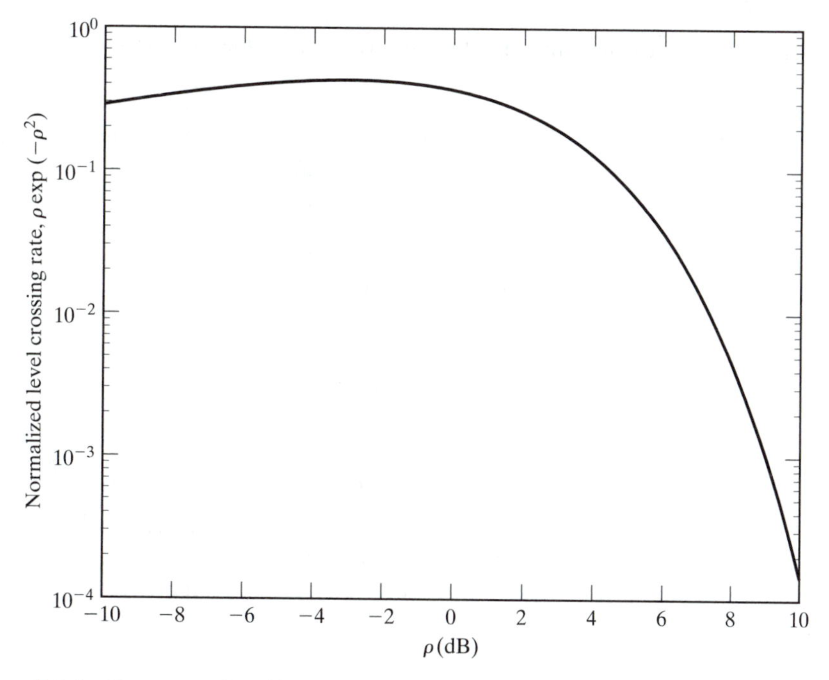

Figure 2.24 The normalized level crossing rate of the flat Rayleigh fading channel.

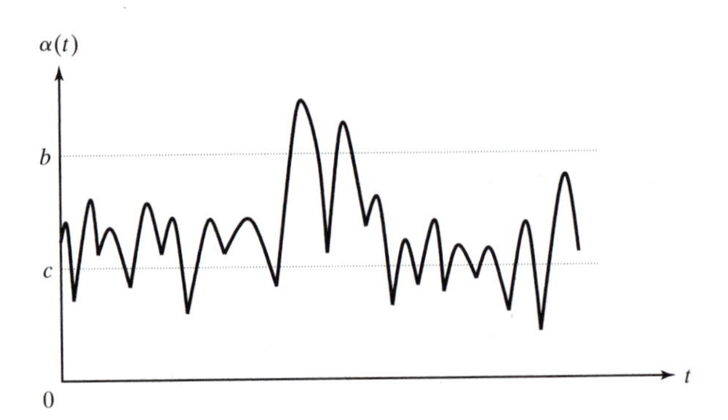

Figure 2.25 An example of amplitude fading level versus time.

observed from Figure 2.24. The maximum LCR is achieved at $\rho = -3$ dB, because the pdf of α is maximized at the threshold.

Average Fade Duration

Definition 2.4 The average fade duration at level R is the average period of time for which the channel amplitude fading level is below the specified threshold R during each fade period.

Let χ_R denote the AFD. It is a statistic closely related to the LCR. Mathematically, the AFD can be represented as

$$\chi_R = \text{E[the period that the amplitude fading level stays below the threshold } R \text{ in each upward crossing]}.$$

For the example shown in Figure 2.23, the AFD is $\sum_{i=1}^{5} t_i/5$. From the definitions of LCR and AFD, we have

$$\begin{aligned} N_R \cdot \chi_R &= \lim_{T\to\infty} \frac{M_T}{T} \cdot \frac{\sum_{i=1}^{M_T} t_i}{M_T} \\ &= \lim_{T\to\infty} \frac{\sum_{i=1}^{M_T} t_i}{T} \\ &= P(\alpha \le R). \end{aligned} \tag{2.5.12}$$

Equation (2.5.12) provides a relationship among the three statistics (i.e., LCR, AFD, and the cumulative distribution function (cdf) of the amplitude fading α). Thus, if any two statistics are known, the third one can also be determined. For the Rayleigh fading environment, the cdf of α is

$$P(\alpha \le x) = \int_0^x f_\alpha(y)\,dy = 1 - \exp\left(-\frac{x^2}{2\sigma_z^2}\right). \tag{2.5.13}$$

By Eqs. (2.5.11)–(2.5.13), the corresponding AFD is

$$\chi_R = \frac{P(A \le R)}{N_R} = \frac{1 - \exp(-R^2/2\sigma_z^2)}{\sqrt{2\pi}\,\nu_m (R/\sqrt{2}\sigma_z)\exp(-R^2/2\sigma_z^2)} = \frac{\exp(\rho^2) - 1}{\sqrt{2\pi}\,\nu_m \rho}. \tag{2.5.14}$$

The AFD is a product of two components. The first component, $1/(\sqrt{2\pi}\nu_m)$, indicates that the AFD is inversely proportional to the mobile speed and the carrier frequency.

The second term, $[\exp(\rho^2) - 1]/\rho$, depends only on the normalized threshold ρ. Figure 2.26 shows how the component changes with the normalized threshold in dB. The value of the AFD increases dramatically as the threshold ρ increases much above the rms value. This can be explained from Figure 2.25. With a large threshold value, it is very unlikely for the amplitude level α to cross the threshold. Therefore, the length of time that α stays below the threshold can be very long.

Knowledge of the AFD value helps to determine the most likely number of bits that may be lost during a fade. This is useful for relating the received signal-to-noise ratio (SNR) during a fade to the instantaneous bit error rate (BER).

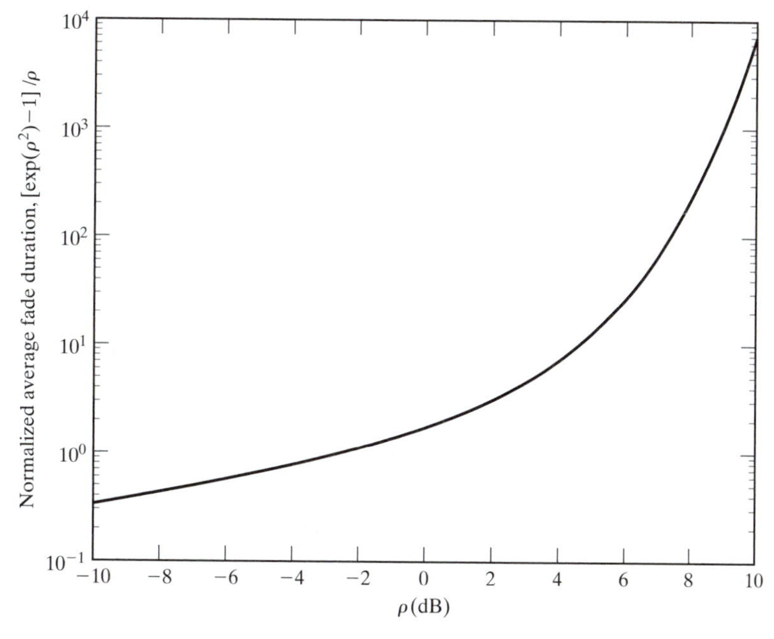

Figure 2.26 The normalized average fade duration of the flat Rayleigh fading channel.

Example 2.6 The LCR N_R

Consider a flat Rayleigh fading channel. Determine the positive-going level crossing rate for $\rho = 1$, when the maximum Doppler frequency $\nu_m = 20$ Hz. Compute the maximum velocity of the mobile if the carrier frequency is 900 MHz.

Solution

$$\begin{aligned} N_R &= \sqrt{2\pi}\nu_m \rho e^{-\rho^2}; \quad \nu_m = 20 \text{ Hz}, \ \rho = 1 \\ &= \sqrt{2\pi} \times 20 \times 1 \times e^{-1} \\ &= 18.44 \text{ crossings per second.} \end{aligned}$$

The maximum velocity is $V_m = \nu_m \cdot \frac{c}{f_c} = 20 \times \frac{3\times10^8}{9\times10^8}$ m/s = 6.66 m/s = 24 km/hr.

Example 2.7 The AFD χ_R

Suppose a flat Rayleigh fading channel exhibits a maximum Doppler frequency $\nu_m = 200$ Hz.

a. Determine the average fade duration for a normalized threshold level $\rho = 0.1$.
b. How does the result in part (a) changes when ρ is increased to 1.0?
c. How does the result in part (a) changes when ρ is reduced to 0.01?

Solution

a. At $\rho = 0.1$,

$$\chi_R = \frac{e^{(0.1)^2} - 1}{\sqrt{2\pi} \times 200 \times 0.1} = 200 \ \mu\text{s}.$$

b. Increasing the threshold level will increase the AFD value. At $\rho = 1$,

$$\chi_R = \frac{e^{1^2} - 1}{\sqrt{2\pi} \times 200 \times 1} = 3.43 \text{ ms}.$$

c. Similarly, a decrease in threshold level will decrease the AFD value. At $\rho = 0.01$,

$$\chi_R = \frac{e^{(0.01)^2} - 1}{\sqrt{2\pi} \times 200 \times 0.01} = 19.9 \ \mu\text{s}.$$

Example 2.8 The LCR N_R and AFD χ_R

Consider a mobile cellular system in which the carrier frequency is $f_c = 900$ MHz and the mobile travels at a speed of 24 km/h. Calculate the AFD and LCR at the normalized level $\rho = 0.294$.

Solution At $f_c = 900$ MHz, the wavelength is $\lambda = \frac{c}{f_c} = \frac{3\times10^8}{900\times10^6} = \frac{1}{3}$ m. The velocity of the mobile is $V = 24$ km/h $= 6.67$ m/s. The maximum Doppler frequency is $\nu_m = V/\lambda = \frac{6.67}{1/3} =$ 20 Hz. The average duration of fades below the normalized level $\rho = 0.294$ is

$$\chi_R = \frac{e^{\rho^2} - 1}{\sqrt{2\pi}\nu_m \rho} = \frac{e^{(0.294)^2} - 1}{\sqrt{2\pi} \times 20 \times 0.294} = 0.0061 \text{ s}.$$

The level crossing rate at $\rho = 0.294$ is

$$
\begin{aligned}
N_R &= \sqrt{2\pi}\, \nu_m \rho e^{-\rho^2} \\
&= \sqrt{2\pi} \times 20 \times 0.294 e^{-(0.294)^2} \\
&= 16 \text{ upcrossings/second.}
\end{aligned}
$$

SUMMARY

Unlike a guided wire, the wireless propagation channel is prone to atmospheric conditions. A consequence of this is that electromagnetic propagation through the wireless channel will suffer different degrees of impairment. To understand the channel disturbance on the transmitted signal, so as to facilitate transmitter and receiver design as a means of combating channel impairments, we have examined commonly used analytical methods for modeling the channel characteristics. The next chapter will study modulation methods commonly used in a mobile radio environment.

ENDNOTES

1. For a general discussion on channel fading and its characterization, see the paper by Sklar [140] and the book by Kennedy [71].
2. For detailed discussion of the LTV channel model, the channel functions, and correlation functions, see the paper by Bello [12].
3. For background on random variables, probability density distributions, and random processes, see the book by Papoulis [108].
4. For detailed derivation of the free-space propagation path loss, see the book by Pratt and Bostian [122].
5. For path loss models in an outdoor environment, see the papers by Okumura *et al.* [106] and Hata [60] for the Okumura–Hata model, and Chapter 2 of the book by Lee [81] for Lee's model. Other models are given in [129], [89], [43], [36], and [156].
6. For cell coverage area, see the papers by Leonardo and Yacoub [83, 84].
7. Characterization of an indoor propagation channel is reviewed in the paper by Hashemin [59]. A statistical model for indoor propagation is given in the paper by Saleh and Valenzula [133].
8. For detailed analysis of the short-term flat fading characteristics (such as Rayleigh and Rician fading), see the papers by Clarke [30] and Lee [78], Chapter 1 of the book edited by Jakes [42], and Chapters 3 and 4 of the book by Lee [82]. For computer simulation of correlated Rayleigh and Rician fading channels, see the book edited by Jakes [42] and the paper by Young and Beaulieu [162].
9. Other references on wireless propagation channels include the textbooks by Griffiths [56], Parsons [109], and Rappaport [128].

PROBLEMS

P2-1 For a linear time-invariant channel with impulse response $h(t)$, given the channel input $x(t)$, the channel output is $r(t) = x(t) \star h(t)$. Verify that this channel input–output relation is a special case of Eq. (2.2.6).

P2-2 A wireless channel is specified by the time-variant channel impulse response

$$h(\tau, t) = \left(1 - \frac{\tau}{T}\right) \cos(\Omega t + \theta_0), \quad 0 \le \tau \le T,$$

where $T = 0.05$ ms, $\Omega = 10\pi$, and $\theta_0 \in [-\pi, +\pi]$ is a constant.

a. Determine the channel time-variant transfer function.
b. Given that the channel input signal is

$$x(t) = \begin{cases} 1, & 0 \le t \le T_s \\ 0, & \text{otherwise} \end{cases},$$

where $0 < T_s < T$, determine the channel output signal.
c. For continuous digital transmission with symbol interval T_s, if the channel fading is frequency selective, specify the relation between T_s and T.

P2-3 Derive the relation given in Eq. (2.2.8).

P2-4 For the channel specified in Example 2.1 and the transmitted signal $x_1(t)$, calculate the received signal by using the relation derived in Problem 2-3.

P2-5 Verify that the channel specified in Example 2.3 is a WSSUS channel.

P2-6 Consider a fading channel with multipath intensity profile

$$\phi_h(\tau) = e^{-\tau/T}, \quad \tau \ge 0.$$

Assuming that $T = 10$ ns, determine

a. the mean propagation delay,
b. the rms delay spread, and
c. the coherence bandwidth of the channel.

P2-7 The time-variant transfer function of a WSSUS channel is given by

$$H(f, t) = \alpha(t) e^{j\theta(t)}.$$

a. Does the channel exhibit frequency-selective fading or flat fading? Explain.
b. Derive the channel impulse response $h(\tau, t)$.
c. Given that the transmitted signal is $x(t)$, derive the received signal $r(t)$ in the absence of background noise. From the relation between $x(t)$ and $r(t)$, find the mean delay and delay spread of the channel.

P2-8 Let W_s be the signal bandwidth and T_m be the multipath delay spread of the channel.

a. Discuss the conditions under which the channel exhibits
   (1) frequency selective fading,
   (2) flat fading.

b. Specify the channel conditions under which the time-variant channel transfer function is

$$H(f,t) = \alpha(t)e^{-j\theta(t)},$$

where $\alpha(t)$ and $\theta(t)$ are real-value time functions. Does the channel introduce Doppler shifts? Explain.

P2-9 The delay psd of a WSSUS channel is given by

$$\phi_h(\tau) = \begin{cases} 0.5[1+\cos(2\pi\tau/T)], & 0 \le \tau \le T/2 \\ 0, & \text{otherwise.} \end{cases}$$

a. Find the channel frequency correlation function.
b. Calculate the mean delay, the multipath delay spread, and the coherence bandwidth.
c. If $T = 0.1$ ms, determine whether the channel exhibits frequency-selective fading to the GSM systems.

P2-10 A WSSUS channel has a multipath delay spread of $T_m = 1$ s and a Doppler spread of $B_d = 0.02$ Hz. The total channel bandwidth at bandpass available for signal transmission is $W_s = 5$ Hz. To reduce the effects of intersymbol interference, the signal designer selects a symbol duration $T_s = 10$ s.

a. Determine the coherence bandwidth and the coherence time.
b. Does this channel exhibit frequency selective fading? Explain.
c. Does this channel exhibit slow or fast fading? Explain.

P2-11 Derive the relation between the time-variant transfer function $H(f,t)$ and the Doppler spread function $H(f,\nu)$.

P2-12 Given the channel input signal $x(t)$ and the channel delay-Doppler spread function $H(\tau,\nu)$, show that the channel output signal is

$$r(t) = \int_{-\infty}^{\infty}\int_{-\infty}^{\infty} x(t-\tau)H(\tau,\nu)e^{j2\pi\nu t}d\nu\, d\tau.$$

How do you interpret the channel function $H(\tau,\nu)$?

P2-13 a. For a WSSUS channel, show that the autocorrelation function of $H(\tau,\nu)$ can be represented in the form of

$$\begin{aligned}\Phi_H(\tau_1,\tau_2,\nu_1,\nu_2) &\triangleq \frac{1}{2}E[H^*(\tau_1,\nu_1)H(\tau_2,\nu_2)] \\ &= \Phi_H(\tau_1,\nu_1)\delta(\tau_1-\tau_2)\delta(\nu_1-\nu_2).\end{aligned}$$

b. Derive the relation between $\Phi_H(\tau,\nu)$ and $\phi_h(\tau,\Delta t)$, where $\phi_h(\tau,\Delta t) = \mathcal{F}_{\Delta\tau}\{\frac{1}{2}E[h^*(\tau,t)h(\tau+\Delta\tau, t+\Delta t)]\}$.
c. $\Phi_H(\tau,\nu)$ is called the channel scattering function. What does the scattering function tell you?

P2-14 Consider a WSSUS channel with scattering function

$$\Phi_H(\tau,\nu) = \Phi_1(\tau)\cdot\Phi_2(\nu),$$

where

$$\Phi_1(\tau) = \begin{cases} 1, & 0 \le \tau \le 100 \text{ ms} \\ 0, & \text{otherwise} \end{cases}$$

$$\Phi_2(\nu) = \begin{cases} \frac{1}{\nu_m}[1 - (\nu/\nu_m)^2], & 0 \le |\nu| \le \nu_m \\ 0, & \text{otherwise} \end{cases}$$

Assume $\nu_m = 10$ Hz. Find

a. the delay psd,
b. the Doppler psd,
c. the mean delay and the rms delay spread,
d. the maximum Doppler shift, the mean Doppler shift, the rms Doppler spread, and
e. the coherence bandwidth and the coherence time of the channel.

P2-15 The shadowing phenomenon can be described by a random variable ϵ(dB) which follows the Gaussian distribution with zero mean and variance σ_ϵ in dB, as given by Eq. (2.4.14). Derive the log-normal distribution, Eq. (2.4.16), for ϵ in the linear scale.

P2-16 In the propagation over smooth plane as shown in Figure 2.17, verify that $d_2 - d_1 \approx 2h_t h_r/d$, given that $d \gg h_t$ and $d \gg h_r$.

P2-17 Consider the cell coverage under the assumption of the free-space propagation model. If it is required that the path loss at the cell border should not be larger than α dB, find an expression for the cell radius.

P2-18 In practice, the cell coverage depends on the propagation environment. Consider a log-normal propagation environment. The received signal power P_r at a distance d ($> d_0$) is

$$P_{r(\text{dBW})} = P_{0(\text{dBW})} - 10\kappa \log_{10}(d/d_0) + \epsilon_{(\text{dB})},$$

where P_0 is the received power at a properly chosen reference distance d_0, κ is the path loss exponent, $\epsilon_{(\text{dB})}$ is a random variable uniformly distributed over $[-a, +a]$. The cell coverage is defined as the service area of a base station over which the signal power (from the base station) received at a mobile station is larger than ξ with a probability of a. Find an expression for evaluating the cell coverage area.

P2-19 For the Rayleigh fading environment studied in Section 2.5,

a. show that Z_c and Z_s are independent, and
b. derive the joint pdf of Z_c and Z_s as given in Eq. (2.5.2).

P2-20 The channel gain of a flat Rayleigh fading channel, $Z_c - jZ_s$, can also be represented as $\alpha \exp(-j\theta)$, where the amplitude fading $\alpha = \sqrt{Z_c^2 + Z_s^2}$ and the phase distortion $\theta = \tan^{-1}[Z_s/Z_c]$.

a. Derive the joint pdf of α and θ.
b. Show that α and θ are independent.
c. Find $E[\alpha]$ and $E[\alpha^2]$.

P2-21 Derive the pdf of the carrier phase distortion in the Rician fading, as given by Eq. (2.5.7).

P2-22 Consider a cellular system with a carrier frequency of 2,000 MHz. Suppose the user is in a vehicle moving at a speed of 60 km/hr. Assuming that the channel exhibits flat Rayleigh fading, find

a. the LCR at the normalized level of -3 dB,
b. the AFD at the normalized level of -3 dB.

P2-23 The LCR at the normalized threshold ρ for a flat Rayleigh fading channel can be represented as

$$N_R = \sqrt{2\pi}\nu_m \cdot n_R,$$

where $n_R = \rho \exp(-\rho^2)$ is the normalized LCR with respect to $\sqrt{2\pi}\nu_m$, and ν_m is the maximum Doppler shift.

a. Find the normalized threshold value ρ_0 at which the LCR n_R achieves the maximum value.
b. Explain why the LCR at ρ decreases as ρ deviates from ρ_0.

P2-24 Figure 2.27 gives a block diagram to generate N uniform samples of the complex gain, $Z_c(t) - jZ_s(t)$, of a flat Rayleigh fading channel [162]. Let T_s denote the sampling period in s, ν_m the maximum Doppler shift in Hz, and σ_z the parameter in the Rayleigh distribution. The two Gaussian random variable generators generate two independent and identically distributed (iid) random sequences $\{X_k\}|_{k=0}^{N}$ and $\{Y_k\}|_{k=0}^{N}$, each sample having zero mean and variance σ_1^2 as follows:

$$E[X_k] = E[Y_k] = 0$$
$$E[X_k^2] = E[Y_k^2] = \sigma_1^2$$
$$E[X_k X_l] = E[Y_k Y_l] = 0, \quad k \neq l$$
$$E[X_k Y_l] = 0, \quad \forall k, l$$

The sequence $\{Y_k\}$ is converted to the imaginary component by multiplication with $-j$. The summed sequence, $\{X_k - jY_k\}$, is a complex Gaussian sequence. The complex

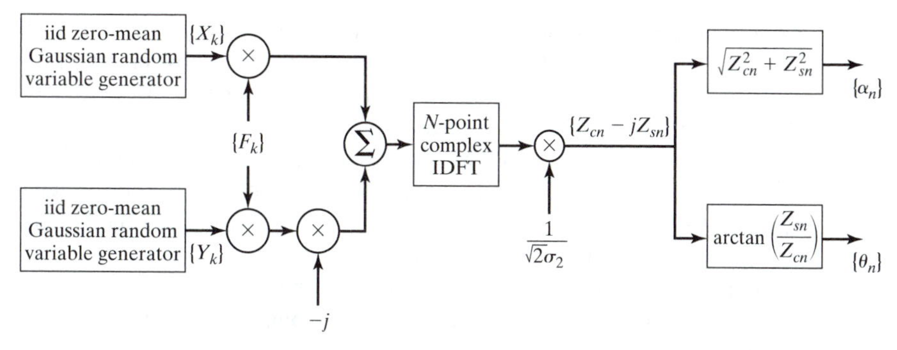

Figure 2.27 Block diagram of a Rayleigh fading channel simulator.

sequence is weighted by a sequence of filter coefficients $\{F_k\}$ given by

$$F_k = \begin{cases} 0, & k = 0 \\ \sqrt{\dfrac{1}{2\sqrt{1-\left(\dfrac{k}{Nf_m}\right)^2}}}, & k = 1, 2, \ldots, k_m - 1 \\ \sqrt{\dfrac{k_m}{2}\left[\dfrac{\pi}{2} - \arctan\left(\dfrac{k_m - 1}{\sqrt{2k_m - 1}}\right)\right]}, & k = k_m \\ 0, & k = k_m + 1, \ldots, N - k_m - 1 \\ \sqrt{\dfrac{k_m}{2}\left[\dfrac{\pi}{2} - \arctan\left(\dfrac{k_m - 1}{\sqrt{2k_m - 1}}\right)\right]}, & k = N - k_m \\ \sqrt{\dfrac{1}{2\sqrt{1-\left(\dfrac{N-k}{Nf_m}\right)^2}}}, & k = N - k_m + 1, \ldots, N-2, N-1 \end{cases},$$

where $f_m = \nu_m T_s$, and $k_m = \lfloor f_m N \rfloor$ is the largest integer less than or equal to $f_m N$. An inverse discrete Fourier transform (IDFT) is then taken of this filtered complex sequence to form the complex time sequence of length N, $\{z_{cn} - jz_{sn}\}|_{n=0}^{N}$, where $z_{cn} = z_c(nT_s)$ and $z_{sn} = z_s(nT_s)$. Both z_{cn} and z_{sn} have zero mean and the same variance σ_2^2 given by

$$\sigma_2^2 = \sigma_1^2 \sum_{k=0}^{N-1} \left(\frac{F_k}{N}\right)^2.$$

As a result, both $Z_{cn} = z_{cn}/(\sqrt{2}\sigma_2)$ and $Z_{sn} = z_{sn}/(\sqrt{2}\sigma_2)$ have zero mean and variance $1/2$. The second moment of the amplitude fading is normalized to unity so that $E[\alpha_n^2] = 1$.

a. Write a program to simulate the complex channel gain of the Rayleigh fading channel. Note that the IDFT block can be implemented by inverse fast Fourier transform (IFFT). The input parameters are the user velocity V, carrier frequency f_c, sampling interval T_s, and sequence length N.

b. Given $V = 50$ km/hr, $f_c = 900$ MHz, $T_s = 1$ ms, and $N = 1000$, plot the empirical pdfs for the amplitude fading sequence $\{\alpha_n\}$ and phase distortion sequence $\{\theta_n\}$, where $\alpha_n = \sqrt{Z_{cn}^2 + Z_{sn}^2}$ and $\theta_n = \arctan(Z_{sn}/Z_{cn})$. Compare the empirical pdfs with the theoretical ones, respectively.

c. Using the same input parameters as in (*b*), find the empirical LCR and AFD values for the amplitude fading sequence $\{\alpha_n\}$ and compare them with the theoretical values.

3

Bandpass Transmission Techniques for Mobile Radio

This chapter is concerned with methods to mitigate the channel impairments identified and characterized in Chapter 2. The focus is on signal representation with built-in noise immunity, and receiver design to compensate for channel disturbance. With properly designed transmitted signals and corresponding receivers, the effective channel disturbance is then characterized as Rayleigh fading and additive white Gaussian noise (AWGN). The design of post-detectors at the receiving end is then directed at reducing the probability of transmission error in the presence of Rayleigh fading and AWGN.

3.1 INTRODUCTION

Elements of a Digital Communications System. In a digital communications system, the source to be transmitted is discrete, both in time and in amplitude. Unless specifically generated, information sources (e.g., voice, video, etc.) are analog in nature. An analog-to-digital (A/D) converter is needed to convert the analog source to a digital representation. At the receiving end, a digital-to-analog (D/A) converter is used to recover the analog waveform before presentation to the end user. In the most basic form, digital sequences are represented in a binary format, with symbols drawn from the set $\{0, 1\}$.

A generic functional block diagram of a binary digital communications system is shown in Figure 3.1. Sources normally contain redundant information. From a bandwidth conservation

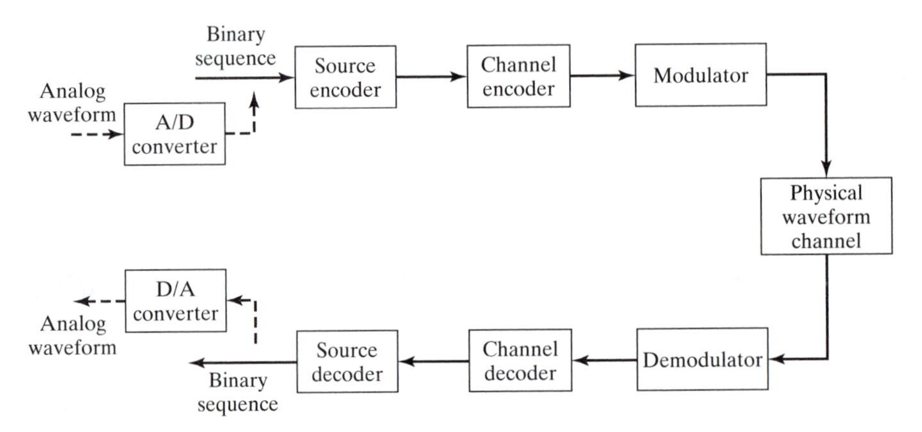

Figure 3.1 Functional block diagram of a binary digital communications system.

point of view, redundant information should be removed prior to transmission. Moreover, the redundant information is normally unknown and cannot be used to combat channel noise. This is the reason the digital communications system shown in Figure 3.1 comprises a source encoder, channel encoder, and digital modulator at the transmitting end and a digital demodulator, channel decoder, and source decoder in tandem at the receiving end. The source encoder removes the unknown redundant information while the channel encoder inserts a controlled amount of known redundancy prior to modulation and transmission. The reason for having both a source encoder and a channel encoder is that the source encoded sequence, lacking redundancy, is very susceptible to channel noise. Channel coding, which introduces known redundancy that can be used at the receiver for decoding, endows the information sequence with noise immunity.

This text is not concerned with source coding or channel coding. As a noise suppression mechanism, channel coding is incapable of combating disturbances due, for example, to frequency selective fading or multipath propagation. Signal design and receiver optimization together have the ability to combat fading and multipath effects. Modulation is an integral component in signal design, and demodulation is the front-end processing in an optimized receiver. To combat channel impairments that cannot be tackled by channel coding, this chapter focuses on modulator/demodulator design, while post processing in receiver optimization is described in the next chapter. The effective communication channel seen by the channel encoder/decoder pair is often referred to as the *coding channel*. The residual disturbance in the *coding channel* should be of the variety that channel coding is capable of mitigating. Thus, properly designed signals and optimized receivers are central to mitigation of channel impairments in the radio propagation.

Bandpass Transmission. Communication channels are analog systems by nature. A signal carrying digital information must first be converted to an analog waveform prior to transmission. At the receiving end, analog signal is converted back to a digital format before presentation to the end user. The conversion process at the transmitting end is known as modulation, and that at the receiving end is demodulation or detection. For wireless communications, the criteria commonly used to evaluate the suitability of a modulation scheme include: (*a*) compact power spectral density for high radio spectral utilization efficiency and for low adjacent channel interference, especially in narrowband systems, (*b*) good transmission performance (i.e., low transmission error rate for a given transmitted power and channel condition), and (*c*) small (or no) envelope fluctuations after bandpass filtering so that the use of a power efficient nonlinear amplifier does not introduce much nonlinear distortion into the transmitted signal, which is very important for battery-powered handsets.

The source information is normally represented as a baseband (low-pass) signal. Because of signal attenuation, it is necessary to move the baseband signal spectrum to reside at a much higher frequency band centered at f_c, called the carrier frequency, in the radio spectrum. In other words, modulation injects a carrier frequency onto the baseband signal. At the receiving end, the demodulation process removes the carrier frequency to recover the baseband information signal. The frequency conversion is required to facilitate efficient radiation and to circumvent severe signal attenuation, as the dimension of both transmitter and receiver antennas is in the order of the wavelength ($\lambda = c/f_c$, where $c \approx 3 \times 10^8$ m/s is the velocity of electromagnetic waves) of the transmitted signal. Furthermore, by choosing different carrier frequencies for different signals to be transmitted, the modulation/demodulation process facilitates channel assignment and reduces interference from other transmissions.

Let $v(t)$ represent the baseband signal waveform carrying the information to be sent. In general, the baseband signal $v(t)$ is complex, and can be represented mathematically as

$$v(t) = a(t) \exp[j\zeta(t)],$$

where $a(t)$ is the amplitude function and $\zeta(t)$ is the phase function. The maximum frequency, f_m, of $v(t)$ is an approximate measure of its bandwidth. The Fourier transform of $v(t)$ is given by

$$V(f) = \int_{-\infty}^{\infty} v(t) \exp(-j2\pi f t)\, dt.$$

$V(f)$ is a Hermitian function of frequency, so that the amplitude spectrum is an even function of frequency and the phase spectrum is an odd function of frequency. An amplitude spectrum is illustrated in Figure 3.2.

A functional block diagram of a generic modulation procedure for signal $v(t)$ is depicted in Figure 3.3. With the modulating signal $v(t)$ being a complex-valued function, the carrier signal is in the form of an exponential, $A_c \exp(j2\pi f_c t)$, where A_c is a constant denoting the amplitude of

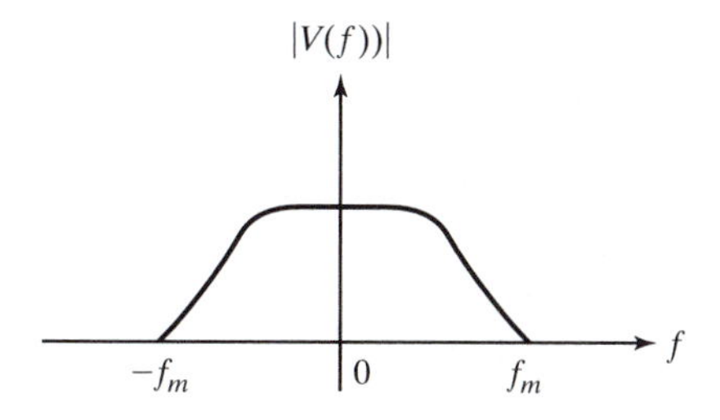

Figure 3.2 Illustration of amplitude spectrum.

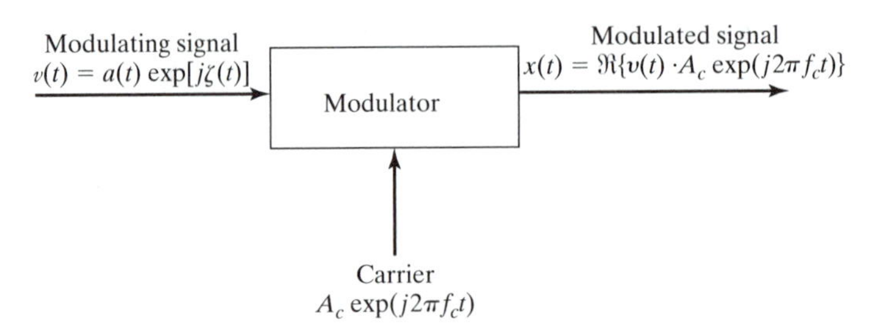

Figure 3.3 Generic modulation procedure.

the carrier. The transmitted signal must be real. Therefore, the modulated signal, $x(t)$, is given by

$$\begin{aligned} x(t) &= \Re\{v(t)A_c \exp(j2\pi f_c t)\} \\ &= A_c a(t)\cos(2\pi f_c t + \zeta(t)) \\ &= A_c a(t)\cos\zeta(t)\cos(2\pi f_c t) - A_c a(t)\sin\zeta(t)\sin(2\pi f_c t). \end{aligned}$$

The modulation can be classified into two categories: linear modulation and nonlinear modulation. A modulation process is *linear* when both $a(t)\cos\zeta(t)$ and $a(t)\sin\zeta(t)$ terms are linearly related to the message information signal. Examples of linear modulation include

a. amplitude modulation, where the modulating signal affects only the amplitude of the modulated signal (i.e., when $\zeta(t)$ is a constant $\forall t$ (for any t)), and
b. phase modulation (with a rectangular phase shaping function) where the modulating signal affects only the phase of the modulated signal (i.e., when $\zeta(t)$ is a constant over each signaling (symbol) interval and $a(t)$ is a constant $\forall t$).

The modulation process is *nonlinear* when the modulating signal, $v(t)$, affects the frequency of the modulated signal. The definition of nonlinear is that superposition does not apply. The modulation process is nonlinear whether or not the amplitude of the modulating signal is a function of time. Consider a frequency modulation process and let $a(t) = a \forall t$. Then, the nonlinearly modulated signal is

$$x(t) = a \cdot A_c \cos(2\pi f_c t + \zeta(t))$$

where the angle, $\zeta(t)$, is the integral of a frequency function.

Example 3.1 Binary Digital Modulation Schemes

Suppose the information to be transmitted is a binary sequence "**10111001**". The binary sequence is to be transmitted at a rate of R_s[1] symbols per second (i.e., the transmission symbol interval for each binary digit is $T_s = 1/R_s$ seconds).

a. *Binary amplitude shift keying* (ASK): This is linear modulation with $\zeta(t) = 0$. The information is carried by the amplitude component $a(t)$: $a(t) = 1$ for symbol "1" and $a(t) = 0$

[1] For binary signaling, the symbol rate R_s equals the bit rate R_b.

for symbol "0". The modulated signal is

$$x(t) = \begin{cases} A_c \cos(2\pi f_c t), & \text{symbol "1"} \\ 0, & \text{symbol "0"} \end{cases}.$$

The binary ASK signaling process is illustrated in Figure 3.4(a).

b. *Binary phase shift keying* (PSK): This is linear modulation with $a(t) = 1$. The information is carried by the phase component $\zeta(t)$: $\zeta(t) = 0$ for symbol "1" and $\zeta(t) = \pi$ for symbol "0". The modulated signal is

$$x(t) = \begin{cases} A_c \cos(2\pi f_c t), & \text{symbol "1"} \\ A_c \cos(2\pi f_c t + \pi) = -A_c \cos(2\pi f_c t), & \text{symbol "0"} \end{cases}.$$

The binary PSK signaling process is illustrated in Figure 3.4(b). In the preceding equation for $x(t)$, it is readily observed that the waveforms corresponding to the symbols "0" and "1" are antipodal. In this regard, binary PSK is equivalent to double sideband suppressed carrier linear modulation.

c. *Binary frequency shift keying* (FSK): This is nonlinear modulation with $a(t) = 1$. The information is carried by the phase component $\zeta(t)$: $\zeta(t) = 2\pi(\Delta/2)t + \phi_1$ for symbol "1" and $\zeta(t) = -2\pi(\Delta/2)t + \phi_2$ for symbol "0", where Δ is the frequency separation between

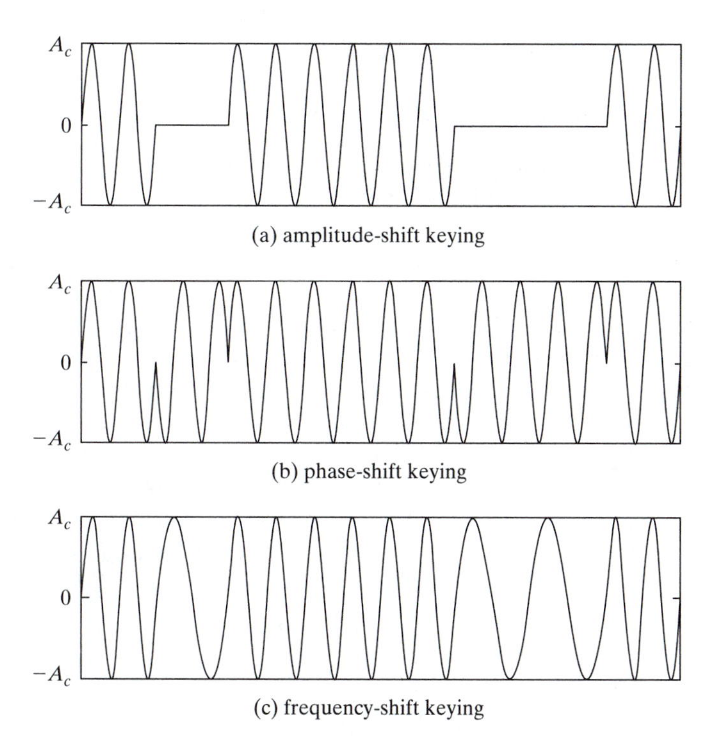

Figure 3.4 Modulated signal waveforms for binary sequence "10111001".

the signals for symbols "1" and "0" respectively, ϕ_1 and ϕ_2 are any constants in $[-\pi, +\pi]$. The modulated signal is

$$x(t) = \begin{cases} A_c \cos[2\pi(f_c + 0.5\Delta)t + \phi_1], & \text{symbol "1"} \\ A_c \cos[2\pi(f_c - 0.5\Delta)t + \phi_2], & \text{symbol "0"} \end{cases}.$$

The binary FSK signaling process is illustrated in Figure 3.4(c) for $\phi_1 = \phi_2 = 0$. In this modulation, the instantaneous frequency of the modulated signal $x(t)$ is $f(t) = f_c + 0.5\Delta$ for symbol "1" and $f(t) = f_c - 0.5\Delta$ for symbol "0". The instantaneous phase of $x(t)$ is obtained as the integral of the instantaneous frequency, and is given by

$$\phi(t) = \int_{-\infty}^{t} 2\pi f(\tau)\, d\tau.$$

This chapter is organized as follows. In Section 3.2, we first study the concept of the signal space and decision regions when using a matched filter to detect the transmitted information symbol. In Section 3.3, we then study various digital modulation schemes, including those used in the second generation wireless systems. In Sections 3.4 and 3.5, we analyze the power spectral density and transmission accuracy, respectively.

3.2 SIGNAL SPACE AND DECISION REGIONS

3.2.1 Vector-Space Representation of M-ary Signals

To understand digital modulation, it is important to know that a signal can be equivalently represented both in the time domain and in the signal space domain.

The digital source generates digital symbols for transmission at a rate of R_s symbols per second. The symbols are taken from an alphabet of size M. Normally, $M = 2^l$, where l is a positive integer. Each symbol can be represented by l binary digits. The transmission rate can be equivalently represented by $R_b = lR_s$ bits per second (bps), where R_b is the bit rate.

Consider the transmission of a piece of information represented by symbols from an alphabet of size M, $\{s_m\}_{m=1}^{M}$. Each symbol is represented by a unique baseband waveform. The information is transmitted at a symbol rate R_s. That is, over each time interval of $T_s = 1/R_s$, one of the M waveforms, $\{v_m(t)\}_{m=1}^{M}$, is selected for transmission. Each of the waveforms, $v_m(t)$, can be represented as a point in an N-dimensional signal space defined by orthonormal basis functions $\{\tilde{\varphi}_n(t)\}_{n=1}^{N}$ over the interval $[0, T_s]$. The constraint on the dimension of the signal space is $N \leq M$.

If $\tilde{\varphi}_i(t)$ and $\tilde{\varphi}_j(t)$ are members of an orthonormal set over $t \in [0, T_s]$, they must satisfy the orthonormal condition

$$\int_0^{T_s} \tilde{\varphi}_i(t)\tilde{\varphi}_j^*(t)\, dt = \delta(i - j),$$

where

$$\delta(i - j) = \begin{cases} 1, & \text{if } i = j \\ 0 & \text{otherwise} \end{cases} \tag{3.2.1}$$

is the Kronecker delta function. $v_m(t)$, $m = 1, 2, \cdots, M$, can be expressed as

$$v_m(t) = \sum_{n=1}^{N} v_{mn}\tilde{\varphi}_n(t),$$

where

$$v_{mn} = \int_0^{T_s} v_m(t)\tilde{\varphi}_n^*(t)\, dt.$$

In the signal space defined by the orthonormal set $\{\tilde{\varphi}_n(t)\}_{n=1}^N$, $v_m(t)$ can be represented by the vector $\vec{v}_m = (v_{m1}, v_{m2}, \cdots, v_{mN})$, $m = 1, \cdots, M$. The orthonormal functions $\{\tilde{\varphi}_n(t)\}_{n=1}^N$ can be constructed from the waveforms $\{v_m(t)\}_{m=1}^M$ (using the Gram–Schmidt orthogonalization procedure described in Appendix A). The corresponding bandpass waveforms $\{x_m(t)\}_{m=1}^M$, $m = 1, \cdots, M$, can also be represented as

$$x_m(t) = \sum_{n=1}^{N} x_{mn}\varphi_n(t), \tag{3.2.2}$$

where

$$x_{mn} = \int_0^{T_s} x_m(t)\varphi_n(t)\, dt \tag{3.2.3}$$

and $\{\varphi_n(t)\}_{n=1}^N$ is a real-valued orthonormal set. In the signal space defined by the orthonormal set $\{\varphi_n(t)\}_{n=1}^N$, the signal $x_m(t)$ can be represented by the vector

$$\vec{x}_m = (x_{m1}, x_{m2}, \cdots, x_{mn}).$$

By Rayleigh's energy theorem, the energy of the bandpass signal $x_m(t)$ is

$$E_m = \int_0^{T_s} x_m^2(t)\, dt = \sum_{n=1}^{N} x_{mn}^2 \triangleq ||\vec{x}_m||^2$$

where $||\cdot||$ denotes the norm of the vector. The relation between two waveforms, $x_m(t)$ and $x_k(t)$, $m, k \in \{1, 2, \ldots, M\}$, are represented by their correlation coefficient ρ_{km}, as defined in the following.

Definition 3.1 The correlation coefficient between real-valued signals $x_m(t)$ and $x_k(t)$ is

$$\rho_{km} \triangleq \frac{1}{\sqrt{E_k E_m}} \int_0^{T_s} x_m(t)x_k(t)\, dt = \frac{\vec{x}_m \cdot \vec{x}_k}{||\vec{x}_m|| \cdot ||\vec{x}_k||}.$$

If $x_k(t) = \alpha x_m(t)$, then $\rho_{km} = 1$ for $\alpha > 0$ and $\rho_{km} = -1$ for $\alpha < 0$; if $x_k(t)$ and $x_m(t)$ are orthogonal, then $\rho_{km} = 0$.

The squared Euclidean distance between $x_k(t)$ and $x_m(t)$ is defined as

$$d_{km}^2 = ||\vec{x}_m - \vec{x}_k||^2 = E_m + E_k - 2\sqrt{E_m E_k}\rho_{km}.$$

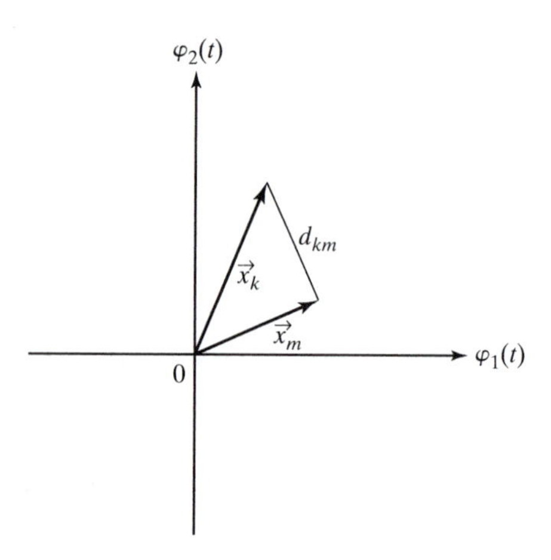

Figure 3.5 Signal constellation for $N = 2$.

If $E_m = E_s \forall m$, then

$$d_{km}^2 = 2E_s(1 - \rho_{km}).$$

The signal constellation for $N = 2$, depicting the Euclidean distance d_{km}, is shown in Figure 3.5.

3.2.2 Signal Detection and Optimal Receiver

For an M-ary signal represented in an N-dimensional signal space, the observation signal space of the received signal can be divided into M decision regions. If the interference or noise causes the received signal vector to lie outside its decision region, an error in the signal detection at the receiver will occur. The signal space representation lends itself conveniently to decision making.

Detection Problem. Consider an additive white Gaussian noise (AWGN) channel, where the received signal is the transmitted signal plus a white Gaussian noise component with zero mean and two-sided power spectral density $N_0/2$ watts/Hz. Figure 3.6 shows the block diagram of the channel and receiver, where the M-ary transmitted signal $x(t) \in \{x_m(t)\}_{m=1}^{M}$ for $t \in [0, T_s]$ is represented in an N-dimensional signal space specified by the orthonormal set $\{\varphi_n(t)\}_{n=1}^{N}$. The

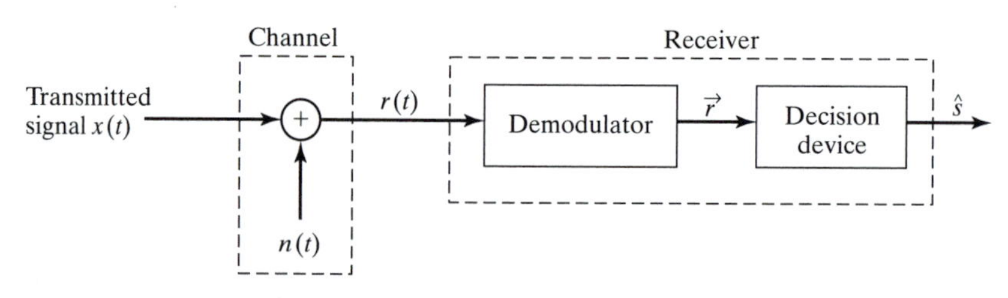

Figure 3.6 Signal detection in AWGN.

received signal is

$$r(t) = x(t) + n(t), \tag{3.2.4}$$

where $n(t)$ is the white Gaussian noise introduced by the channel. In the N-dimensional signal space, the received signal can be represented by a vector $\vec{r} = (r_1, r_2, \ldots, r_N)$, where

$$r_n = \int_0^{T_s} r(t)\varphi_n(t)\,dt, \quad n = 1, 2, \ldots, N. \tag{3.2.5}$$

Once the signal space is defined, $\vec{r}$ and $r(t)$ are equivalent; knowing one will yield the other. This means that $\vec{r}$ contains all the information of $r(t)$. On the other hand, it is much easier to deal with $\vec{r}$ than with $r(t)$ as $\vec{r}$ is not a function of time. In general, in an uncoded system, as shown in Figure 3.6, the receiver can be divided into two components: one is the demodulator (also called detector) which maps the received signal $r(t)$ to the received vector $\vec{r}$, and the other is the decision device which estimates the information carried in the transmitted signal $x(t)$ based on the received vector $\vec{r}$.

Matched Filter. Based on Eq. (3.2.5), Figure 3.7(a) shows the structure of the demodulator. It consists of N parallel correlators, each correlating the received signal $r(t)$ with a unique function $\varphi_n(t)$ from the orthonormal set over the symbol interval. As the multipliers and integrators in the correlators are expensive to implement in practice, each correlator can also be implemented by using a matched filter and sampling the filter output at the end of the symbol interval, as shown in Figure 3.7(b). The impulse response of the matched filter at the nth branch is defined as

$$h_n(t) = \varphi_n(T_s - t).$$

The output of the matched filter is $y_n(t) = r(t) \star h_n(t)$. At $t = T_s$, the output is

$$\begin{aligned} y_n(T_s) &= \int_{-\infty}^{\infty} r(\tau)h_n(T_s - \tau)\,d\tau \\ &= \int_{-\infty}^{\infty} r(\tau)\varphi_n(\tau)\,d\tau \\ &= \int_0^{T_s} r(\tau)\varphi_n(\tau)\,d\tau \\ &= r_n, \end{aligned}$$

which is exactly the output of the nth correlator in Figure 3.7(a). That is, the filter extracts the signal component in the input $r(t)$, where the component is proportional to $\varphi_n(t)$. To do so, the impulse response of the filter is matched to the waveform $\varphi_n(t)$, hence the name *matched filter*.

Decision Regions. Given the demodulator output $\vec{r}$, the decision device is to perform a mapping from $\vec{r}$ to an estimate $\hat{s}$ of the transmitted symbol in a way that will minimize the probability of error in the decision making process. As shown in Appendix B, in the case that all transmitted

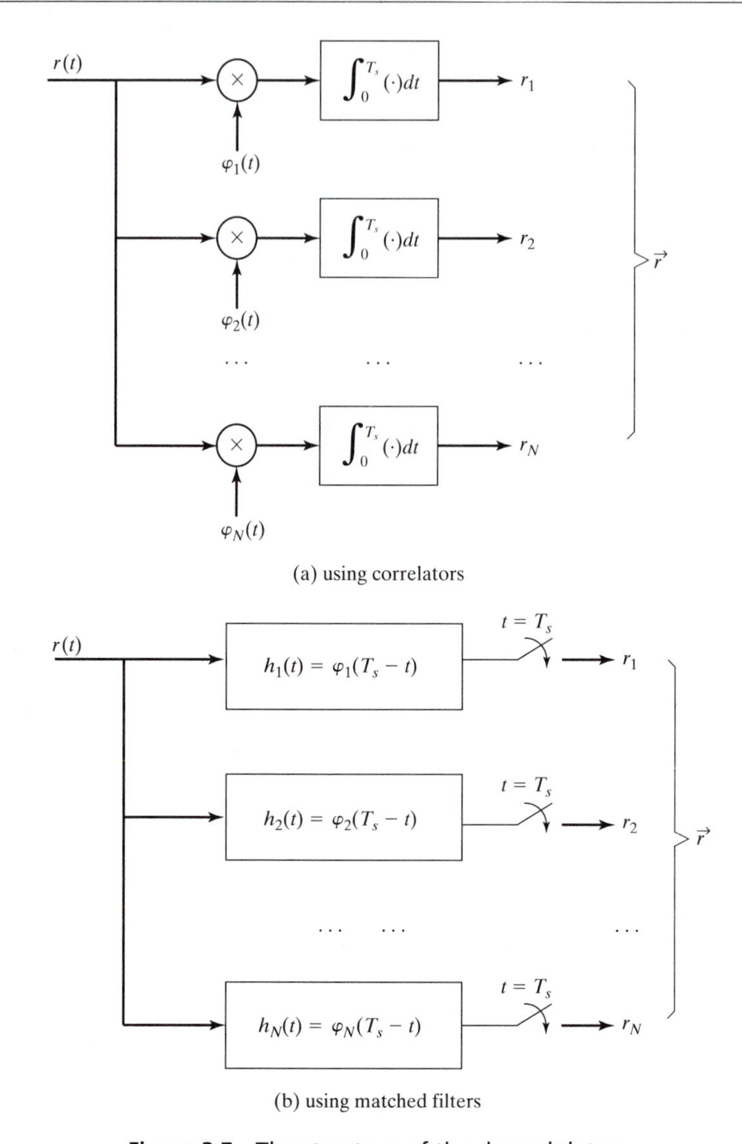

Figure 3.7 The structure of the demodulator.

symbols, $\{s_m\}_{m=1}^{M}$, are equally likely, the following *maximum likelihood (ML) decision rule* is an optimum decision rule for an AWGN channel.

ML Decision Rule: Set $\hat{s} = s_m$ if $||\vec{r} - \vec{x}_m|| \leq ||\vec{r} - \vec{x}_k||$ for all $k = 1, 2, \ldots, M$ and $k \neq m$.

In other words, the decision rule is to choose the message point closest to the received signal point, which is intuitively satisfying. For an N-dimensional observation space, the optimum receiver is simply a partitioning of the signal space into regions, the point of which is closest to the corresponding transmitted signal vector $\vec{x}_m$. These regions are called decision regions. That is, the observation space D is partitioned into subspaces $D_1, D_2, \ldots, D_M$ corresponding to the message symbols $s_1, s_2, \ldots, s_M$, respectively. Each subspace D_m is a decision region. If $\vec{r} \in D_m$, then set $\hat{s} = s_m$. The decision regions are partitioned in this way because the channel noise component is zero-mean Gaussian which is more likely to take a smaller value than a larger value.

The receiver consisting of the matched-filter demodulator and the decision device using the ML decision rule is commonly referred to as matched filter receiver. The receiver is optimum in the sense that it minimizes the probability of transmission error in an AWGN channel.

Example 3.2 Matched Filter

The transmitted signal is given by

$$g(t) = e^{-\alpha t}, \quad 0 \le t \le T$$

where T is the symbol interval. The signal is corrupted by additive white Gaussian noise with zero mean and two-sided power spectral density $N_0/2$ watts/Hz. The optimum detector is the matched filter.

(*a*) Find and sketch the impulse response of the filter which is matched to the signal $g(t)$.
(*b*) Derive the output of the filter and determine its maximum value.

Solution

a. The impulse response of the filter matched to $g(t)$ is

$$h(t) = g(T - t) = \begin{cases} e^{-\alpha(T-t)}, & 0 \le t \le T \\ 0, & \text{otherwise} \end{cases},$$

which is shown in Figure 3.8.

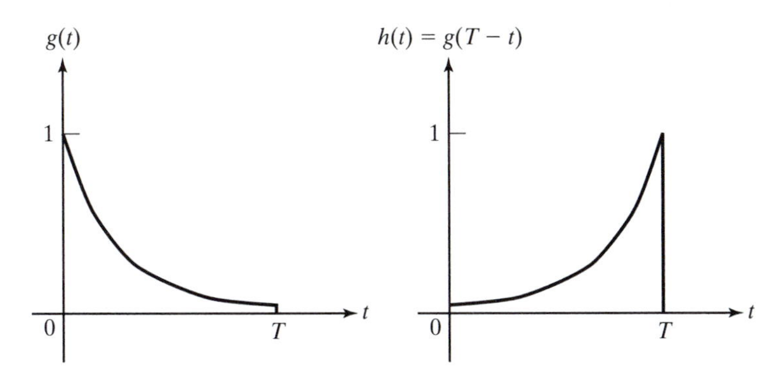

Figure 3.8 The input signal and impulse response of the matched filter.

b. The output of the matched filter in the absence of noise is

$$
\begin{aligned}
y(t) &= g(t) \star h(t) \\
&= \int_{-\infty}^{\infty} g(\tau) h(t-\tau)\, d\tau \\
&= \begin{cases} \int_0^t e^{-\alpha\tau} e^{-\alpha[T-(t-\tau)]}\, d\tau, & t \in [0, T] \\ \int_{t-T}^{T} e^{-\alpha\tau} e^{-\alpha[T-(t-\tau)]}\, d\tau, & t \in [T, 2T] \\ 0, & \text{elsewhere} \end{cases} \\
&= \begin{cases} \frac{1}{2\alpha}[e^{-\alpha(T-t)} - e^{-\alpha(t+T)}], & t \in [0, T] \\ \frac{1}{2\alpha}[e^{-\alpha(t-T)} - e^{-\alpha(3T-t)}], & t \in [T, 2T]. \\ 0, & \text{elsewhere} \end{cases}
\end{aligned}
$$

Figure 3.9 plots $y(t)$ for $\alpha = 2$ and $T = 5$. As shown in Figure 3.9, $y(t)$ achieves the maximum value at $t = T$

$$y(T) = \frac{1}{2\alpha}[1 - e^{-2\alpha T}].$$

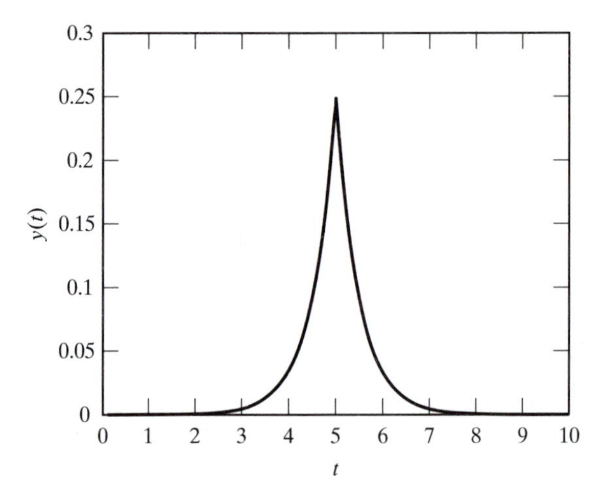

Figure 3.9 The output of the matched filter for $\alpha = 2$ and $T = 5$.

Example 3.3 Decision Region of Binary PSK (BPSK) Signals

As given in Example 3.1, the two BPSK signals, $x_m(t)$, $m = 1, 2$, are

$$x_1(t) = \sqrt{\frac{2E_b}{T_b}} \cos(2\pi f_c t), \quad 0 \le t \le T_b$$

and

$$x_2(t) = -\sqrt{\frac{2E_b}{T_b}} \cos(2\pi f_c t), \quad 0 \le t \le T_b,$$

where $E_b = \int_{-\infty}^{\infty} x_m^2(t)\, dt = A_c^2 T_b/2$ is the bit energy and T_b is the bit duration. The correlation coefficient between the two signals is $\rho_{12} = -1$. Binary waveforms that are negative of each other, as in BPSK where $x_2(t) = -x_1(t)$, are called antipodal signals. Only one basis function, $\varphi_1(t)$, is needed to represent the BPSK signals. We can choose

$$\varphi_1(t) = \begin{cases} \sqrt{\frac{2}{T_b}} \cos(2\pi f_c t), & 0 \le t \le T_b \\ 0, & \text{otherwise} \end{cases} \tag{3.2.6}$$

so that the BPSK signals can be represented as

$$x_1(t) = \sqrt{E_b}\varphi_1(t), \quad 0 \le t \le T_b$$

and

$$x_2(t) = -\sqrt{E_b}\varphi_1(t), \quad 0 \le t \le T_b.$$

The signal constellation of BPSK is shown in Figure 3.10. In the one-dimensional signal space, the optimum decision regions for the equally likely symbols are $D_1 = (0, \infty)$ for symbol "1" and $D_2 = (-\infty, 0)$ for symbol "0". The boundary between the two decision regions is the point with coordinate 0. The decision rule is

$$\text{set } \hat{s} = \text{"1"}, \quad \text{if } \vec{r} \in D_1,$$
$$\hat{s} = \text{"0"}, \quad \text{if } \vec{r} \in D_2.$$

At the boundary (i.e., at $\vec{r} = 0$), set $\hat{s}$ to be "1" or "0" arbitrarily. A detection error occurs if $\vec{r} \in D_2$ when $x_1(t)$ was sent or if $\vec{r} \in D_1$ when $x_2(t)$ was sent.

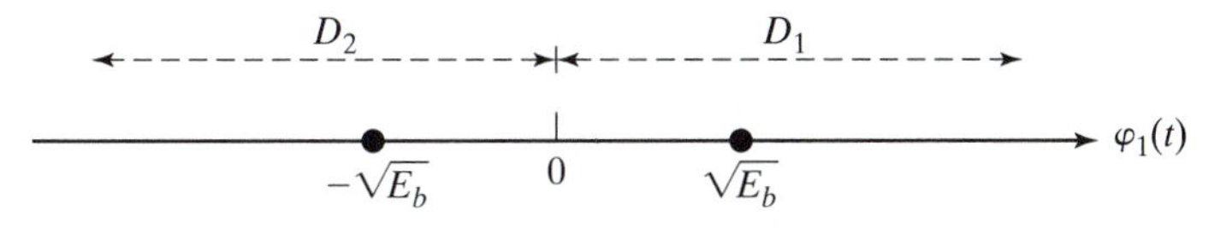

Figure 3.10 BPSK signal constellation and the decision regions.

3.3 DIGITAL MODULATION

The most commonly used digital modulation techniques are M-ary phase shift keying (MPSK) such as differential quadrature phase shift keying (DQPSK), and minimum shift keying (MSK). In the second generation cellular systems, North American IS-54 and Japanese PDC use $\pi/4$ shifted DQPSK ($\pi/4$-DQPSK), European Global System for Mobile Communications (GSM) uses Gaussian filtered MSK (GMSK), and Motorola Integrated Radio System (MIRS) uses orthogonal frequency division multiplexing (OFDM).

3.3.1 M-ary Phase Shift Keying (MPSK)

In M-ary signaling, the source symbols, s_m, can be thought of as taking on integer values (i.e., $s_m \in [\pm 1, \pm 3, \cdots, \pm(M-1)]$). There will thus be M possible waveforms. During a signaling interval T_s, one of the waveforms is selected for transmission. The mth waveform is represented as

$$x_m(t) = \sqrt{\frac{2E_s}{T_s}} \cos\left(2\pi f_c t + \frac{2\pi(m-1)}{M} + \zeta_0\right), \quad 0 \le t \le T_s, \quad m = 1, 2, \cdots, M,$$

where $\zeta_0 = 0$ or π/M, $E_s = \int_{-\infty}^{\infty} x_m^2(t)\, dt$ is the symbol energy, and T_s is the symbol duration. As each waveform represents l binary digits, we have $E_s = lE_b$ and $T_s = lT_b$, where E_b and T_b are the bit energy and bit duration, respectively. The most commonly used are $M = 2$ with $\zeta_0 = 0$ (BPSK) as studied in Example 3.3, and $M = 4$ (QPSK) with $\zeta_0 = 0$ or $\pi/4$.

Quadrature Phase Shift Keying (QPSK). With QPSK signaling, the symbols can take on four possible values, each corresponding to two bits. For $\zeta_0 = \pi/4$, the modulated signal can be represented as

$$\begin{aligned} x_m(t) = &\sqrt{\frac{2E_s}{T_s}} \cos\left[(2m-1)\frac{\pi}{4}\right] \cos(2\pi f_c t) \\ &- \sqrt{\frac{2E_s}{T_s}} \sin\left[(2m-1)\frac{\pi}{4}\right] \sin(2\pi f_c t), \quad 0 \le t \le T_s, \quad m = 1, 2, 3, 4. \end{aligned}$$

Letting

$$\begin{cases} \varphi_1(t) = \sqrt{\dfrac{2}{T_s}} \cos(2\pi f_c t), & 0 \le t \le T_s \\ \varphi_2(t) = \sqrt{\dfrac{2}{T_s}} \sin(2\pi f_c t), & 0 \le t \le T_s \end{cases},$$

we have

$$x_m(t) = \sqrt{E_s} \cos\left[(2m-1)\frac{\pi}{4}\right] \varphi_1(t) - \sqrt{E_s} \sin\left[(2m-1)\frac{\pi}{4}\right] \varphi_2(t).$$

In the two-dimensional signal space defined by the orthonormal set $\{\varphi_1(t), \varphi_2(t)\}$, the QPSK signal $x_m(t)$ can be represented by a point

$$\vec{x}_m = \left(\sqrt{E_s}\cos\left[(2m-1)\frac{\pi}{4}\right], -\sqrt{E_s}\sin\left[(2m-1)\frac{\pi}{4}\right]\right).$$

The signal constellation of QPSK for $\zeta_0 = \pi/4$ is shown in Figure 3.11(a), where $E_b = E_s/2$. The boundaries of the decision regions are the $\varphi_1(t)$ and $\varphi_2(t)$ axes. That is, the optimum decision regions for the equally likely symbols are just the four quadrants. The decision rule is:

$$\begin{aligned} \text{set } \hat{s} &= \text{“00”}, \quad \text{if } \vec{r} \in D_1; \\ \hat{s} &= \text{“01”}, \quad \text{if } \vec{r} \in D_2; \\ \hat{s} &= \text{“11”}, \quad \text{if } \vec{r} \in D_3; \\ \hat{s} &= \text{“10”}, \quad \text{if } \vec{r} \in D_4. \end{aligned}$$

For example, if the received signal vector ($\vec{r}_1$) corresponding to the sent signal vector ($\vec{x}_1$) falls in D_4, then a detection error occurs as the received signal is decoded as $\vec{x}_4$ being sent. In QPSK, each symbol represents two bits; a symbol detection error may correspond to a one-bit error or a two-bit error. In an AWGN channel, a small noise component is more likely to happen than a large noise component. As a result, when a detection error happens, it is more likely that the symbol was detected as one of the two adjacent symbols in the signal constellation. In order to minimize transmission bit error rate, the binary digits are assigned to the symbols in such a way that the adjacent symbols differ by only one binary digit. This way of mapping binary digits to the symbols is called *Gray encoding*. Figure 3.11(a) also shows an example of Gray encoding for the QPSK signals. Figure 3.11(b) shows the signal constellation, decision regions, and Gray mapping for QPSK with $\zeta_0 = 0$.

The generic QPSK transmitter and receiver are shown in Figure 3.12. At the transmitter, the input to the QPSK transmitter is a non-return-to-zero (NRZ) binary data stream $\{a_n\}$ at the bit rate R_b. The input data stream is converted to an in-phase waveform $I(t)$ and a quadrature waveform $Q(t)$, each at the symbol rate $R_s = R_b/2$. Based on the QPSK signal constellation used (as shown in Figure 3.11), at each symbol interval, the functions $I(t)$ and $Q(t)$ take on the values of the signal point projections on the $\varphi_1(t)$ and $\varphi_2(t)$ axes, respectively. The in-phase and quadrature components are multiplied with their corresponding carriers, and then combined to produce the QPSK signal. At the receiver, the received signal is the transmitted QPSK signal plus a white Gaussian noise component for an AWGN channel. For coherent detection, it is necessary to use the carrier recovery circuit to recover the in-phase and quadrature carriers from the received signal. The product modulators (i.e., the multipliers) and matched filters (matching to the baseband waveform) at the in-phase and quadrature branches retrieve the noisy in-phase and quadrature baseband signal components, respectively. Depending on the decision region in which the baseband components fall, the decision device(s) makes a decision on which 2-bit symbol was sent. For QPSK with $\zeta = \pi/4$, given the in-phase and quadrature baseband components, the detection on the even numbered binary digit is uncorrelated to the detection on the odd numbered digit, as indicated by the decision regions. As a result, in Figure 3.12(b), each of the in-phase and quadrature branches has its own decision device for detecting the bit in the QPSK symbol. The

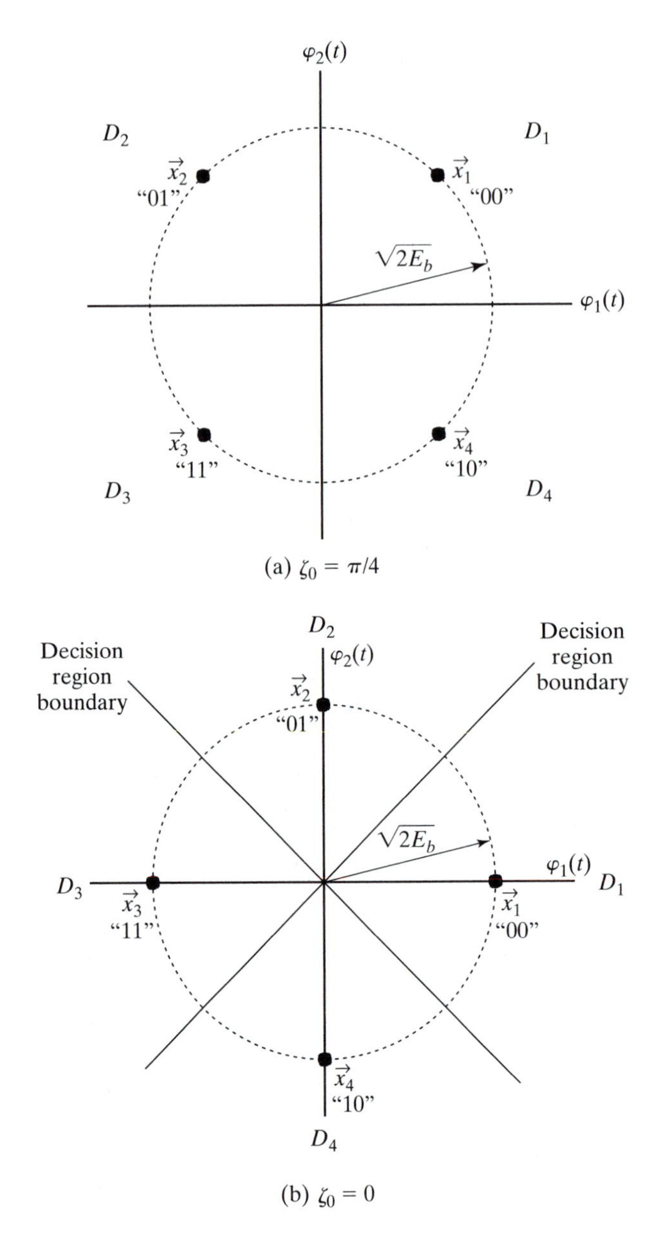

Figure 3.11 QPSK signal constellation, decision regions, and Gray encoding.

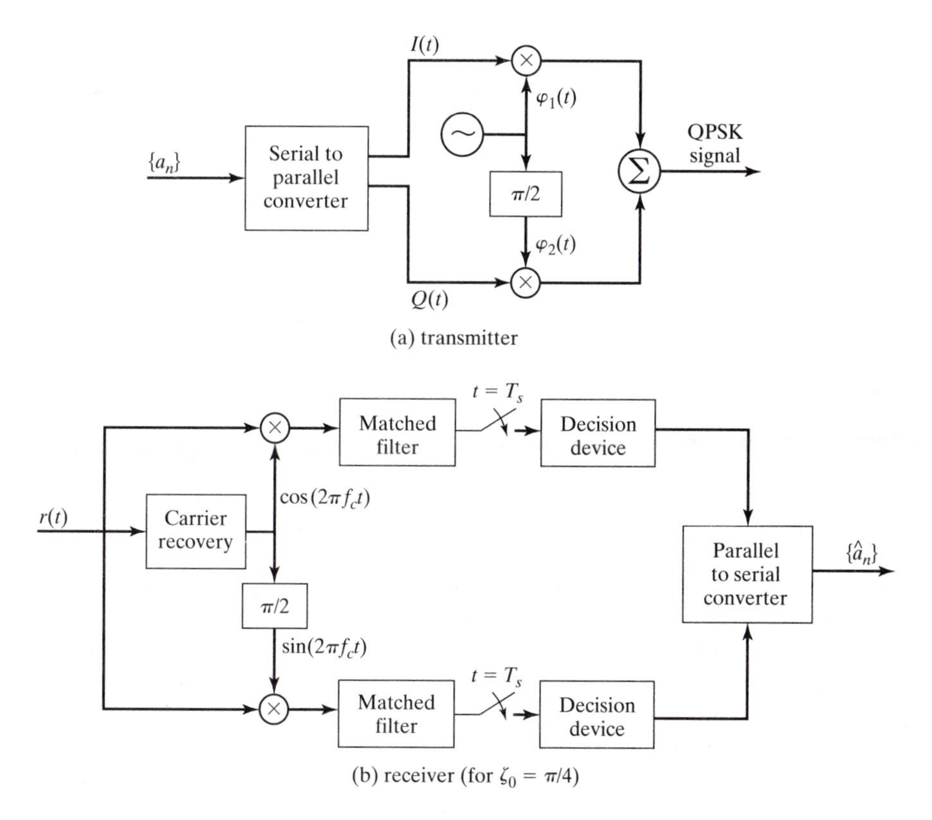

Figure 3.12 QPSK transmitter and receiver.

parallel to serial converter puts the odd numbered and even numbered detected binary digits in order and outputs the estimated binary sequence $\{\hat{a}_n\}$.

$\pi/4$-DQPSK. $\pi/4$-DQPSK is the modulation scheme used in the North American second generation IS-54 digital cellular systems. It is more spectrally efficient than MSK and has less envelope variations than QPSK with the same bandpass filtering. To introduce the modulation, first we neglect the differential encoding and consider $\pi/4$ shifted QPSK. It is a superposition of two QPSK signal constellations offset by $\pi/4$ relative to each other, as shown in Figure 3.13, one QPSK with $\zeta_0 = 0$ for odd-numbered symbols and the other with $\zeta_0 = \pi/4$ for even-numbered symbols. The two signal constellations are used alternately from symbol to symbol. At any symbol interval, a signal point is chosen from the QPSK constellation to represent the two information bits. There are a total of eight possible signal points, and the phase change between adjacent QPSK symbols is limited to the set $\{\pi/4, 3\pi/4, 5\pi/4, 7\pi/4\}$. As a result, the modulation has the following advantages:

a. Unlike QPSK, there is no phase change of π between adjacent symbols. This is the reason that, after bandpass filtering, $\pi/4$ shifted QPSK has a smaller envelope variation than QPSK.

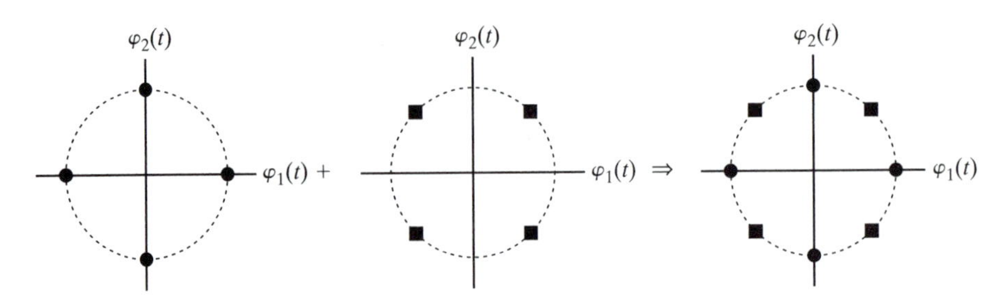

Figure 3.13 Signal constellation of $\pi/4$ shifted QPSK.

b. Unlike QPSK, there must be a phase difference between adjacent symbols. This makes symbol synchronization in the receiver easier.

c. With the four possible phase changes $\{\pi/4, 3\pi/4, 5\pi/4, 7\pi/4\}$, noncoherent detection of $\pi/4$ shifted QPSK signals is possible. By using a frequency discriminator for detection, the receiver structure can be very simple. This is very important for battery-powered handsets.

As discussed in Chapter 2, a wireless channel often introduces a carrier phase distortion to the signal transmitted over it. To combat the effect of the phase distortion on signal detection, differential encoding can be used. $\pi/4$-DQPSK is $\pi/4$ shifted QPSK combined with differential encoding, where the information is carried by the phase difference of adjacent transmitted symbols. If the phase distortion introduced by a wireless channel does not change much over an interval of two symbol durations (which is normally true), the phase difference between two adjacent symbols is almost independent of the channel phase distortion. To combine the differential encoding with $\pi/4$-QPSK, each information symbol should be mapped to one of the four possible phase changes (i.e., $\pi/4$, $3\pi/4$, $5\pi/4$, and $7\pi/4$) between adjacent $\pi/4$-QPSK symbols. The steps to generate a $\pi/4$-DQPSK signal from a binary information sequence $\{a_n\}$ are described mathematically in the following:

Step 1: Convert the input sequence $\{a_n\}$ at rate R_b into an odd-numbered digit sequence $\{a_{1n}\}$ and an even-numbered digit sequence $\{a_{2n}\}$ both at rate $R_b/2$ (by using a serial-to-parallel converter).

Step 2: According to Table 3.1, map each of the two-bit symbols (a_{1n}, a_{2n}) to the phase difference ϕ_n between the $\pi/4$-DQPSK signals transmitted over the current symbol interval $[nT_s, nT_s + T_s]$ and the previous symbol interval $[nT_s - T_s, nT_s]$.

Table 3.1 Mapping Between the Information Symbol and Phase Difference in $\pi/4$-DQPSK

(a_{1n}, a_{2n})	ϕ_n
00	$\pi/4$
01	$3\pi/4$
11	$5\pi/4$
10	$7\pi/4$

Step 3: Generate the baseband in-phase and quadrature signal components as

$$\begin{cases} I_n = I_{n-1}\cos\phi_n - Q_{n-1}\sin\phi_n & \in \{0, \pm 1, \pm 1/\sqrt{2}\} \\ Q_n = I_{n-1}\sin\phi_n + Q_{n-1}\cos\phi_n & \in \{0, \pm 1, \pm 1/\sqrt{2}\} \end{cases}. \tag{3.3.1}$$

In general, the baseband in-phase and quadrature components can be represented in terms of the carrier phase of a $\pi/4$-DQPSK signal for the symbol by

$$\begin{cases} I_n &= \cos\Phi_n \\ Q_n &= \sin\Phi_n \end{cases}, \tag{3.3.2}$$

where Φ_n is the carrier phase for the nth symbol. Using Eq. (3.3.2) to represent the I_{n-1} and Q_{n-1} components in Eq. (3.3.1), we have

$$\begin{cases} I_n &= \cos\Phi_{n-1}\cos\phi_n - \sin\Phi_{n-1}\sin\phi_n = \cos(\Phi_{n-1}+\phi_n) \\ Q_n &= \cos\Phi_{n-1}\sin\phi_n + \sin\Phi_{n-1}\cos\phi_n = \sin(\Phi_{n-1}+\phi_n) \end{cases}. \tag{3.3.3}$$

From Eqs. (3.3.2) and (3.3.3), we observe that

$$\Phi_n = \Phi_{n-1} + \phi_n \quad \text{or} \quad \phi_n = \Phi_n - \Phi_{n-1}. \tag{3.3.4}$$

That is, the information symbol (a_{1n}, a_{2n}) indeed is represented by the phase difference (ϕ_n) between the carrier phase of the transmitted signal over the current symbol interval and that over the previous symbol interval.

Step 4: Generate the transmitted signal over the current (nth) symbol interval as

$$\begin{aligned} x(t) &= \sqrt{\frac{2E_s}{T_s}}\cos(2\pi f_c t + \Phi_n) \\ &= \sqrt{E_s}\cos\Phi_n \underbrace{\sqrt{\frac{2}{T_s}}\cos(2\pi f_c t)}_{\varphi_1(t-nT_s)} - \sqrt{E_s}\sin\Phi_n \underbrace{\sqrt{\frac{2}{T_s}}\sin(2\pi f_c t)}_{\varphi_2(t-nT_s)}, \end{aligned} \tag{3.3.5}$$

where the functions

$$\varphi_1(t) = \begin{cases} \sqrt{\frac{2}{T_s}}\cos(2\pi f_c t), & 0 \le t \le T_s \\ 0, & \text{otherwise} \end{cases}$$

and

$$\varphi_2(t) = \begin{cases} \sqrt{\frac{2}{T_s}}\sin(2\pi f_c t), & 0 \le t \le T_s \\ 0, & \text{otherwise} \end{cases}$$

form the orthonormal set which defines the signal space for the modulation, as shown in Figure 3.13. Here, it is assumed that $f_c T_s$ is an integer.

Figure 3.14 shows a generic $\pi/4$-DQPSK transmitter and receiver. The information sequence $\{a_n\}$ is first divided into an odd-numbered bit stream $\{a_{1n}\}$ and an even-numbered bit stream

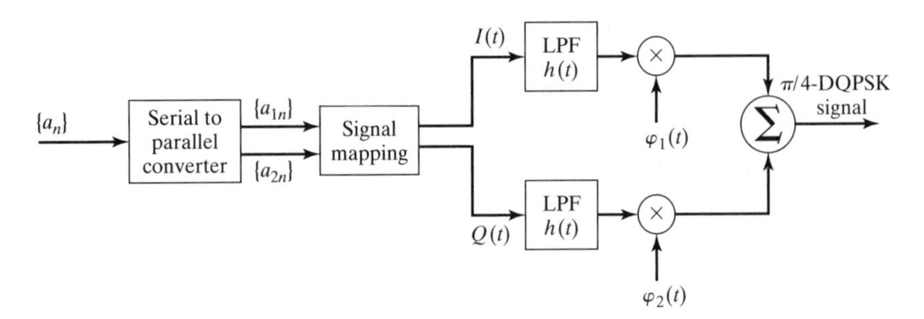

Figure 3.14 $\pi/4$-DQPSK transmitter.

$\{a_{2n}\}$. The two-bit symbol sequence $\{a_{1n}, a_{2n}\}$ is then mapped to an in-phase waveform $I(t)$ and a quadrature waveform $Q(t)$ which take on constant values $\sqrt{E_s}I_n = \sqrt{E_s}\cos(\Phi_n)$ and $\sqrt{E_s}Q_n = \sqrt{E_s}\sin(\Phi_n)$, respectively, over the nth symbol interval, according to Table 3.1 and Eq. (3.3.4). The LPFs (low-pass filters) at both in-phase and quadrature branches with the impulse response $h(t)$ are for baseband pulse shaping in order to suppress the out-of-band radiation. In the IS-54 standard, it is specified that the LPFs shall have linear phase and square root raised cosine frequency response of the form

$$|H(f)| = \begin{cases} 1, & 0 \le f \le \dfrac{1-\alpha}{2T_s} \\ \sqrt{\dfrac{1}{2}\left\{1 - \sin\left[\dfrac{\pi(2fT_s - 1)}{2\alpha}\right]\right\}}, & \dfrac{(1-\alpha)}{2T_s} \le f \le \dfrac{(1+\alpha)}{2T_s} \\ 0, & f > \dfrac{(1+\alpha)}{2T_s} \end{cases},$$

where $\alpha = 0.35$ is the roll-off factor which determines the width of the transition band. The bandpass $\pi/4$-DQPSK signal in the nth symbol interval is given by

$$x_n(t) = \sqrt{E_s}I_n h(t - nT_s)\varphi_1(t - nT_s) - \sqrt{E_s}Q_n h(t - nT_s)\varphi_2(t - nT_s),$$

where $nT_s \le t \le (n+1)T_s$ and $\int_{-\infty}^{\infty} h^2(t)\,dt = 1$.

Figure 3.15 shows the $\pi/4$-DQPSK receiver structure using a baseband differential detector. The receiver structure is very similar to that of QPSK in Figure 3.12(b), but has two differences: (a) no accurate carrier phase synchronization is required here due to differential detection, and (b) a differential phase detector is added at the baseband to extract the phase difference between adjacent symbols which carries information as given in Table 3.1. In the absence of additive channel noise, we have $\hat{I}_n = I_n$ and $\hat{Q}_n = Q_n$. In this case, it can be easily verified that $x_n = \cos(\phi_n)$ and $y_n = \sin(\phi_n)$. In the presence of additive channel noise, the decision device makes the decision that $\hat{a}_{1n}$ ($\hat{a}_{2n}$) is symbol "1" if $x_n > 0$ ($y_n > 0$) and that $\hat{a}_{1n}$ ($\hat{a}_{2n}$) is symbol "0" if $x_n < 0$ ($y_n < 0$).

The $\pi/4$-DQPSK signals can also be demodulated noncoherently using a frequency discriminator, as shown in Figure 3.16(a). The front-end bandpass filter centered at the carrier frequency is

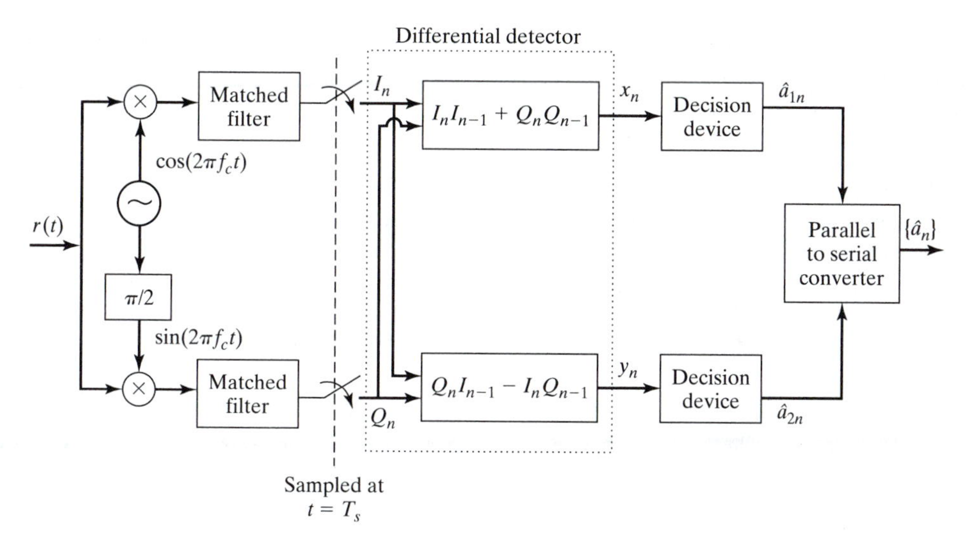

Figure 3.15 $\pi/4$-DQPSK receiver using a baseband differential detector.

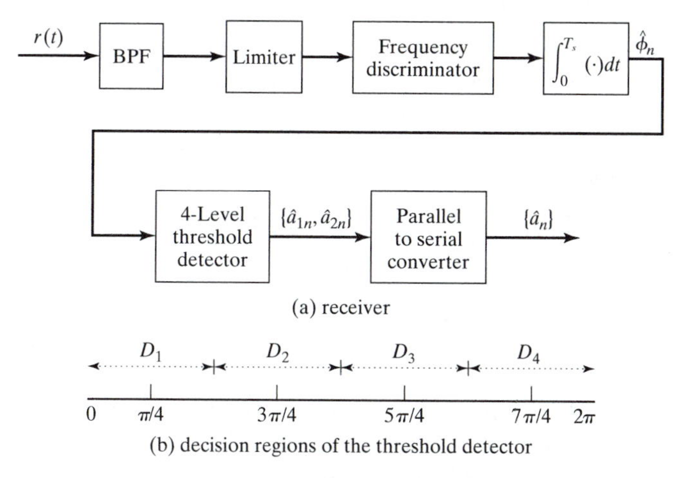

Figure 3.16 $\pi/4$-DQPSK noncoherent receiver using frequency discriminator.

to pass the desired received signal without distortion and to remove out-of-band interference and noise. As the information is carried in the phase component of the received signal, the limiter is used to remove all the amplitude variations introduced by the channel. The frequency discriminator extracts the instantaneous frequency deviation of the received signal from the carrier frequency f_c. The phase difference between two adjacent symbols is obtained by integrating the frequency deviation over one symbol interval. The reconstructed differential phase is passed through a four-level threshold detector. Figure 3.16(b) shows the decision regions for the carrier phase difference $\hat{\phi}_n$ as D_1 for the information symbol "00", D_2 for "01", D_3 for "11", and D_4 for "10". The parallel-to-serial converter outputs $\{\hat{a}_n\}$, which is an estimate of the original binary information sequence $\{a_n\}$.

3.3.2 Minimum Shift Keying (MSK)

The MSK signal is a binary continuous phase frequency shift keying (CPFSK) signal with modulation index of 1/2. The modulation index of 1/2 corresponds to the minimum frequency spacing that makes two FSK signals orthogonal. Let f_1 and f_2 denote the frequencies of the two FSK signals for symbols "0" and "1", respectively. The frequency separation between the two frequencies is $\Delta = f_2 - f_1$. The modulation index, denoted by h, is defined as the product of the frequency separation and the symbol interval. With $h = \Delta \cdot T_b = 1/2$, we have

$$f_2 = f_c + \frac{1}{2}\Delta = f_c + \frac{1}{4T_b}, \quad \text{for symbol "1"}$$

and

$$f_1 = f_c - \frac{1}{2}\Delta = f_c - \frac{1}{4T_b}, \quad \text{for symbol "0"},$$

where $f_c = (f_1 + f_2)/2$ is the center of the two signal frequencies. With respect to the "carrier" phase $2\pi f_c t$, the phase change of the MSK signals over each symbol interval T_b is

$$\begin{cases} 2\pi(f_2 - f_c)T_b = \pi/2, & \text{for symbol "1"} \\ 2\pi(f_1 - f_c)T_b = -\pi/2, & \text{for symbol "0"} \end{cases}.$$

As a result, over the symbol interval $[0,\ T_b]$ the MSK signal can be represented as

$$x(t) = \sqrt{\frac{2E_b}{T_b}} \cos[2\pi f_c t + \phi(t)], \quad 0 \le t \le T_b, \tag{3.3.6}$$

where

$$\phi(t) = \phi(0) + 2\pi\left(\pm\frac{1}{2}\Delta\right)t = \phi(0) \pm \frac{\pi t}{2T_b}, \quad 0 \le t \le T_b,$$

with "+" sign for symbol "1" and "−" sign for symbol "0", and where $\phi(0)$ is the phase at $t = 0$. The phase continuity in MSK can be characterized by its phase trajectory. As the phase change over each symbol interval is either $\pi/2$ or $-\pi/2$, with $\phi(0) = 0$, we have $\phi(t) = 0$ or π for $t = 2nT_b$, and $\phi(t) = \pi/2$ or $-\pi/2$ for $t = (2n+1)T_b$, where n is a positive integer. Within each symbol interval, the phase changes linearly with t. Figure 3.17 shows the MSK phase trajectory for binary information sequence "10111001", under the assumption that the initial phase $\phi(0) = 0$.

From Eq. (3.3.6), the MSK signal can be constructed as (see Appendix C)

$$x(t) = \sqrt{E_b}a_I\varphi_1(t) + \sqrt{E_b}a_Q\varphi_2(t), \quad 0 \le t \le T_b, \tag{3.3.7}$$

where

$$\begin{cases} a_I = \cos[\phi(0)] \\ a_Q = -\sin[\phi(T_b)] \end{cases}$$

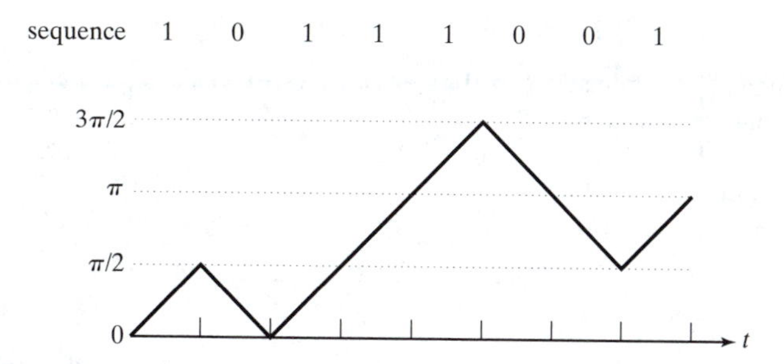

Figure 3.17 MSK phase trajectory for binary sequence "10111001" with $\phi(0) = 0$.

and

$$\begin{cases} \varphi_1(t) = \sqrt{\frac{2}{T_b}} \cos\left(\frac{\pi t}{2T_b}\right) \cos(2\pi f_c t), & -T_b \le t \le T_b \\ \varphi_2(t) = \sqrt{\frac{2}{T_b}} \sin\left(\frac{\pi t}{2T_b}\right) \sin(2\pi f_c t), & 0 \le t \le 2T_b. \end{cases}$$

The signals $\varphi_1(t)$ and $\varphi_2(t)$ form an orthonormal set, which defines the two-dimensional signal space for the MSK signals. Note that MSK signals (different from BPSK signals) cannot be accurately represented by a one-dimensional signal space, even though it is a binary modulation scheme. Figure 3.18 shows the two-dimensional signal space and the decision regions for MSK, based on Eq. (3.3.7), where the details of the four signal points are given in Table 3.2. The axes defined by the orthonormal functions $\varphi_1(t)$ and $\varphi_2(t)$ form the boundaries of the decision regions for the four signal points. That is, the decision regions are the four quadrants, respectively. The signal space of MSK is very similar to that of QPSK with $\zeta_0 = \pi/4$ as given in Figure 3.11(a). However, there are differences between the two signal spaces: (*a*) each signal point in QPSK represents two information bits, whereas it represents one information bit in MSK; (*b*) each decision region in QPSK corresponds to a unique two-bit symbol, whereas in MSK both decision regions D_1 and D_3 are for symbol "0" and D_2 and D_4 for symbol "1"; and (*c*) in QPSK, there

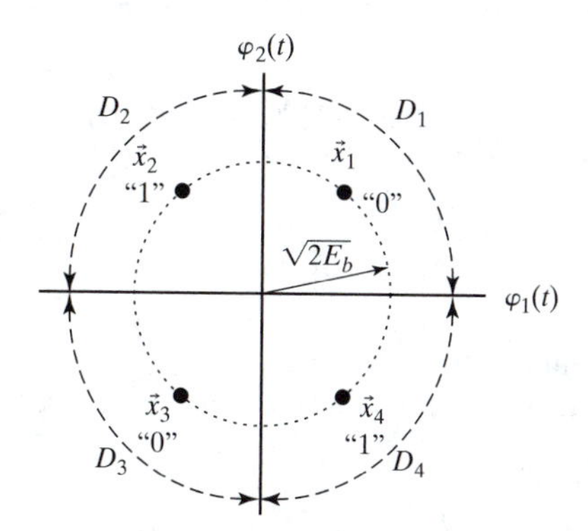

Figure 3.18 The signal space and decision regions of MSK.

Table 3.2 Signal Points in the MSK Signal Space

Signal point	Coordinates (a_I, a_Q)	$\phi(0)$	$\phi(T_b)$	Phase change $\phi(T_b) - \phi(0)$	Information symbol
$\vec{x}_1$	(1, 1)	0	$-\pi/2$	$-\pi/2$	"0"
$\vec{x}_2$	(−1, 1)	π	$-\pi/2$	$\pi/2$	"1"
$\vec{x}_3$	(−1, −1)	π	$\pi/2$	$-\pi/2$	"0"
$\vec{x}_4$	(1, −1)	0	$\pi/2$	$\pi/2$	"1"

is a direct mapping between the symbol phase (which carries the information) and the phase of the signal vector $\vec{x}_m$ $(m = 1, 2, 3, 4)$ in the signal space, whereas such a mapping does not exist for MSK.

Figure 3.19(a) shows the structure of a generic MSK transmitter corresponding to the MSK signal represented by Eq. (3.3.7). Using a serial-to-parallel converter, the information sequence $\{a_n\}$ with bit interval T_b is converted to a waveform $a_{1n}(t)$ for odd-numbered bits and a waveform $a_{2n}(t)$ for even-numbered bits. The waveforms can take on values +1 or −1 and have a duration of $2T_b$ for each bit. The two waveforms are staggered by an interval of T_b. The MSK signal can be detected noncoherently or coherently. In noncoherent detection, the MSK signal is detected in the same way as an FSK signal by using a frequency discriminator, where the constraint of continuous phase is not used. Figure 3.19(b) shows the structure of an MSK receiver with coherent detection which exploits the continuous phase constraint of the signal. The decision device of the in-phase branch estimates the phase of the input signal at $t = 2nT_b$ based on the estimate $\hat{a}_{2n}$, and that of the quadrature branch estimates the phase of the input signal at $t = (2n + 1)T_b$ based on the estimate $\hat{a}_{1n}$, where $n = 0, 1, 2, \ldots$. The final decision device generates the estimate $\{\hat{a}_n\}$ of the binary information sequence based on the phase change over the nth bit interval.

Example 3.4 Orthogonality between the MSK Signals

Verify that the minimum frequency separation between the two MSK signals is $1/2T_b$ in order to achieve orthogonal signaling.

Solution Without loss of generality, consider the transmission of a symbol over the time interval $[0, T_b]$. The MSK signal is

$$x_1(t) = A\cos[2\pi f_1 t + \phi(0)]$$

for symbol "0" and

$$x_2(t) = A\cos[2\pi f_2 t + \phi(0)]$$

for symbol "1", where the initial phase at $t = 0$, $\phi(0)$, is the same for both cases due to the constraint of continuous phase. For $x_1(t)$ and $x_2(t)$ to be orthogonal over the time interval $t \in [0, T_b]$, it is required that

$$\int_0^{T_b} x_1(t)x_2(t)\,dt = 0.$$

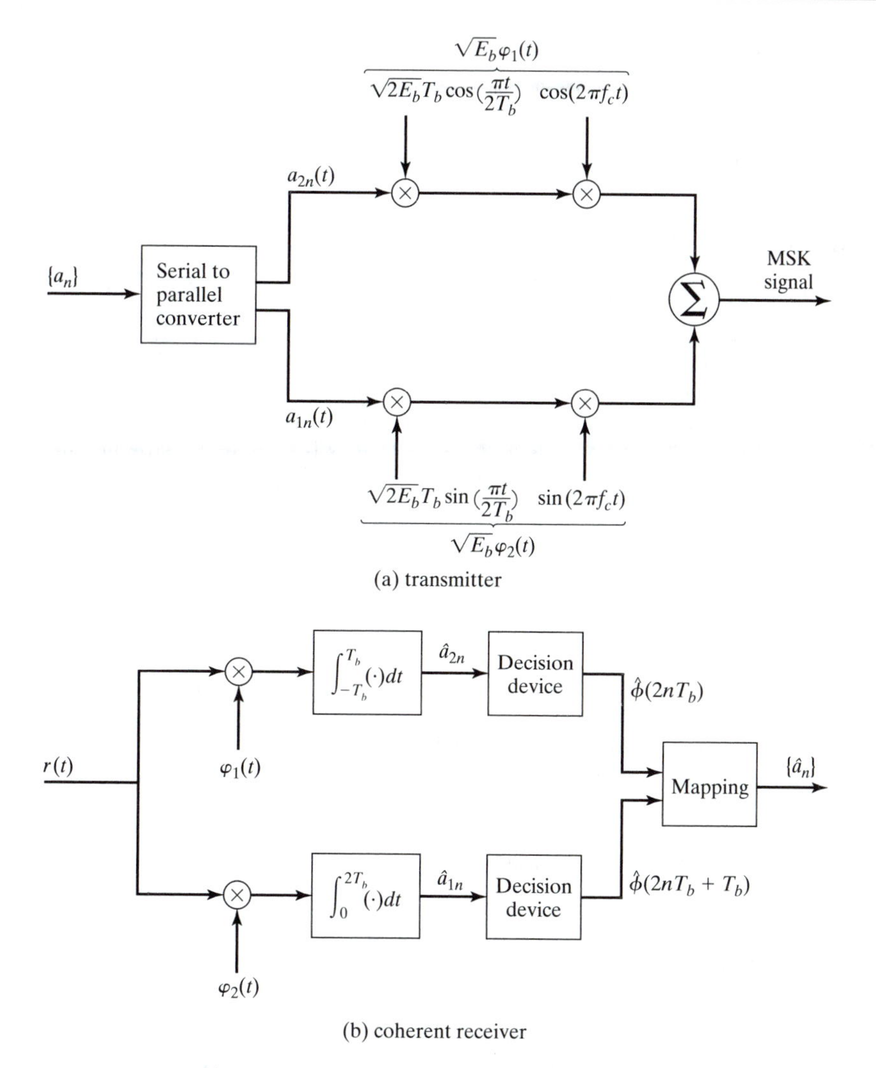

Figure 3.19 MSK transmitter and receiver.

This requirement means that

$$\int_0^{T_b} \cos(2\pi f_1 t + \phi(0)) \cos(2\pi f_2 t + \phi(0))\, dt = 0,$$

$$\Longrightarrow \int_0^{T_b} \cos[2\pi(f_1 + f_2)t + 2\phi(0)]\, dt + \int_0^{T_b} \cos[2\pi(f_2 - f_1)t]\, dt = 0,$$

$$\Longrightarrow \int_0^{T_b} \cos[2\pi(f_2 - f_1)t]\, dt = 0,$$

$$\Longrightarrow \sin[2\pi(f_2 - f_1)T_b] = 0,$$

$$\Longrightarrow \min\{(f_2 - f_1)\} = 1/2T_b,$$

where $\int_0^{T_b} \cos[2\pi(f_1 + f_2)t + 2\phi(0)]\,dt \approx 0$ because $(f_1 + f_2) \gg 1/T_b$.

3.3.3 Gaussian MSK (GMSK)

As mentioned in Section 3.1, for mobile radio transmission, it is desired that the modulation scheme have compact output power spectral density, small or no envelope fluctuations, and high transmission accuracy performance. The three desired characteristics contradict with each other and a certain compromise has to be made, depending on the main design objective. One way to achieve a compact output power spectral density is to properly filter the baseband signal before modulation. If the source signal is prefiltered using a Gaussian shaping function before performing minimum shift keying, as shown in Figure 3.20, the resultant modulated signal is referred to as a Gaussian MSK signal. GMSK is used in European second generation wireless systems such as GSM, because it provides a good compromise among the three desired characteristics.

Let $H(f)$ be the transfer function of the prefilter, which is bell shaped about $f = 0$. $H(f)$ can be expressed as

$$H(f) = \exp\left[-\left(\frac{f^2}{B}\right)\frac{\ln 2}{2}\right], \tag{3.3.8}$$

where B is the bandwidth of the filter. The impulse response of the filter, $h(t)$, is given by the inverse Fourier transform of $H(f)$

$$h(t) = \mathcal{F}^{-1}[H(f)] = \sqrt{\frac{2\pi}{\ln 2}}\,B \exp\left[-\frac{2\pi^2 B^2 t^2}{\ln 2}\right]. \tag{3.3.9}$$

With the low-pass filter, the output of the MSK modulator in Figure 3.20 (i.e., the GMSK signal) can be represented by

$$x(t) = \sqrt{\frac{2E_b}{T_b}} \cos[2\pi f_c t + \phi(t)]. \tag{3.3.10}$$

To describe the phase component $\phi(t)$, we start from the input signal in the transmitter shown in Figure 3.20. The NRZ polar format input is

$$a(t) = \sum_{n=-\infty}^{\infty} a_n \Pi\left(\frac{t - nT_b}{T_b}\right), \tag{3.3.11}$$

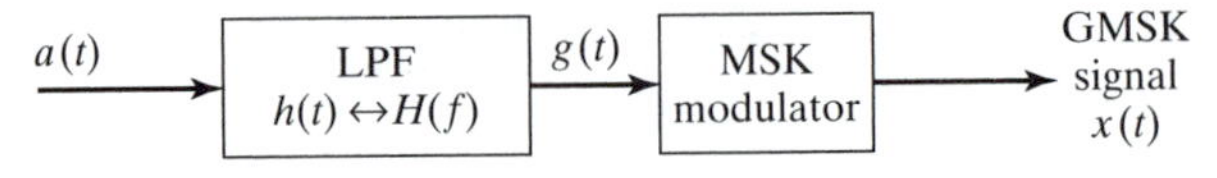

Figure 3.20 Gaussian MSK transmitter.

where

$$a_n = \begin{cases} 1, & \text{if the } n\text{th symbol is "1"} \\ -1, & \text{if the } n\text{th symbol is "0"} \end{cases}$$

and

$$\Pi\left(\frac{t}{T_b}\right) = \begin{cases} 1, & 0 \le t \le T_b \\ 0, & \text{otherwise} \end{cases}.$$

With the impulse response $h(t)$, the output of the LPF is

$$g(t) = a(t) \star h(t) = \sum_{n=-\infty}^{\infty} a_n y(t - nT_b), \tag{3.3.12}$$

where

$$\begin{aligned} y(t) &= \Pi\left(\frac{t}{T_b}\right) \star h(t) \\ &= \int_{t-T_b}^{t} h(z)\, dz \\ &= \frac{1}{2}\left\{\mathrm{erf}\left[\sqrt{\frac{2}{\ln 2}}\pi Bt\right] + \mathrm{erf}\left[-\sqrt{\frac{2}{\ln 2}}\pi B(t - T_b)\right]\right\}, \end{aligned}$$

with $\mathrm{erf}(t) = \frac{2}{\sqrt{\pi}} \int_0^t \exp(-z^2)\, dz$. The phase of the GMSK signal is then

$$\phi(t) = \frac{\pi}{2T_b} \int_{-\infty}^{t} g(z)\, dz. \tag{3.3.13}$$

Over the nth bit interval, $t \in [nT_b, (n+1)T_b]$, the phase can be represented as

$$\phi(t) = \phi(nT_b) \pm \frac{\pi}{2T_b} \int_{nT_b}^{t} y(z - nT_b)\, dz, \tag{3.3.14}$$

where the "+" sign is for $a_n = 1$ and the "−" sign is for $a_n = -1$. With the phase $\phi(t)$, the instantaneous frequency of the GMSK signal is

$$f(t) = f_c + \frac{1}{2\pi}\frac{d\phi(t)}{dt} = f_c \pm \frac{1}{4T_b} y(t - nT_b), t \in [nT_b, (n+1)T_b]. \tag{3.3.15}$$

Figure 3.21 shows the output of the Gaussian filter when the input is a single rectangular pulse $\Pi(t/T_b)$, with BT_b as a parameter. It is observed that the pulse duration is larger than the symbol duration except for $BT_b = \infty$ corresponding to MSK. As the BT_b value decreases, the frequency spectral efficiency increases; however, the pulse interval increases, leading to more severe intersymbol interference which degrades transmission accuracy performance. In GSM, $BT_b = 0.3$ is used. It should be mentioned that the Gaussian filter is noncausal and the impulse response $h(t)$ has to be truncated and time shifted in practical implementation. For example, if the duration of the impulse response is limited to $4T_b$, then it should also be time

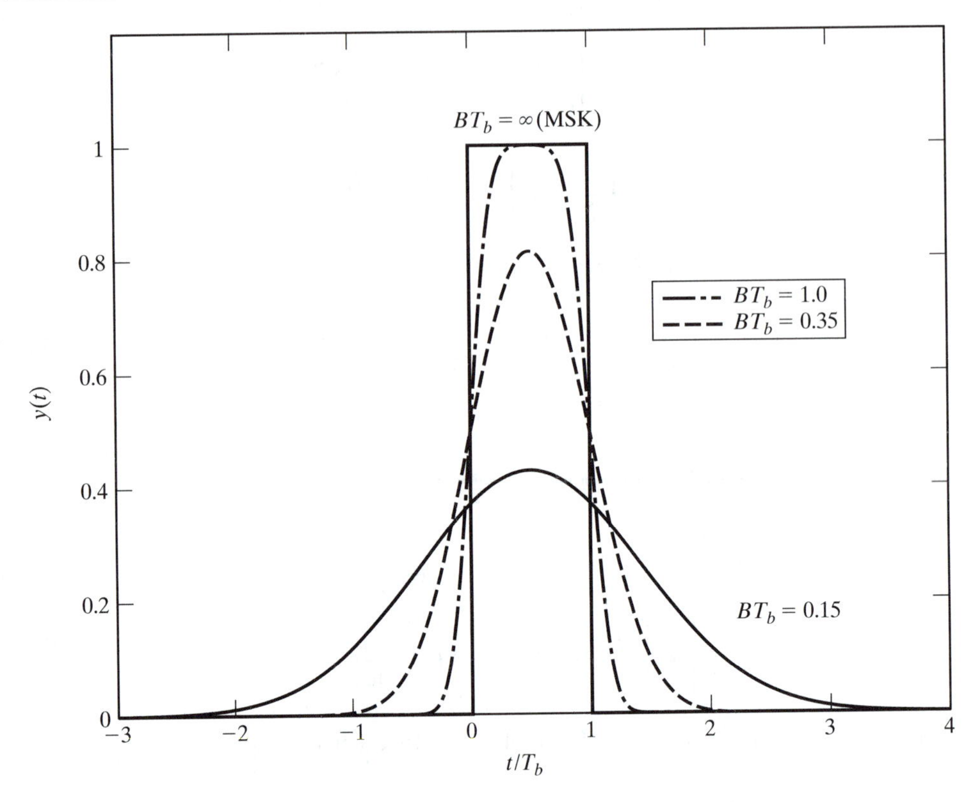

Figure 3.21 Characteristics of Gaussian shaping filter.

shifted by $2T_b$ so that it is physically implementable. The GMSK scheme offers flexibility in the receiver design. Both coherent and differentially coherent demodulation techniques can be used to demodulate GMSK signals [102, 139]. In addition, since GMSK is a kind of FM modulation scheme, the receiver can employ a frequency discriminator detector to detect the transmitted information [45].

3.3.4 Orthogonal Frequency Division Multiplexing (OFDM)

OFDM is a block modulation scheme. Let T_s be the symbol intervals of the input source sequence. A block of N serial symbols is converted into a block of N parallel modulated symbols, each of duration $T = NT_s$. If the rms delay spread of the channel is σ_τ, N is chosen so that $NT_s \gg \sigma_\tau$. This is a means to combat intersymbol interference due to channel time dispersion. In the frequency domain, the bandwidth of each subband signal is $1/N$ times that of the original signal. If the bandwidth of the original signal is large compared with the channel coherence bandwidth, so that the channel exhibits frequency-selective fading, N can be chosen appropriately so that the channel seen by each of the subbands in OFDM exhibits flat fading. Thus, OFDM has the property of mitigating frequency-selective fading.

The complex envelope of OFDM is given by

$$\begin{aligned} v(t) &= \sqrt{\frac{2E_s}{T_s}} \sum_{k=0}^{\infty} \sum_{n=0}^{N-1} a_{k,n} \tilde{\varphi}_n(t-kT) \\ &= \sum_{n=0}^{N-1} v_n(t), \end{aligned} \tag{3.3.16}$$

where $a_{k,n}$ carries the information to be sent over the kth symbol interval $t \in [kT, kT+T]$ and the nth subband ($n = 0, 1, 2, \ldots, N-1$). $v_n(t)$ is the complex envelope of the signal transmitted in the nth subband and is given by

$$v_n(t) = \sqrt{\frac{2E_s}{T_s}} \sum_{k=0}^{\infty} a_{k,n} \tilde{\varphi}_n(t-kT), \tag{3.3.17}$$

where $\{\tilde{\varphi}_n(t)\}_{n=0}^{N-1}$ is a set of complex orthonormal waveforms and is given by

$$\tilde{\varphi}_n(t) = \begin{cases} \exp\left[j2\pi\left(n - \frac{N-1}{2}\right)t/T\right], & t \in [0, T] \\ 0, & t \notin [0, T] \end{cases}. \tag{3.3.18}$$

Each waveform in the set $\{v_n(t)\}_{n=0}^{N-1}$ corresponds to a distinct (nth) subcarrier with frequency $f_c + \frac{2n-(N-1)}{2T}$.

For simplicity, consider $k = 0$. We have

$$v(t) = \sqrt{\frac{2E_s}{T_s}} \sum_{n=0}^{N-1} a_{0,n} \exp\left(j\frac{2\pi nt}{NT_s}\right) \exp\left[-j\frac{\pi(N-1)t}{NT_s}\right], \quad 0 \le t \le NT_s.$$

Note that the term $\exp[-j\frac{\pi(N-1)t}{NT_s}]$ is not a function of n and can be combined with the carrier term $\exp(j2\pi f_c t)$. The complex envelope can thus be written as

$$v(t) = \sqrt{\frac{2E_s}{T_s}} \sum_{n=0}^{N-1} a_{0,n} \exp\left(j\frac{2\pi nt}{NT_s}\right), \quad 0 \le t \le NT_s,$$

with corresponding carrier $\exp[j2\pi(f_c - \frac{N-1}{2NT_s})t]$. Sampling $v(t)$ at $t = \ell T_s$ yields

$$A_{0,\ell} \triangleq v(\ell T_s) = \sqrt{\frac{2E_s}{T_s}} \sum_{n=0}^{N-1} a_{0,n} \exp\left(j\frac{2\pi n\ell}{N}\right), \quad \ell = 0, 1, 2, \cdots, N-1,$$

which is actually proportional to the inverse discrete Fourier transform (IDFT) of $\{a_{0,n}\}$. The samples, $\vec{A}_0 = (A_{0,1}, A_{0,2}, \ldots, A_{0,N-1})$, are then passed through a digital-to-analog (D/A) converter and are used to modulate the carrier which produces the OFDM signal $x(t) = \Re\{v(t) \exp$

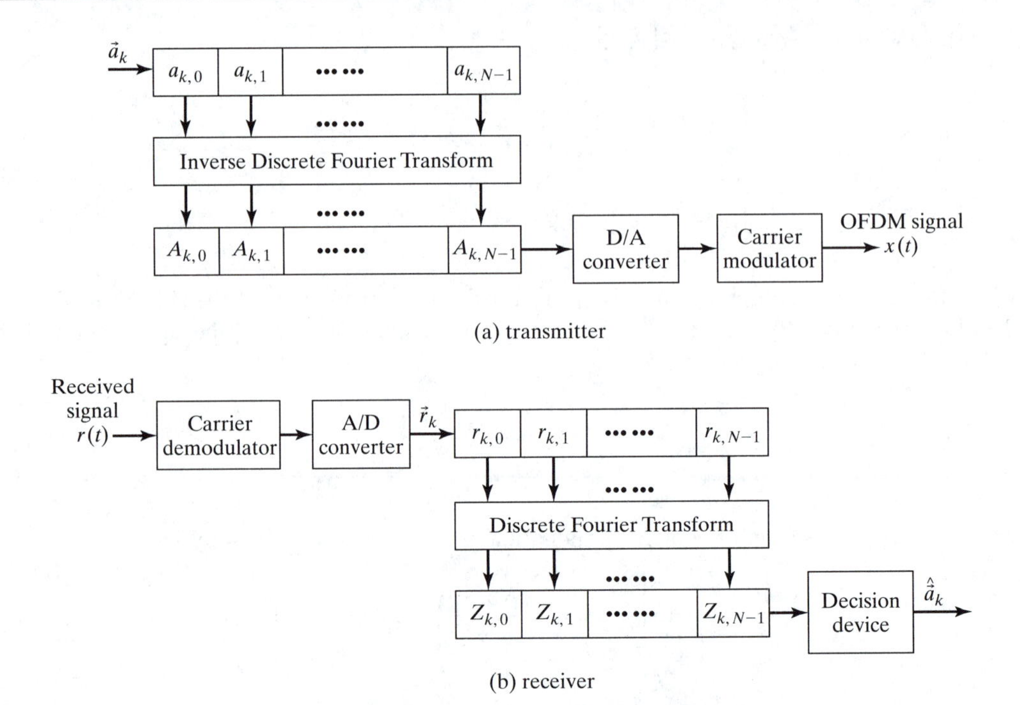

Figure 3.22 OFDM transmitter and receiver.

$[j2\pi(f_c - \frac{N-1}{2NT_s})t]\}$. The OFDM transmitter is shown in Figure 3.22(a), where the input information sequence is partitioned into blocks of length N, each block being represented by $\vec{a}_k$, $k = 0, 1, 2, \ldots$. At the receiver, the received signal frequency is first down converted to the baseband and then converted to digital format using an A/D converter. The output sequence of the A/D converter is also partitioned into blocks of length N and is demodulated block by block. The corresponding block for $k = 0$ has the following N received samples

$$\vec{r}_0 = (r_{0,0}, r_{0,1}, \cdots, r_{0,N-1}).$$

Demodulation is performed by computing the discrete Fourier transform (DFT) on the block $\vec{r}_0$ to yield N decision variables

$$Z_{0,n} = \frac{1}{N}\sum_{\ell=0}^{N-1} r_{0,\ell} \exp\left(-j\frac{2\pi n\ell}{N}\right), \quad n = 0, 1, \cdots, N-1.$$

The OFDM receiver is shown in Figure 3.22(b), where the decision device estimates the transmitted information, $\hat{\vec{a}}_k = (\hat{a}_{k,0}, \hat{a}_{k,1}, \ldots, \hat{a}_{k,N-1})$, based on the N decision variables $Z_{k,n}$, $n = 0, 1, \ldots, N-1$.

3.4 POWER SPECTRAL DENSITY

Power spectral density (psd) describes the distribution of the signal power in the frequency domain and is very important (*a*) in characterizing whether a modulation scheme is frequency efficient or not, (*b*) in channel assignment, and (*c*) in transmitter and receiver design. Consider a narrowband signal with carrier frequency f_c.

$$x(t) = \Re\{v(t)\exp[j2\pi f_c t]\}, \tag{3.4.1}$$

where $v(t) = v_I(t) + jv_Q(t)$ is the complex envelope at baseband, and we assume that the amplitude of the carrier A_c equals 1. It can be easily shown that $x(t)$ is a random process (signal) having a periodic mean and a periodic autocorrelation function with period $T_c = 1/f_c$. Such a process is called a cyclostationary process, whose psd is defined as the Fourier transform of the time-averaged autocorrelation function over each period. The time-averaged correlation function of $x(t)$ is

$$\begin{aligned}
\bar{R}_x(\tau) &= \frac{1}{T_c}\int_0^{T_c} E[x(t)x(t-\tau)]\,dt \\
&= \frac{1}{T_c}\int_0^{T_c} E\{\Re[v(t)e^{j2\pi f_c t}]\Re[v(t-\tau)e^{j2\pi f_c(t-\tau)}]\}\,dt \\
&= \frac{1}{2T_c}\int_0^{T_c} \{\Re[R_v(\tau)e^{j2\pi f_c \tau}] + \Re[E(v(t)v(t-\tau))e^{j2\pi f_c(2t-\tau)}]\}\,dt \\
&\approx \frac{1}{2}\Re[R_v(\tau)e^{j2\pi f_c \tau}],
\end{aligned}$$

where $R_v(\tau) = E[v(t)v^*(t-\tau)]$ is the correlation function of the baseband envelope $v(t)$, and

$$\int_0^{T_c} \Re[E(v(t)v(t-\tau))e^{j2\pi f_c(2t-\tau)}]\}\,dt \approx 0$$

as $v(t)v(t-\tau)$ is approximately a constant over the time interval $[0, T_c]$. As a result, the psd of the narrowband signal $x(t)$ is

$$\begin{aligned}
\Phi_x(f) &= \mathcal{F}[\bar{R}_x(\tau)] \\
&= \mathcal{F}\left\{\frac{1}{2}\Re[R_v(\tau)e^{j2\pi f_c \tau}]\right\} \\
&= \mathcal{F}\left\{\frac{1}{4}[R_v(\tau)e^{j2\pi f_c \tau} + R_v^*(\tau)e^{-j2\pi f_c \tau}]\right\} \\
&= \frac{1}{4}[\Phi_v(f-f_c) + \Phi_v(-f-f_c)],
\end{aligned} \tag{3.4.2}$$

where $\Phi_v(f) = \mathcal{F}[R_v(\tau)]$ is the psd of the baseband envelope $v(t)$. If $\Phi_v(f)$ is an even function, i.e., $\Phi_v(f) = \Phi_v(-f)$, then we have

$$\Phi_x(f) = \frac{1}{4}[\Phi_v(f - f_c) + \Phi_v(f + f_c)].$$

As a result, it is sufficient to determine the psd $\Phi_v(f)$. In digital transmission, the baseband envelope signal $v(t)$ can be represented by

$$v(t) = \sum_{n=-\infty}^{\infty} a_n g(t - nT_s), \tag{3.4.3}$$

where a_n is the discrete random variable carrying information, $g(t)$ describes the basic pulse shape, and T_s is the symbol interval. It can be shown that the psd of $v(t)$ is [123]

$$\Phi_v(f) = \frac{1}{T_s}|G(f)|^2 \sum_{k=-\infty}^{\infty} R_a(k) \exp(-j2\pi k f T_s), \tag{3.4.4}$$

where $G(f) = \mathcal{F}[g(t)]$ and $R_a(k) = E[a_n a_{n-k}^*]$. Note that the psd of $v(t)$ is proportional to $|G(f)|^2$. In the following derivations of psd, it is assumed that

a. symbols "1" and "0" are equally likely,
b. all the symbols are independent of each other, and
c. a rectangular baseband pulse is used.

BPSK. The baseband envelope is given by

$$v(t) = \sum_{n=-\infty}^{\infty} a_n g(t - nT_b), \tag{3.4.5}$$

where

$$a_n = \begin{cases} +1, & \text{if the } n\text{th symbol is "1"} \\ -1, & \text{if the } n\text{th symbol is "0"} \end{cases}$$

and $g(t) = \sqrt{\frac{2E_b}{T_b}}\Pi(\frac{t}{T_b})$ with Fourier transform

$$G(f) = \mathcal{F}[g(t)] = \sqrt{\frac{2E_b}{T_b}} \cdot T_b \text{sinc}(f T_b) \exp\left[-j2\pi f \frac{T_b}{2}\right], \tag{3.4.6}$$

where $\text{sinc}(x) \stackrel{\Delta}{=} \frac{\sin(\pi x)}{\pi x}$. The autocorrelation function of a_n is

$$R_a(k) = E[a_n a_{n-k}^*] = \begin{cases} E[a_n^2], & \text{if } k = 0 \\ E[a_n]E[a_{n-k}], & \text{if } k \neq 0 \end{cases} = \delta(k).$$

As a result, the baseband psd is

$$\Phi_v(f) = \frac{1}{T_b}|G(f)|^2 \sum_{k=-\infty}^{\infty} R_a(k)\exp(-j2\pi k f T_b)$$

$$= \frac{1}{T_b}\left|\sqrt{\frac{2E_b}{T_b}} \cdot T_b \mathrm{sinc}(fT_b)\exp\left[-j2\pi f\frac{T_b}{2}\right]\right|^2$$

$$= 2E_b \mathrm{sinc}^2(fT_b).$$

The psd of BPSK is then given by

$$\Phi_x(f) = \frac{E_b}{2}\{\mathrm{sinc}^2[(f - f_c)T_b] + \mathrm{sinc}^2[(f + f_c)T_b]\}. \tag{3.4.7}$$

QPSK and $\pi/4$-DQPSK. As a constant phase shift of a signal does not affect the psd of the signal, the psd of the QPSK signal with $\zeta = 0$ is the same as that of QPSK with $\zeta = \pi/4$. Considering the case of $\zeta = \pi/4$, the baseband envelope signal is

$$v(t) = \sum_{n=-\infty}^{\infty} a_n g(t - nT_s),$$

where $a_n = a_{1n} + ja_{2n}$ and

$$a_{1n} = \begin{cases} +1, & \text{if the odd-numbered digit of the } n\text{th symbol is "1"} \\ -1, & \text{otherwise} \end{cases},$$

$$a_{2n} = \begin{cases} +1, & \text{if the even-numbered digit of the } n\text{th symbol is "1"} \\ -1, & \text{otherwise} \end{cases},$$

and the basic pulse is $g(t) = \sqrt{\frac{E_s}{T_s}}\Pi(t/T_s)$ with Fourier transform

$$G(f) = \mathcal{F}[g(t)] = \sqrt{\frac{E_s}{T_s}} \cdot T_s \mathrm{sinc}(fT_s)\exp\left[-j2\pi f\frac{T_s}{2}\right].$$

The autocorrelation function of a_n is

$$R_a(k) = E\{[a_{1n} + ja_{2n}][a_{1(n-k)} - ja_{2(n-k)}]\} = 2\delta(k).$$

Substituting $R_a(k)$ and $G(f)$ into Eq. (3.4.4), the psd of the baseband envelope $v(t)$ is

$$\Phi_v(f) = 4E_b \mathrm{sinc}^2(2fT_b). \tag{3.4.8}$$

In deriving Eq. (3.4.8), $E_s = 2E_b$ and $T_s = 2T_b$ are used. The psd of QPSK is then

$$\Phi_x(f) = E_b\{\mathrm{sinc}^2[2(f - f_c)T_b] + \mathrm{sinc}^2[2(f + f_c)T_b]\}. \tag{3.4.9}$$

For $\pi/4$-DQPSK, the signal representation given in Eq. (3.3.5) and Figure 3.13 indicates that (a) the basic pulse $g(t)$ is the same as that in QPSK and (b) the correlation function $R_a(k) = 2\delta(k)$. As a result, the psd of $\pi/4$-DQPSK is the same as that of QPSK.

MSK and GMSK. Recall that the MSK signal in $t \in [0, T_b]$ can be represented by

$$x(t) = \sqrt{\frac{2E_b}{T_b}}\left[a_I \cos\frac{\pi t}{2T_b}\cos(2\pi f_c t) + a_Q \sin\frac{\pi t}{2T_b}\sin(2\pi f_c t)\right], \quad 0 \le t \le T_b, \qquad (3.4.10)$$

which has the complex envelope at baseband

$$v(t) = a_I\left[\sqrt{\frac{2E_b}{T_b}}\cos\left(\frac{\pi t}{2T_b}\right)\right] + ja_Q\left[\sqrt{\frac{2E_b}{T_b}}\sin\left(\frac{\pi t}{2T_b}\right)\right], \quad 0 \le t \le T_b, \qquad (3.4.11)$$

where a_I and a_Q are independent random variables, each taking on the values "+1" and "−1" with equal likelihood depending on the information sequence. Let the impulse response of the symbol shaping filter be

$$g(t) = \sqrt{\frac{2E_b}{T_b}}\cos\left(\frac{\pi t}{2T_b}\right) - T_b \le t \le T_b,$$

which has Fourier transform

$$G(f) = \mathcal{F}[g(t)] = -\frac{1}{4\pi T_b}\sqrt{\frac{2E_b}{T_b}}\frac{\cos(2\pi f T_b)}{f^2 - \dfrac{1}{16T_b^2}}.$$

In general, the complex envelope can be written as

$$v(t) = \sum_{n=-\infty}^{\infty}[a_{I,n}g(t - nT_b) + ja_{Q,n}(t)g(t - nT_b - T_b)],$$

where $a_{I,n}$ and $a_{Q,n}$ are a_I and a_Q, respectively, over the nth symbol interval. As $g(t - nT_b)$ and $g(t - nT_b - T_b)$ are orthogonal, the psd of $v(t)$ is the summation of the psd of $\sum_{n=-\infty}^{\infty} a_{I,n}g(t - nT_b)$ and psd of $\sum_{n=-\infty}^{\infty} a_{Q,n}(t)g(t - nT_b - T_b)$. The psd of the complex envelope is given by

$$\begin{aligned}\Phi_v(f) &= \frac{1}{2T_b}E[(a_{I,n})^2]|\mathcal{F}[g(t)]|^2 + \frac{1}{2T_b}E[(a_{Q,n})^2]|\mathcal{F}[g(t - T_b)]|^2\\ &= \frac{1}{2T_b}|G(f)|^2 + \frac{1}{2T_b}|G(f)\exp[-j2\pi f T_b]|^2\\ &= \frac{32E_b}{\pi^2}\left[\frac{\cos(2\pi f T_b)}{1 - (4fT_b)^2}\right]^2\end{aligned}$$

and the psd of the MSK signal is

$$\Phi_x(f) = \frac{8E_b}{\pi^2}\left\{\frac{\cos[2\pi(f-f_c)T_b]}{1-[4(f-f_c)T_b]^2}\right\}^2 + \frac{8E_b}{\pi^2}\left\{\frac{\cos[2\pi(f+f_c)T_b]}{1-[4(f+f_c)T_b]^2}\right\}^2. \tag{3.4.12}$$

Figure 3.23 shows the normalized (with respect to $4E_b$) psd of MSK, together with those of BPSK and QPSK, versus the normalized frequency difference from the carrier center frequency, $(f - f_c)/R_b$. It can be observed that:

a. compared with QPSK, MSK has smaller sidelobe radiation due to its continuous phase constraint, which leads to less adjacent channel interference;
b. QPSK has a narrower main lobe than MSK, which can lead to higher frequency spectral efficiency, as QPSK is quaternary modulation while MSK is binary modulation; and
c. the spectral efficiency of QPSK is two times that of BPSK, which is achieved by using the quadrature carriers in QPSK.

The psd of GMSK is difficult to derive. Computer simulation can be used to generate the density. In general, GMSK has a more compact psd than MSK, with lower sidelobes. An increase in the BT_b value means an increase in the out-of-band radiation. When the BT_b value approaches infinity, the psd of GMSK approaches that of MSK. On the other hand, as observed in Figure 3.21, a decrease in the BT_b value means more intersymbol interference. As a result, the BT_b value

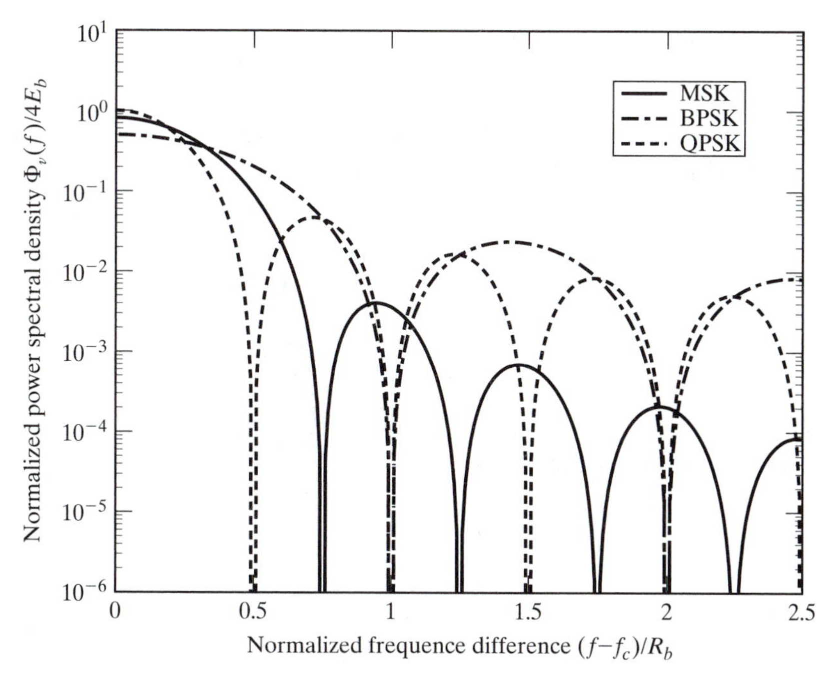

Figure 3.23 Normalized power spectral density of MSK, BPSK, and QPSK.

can be chosen to achieve a compromise between the compact power spectrum of the transmitted signal and the induced intersymbol interference.

OFDM. From Eqs. (3.3.16)–(3.3.18), OFDM can be viewed as independent modulation on orthogonal subcarriers separated by $\frac{1}{T} = \frac{1}{NT_s}$ Hz. For a source symbol sequence $\{a_{k,n}\}$ with zero mean and a symbol shaping pulse $g(t)$ having a Fourier transform $G(f) = \mathcal{F}[g(t)]$, the psd of the complex envelope is given by

$$\Phi_v(f) = \frac{1}{T}\sum_{n=0}^{N-1}\left\{\left|G\left(f - \frac{1}{T}\left(n - \frac{N-1}{2}\right)\right)\right|^2 \sum_{l=-\infty}^{\infty} R_a(l,n)\exp(-j2\pi lfT)\right\},$$

where $R_a(l,n) = E[a_{k,n}a^*_{k-l,n}]$, $T = N/R_s$, and R_s is the information symbol rate. The normalized psd of OFDM (with respect to the peak value) for $N = 4$ and $N = 32$ are shown in Figure 3.24, where it is assumed that $g(t) = \sqrt{\frac{E_s}{T_s}}\Pi(t/T)$ with Fourier transform $G(f) = \sqrt{\frac{E_s}{T_s}} \cdot T\text{sinc}(fT)\exp(-j2\pi f\frac{T}{2})$.

Example 3.5 Calculation of PSD

Determine an expression for the psd of the random pulse train

$$Y(t) = \sum_{n=-\infty}^{\infty} Y_n g(t - nT),$$

where $g(t)$ is a pulse shape defined on $[0, T]$, $Y_n = X_n + \frac{1}{2}X_{n-1}$, and the real-valued random variables X_n are independent and identically distributed, each following a Gaussian distribution with zero mean and variance σ^2.

Solution The psd of $Y(t)$ is given by

$$S_Y(f) = \frac{1}{T}|G(f)|^2 \sum_{k=-\infty}^{\infty} R_Y(k)\exp(-j2\pi kfT),$$

where $G(f) = \mathcal{F}[g(t)]$ and $R_Y(k) = E[Y_nY_{n-k}]$.

The autocorrelation function of $\{X_n\}$ is

$$\begin{aligned} R_X(k) &= E[X_nX_{n-k}] \\ &= \begin{cases} E[X_n^2], & \text{if } k = 0 \\ E(X_n)E(X_{n-k}), & \text{if } k \neq 0 \end{cases} \\ &= \begin{cases} \sigma^2, & \text{if } k = 0 \\ 0, & \text{if } k \neq 0 \end{cases}, \end{aligned}$$

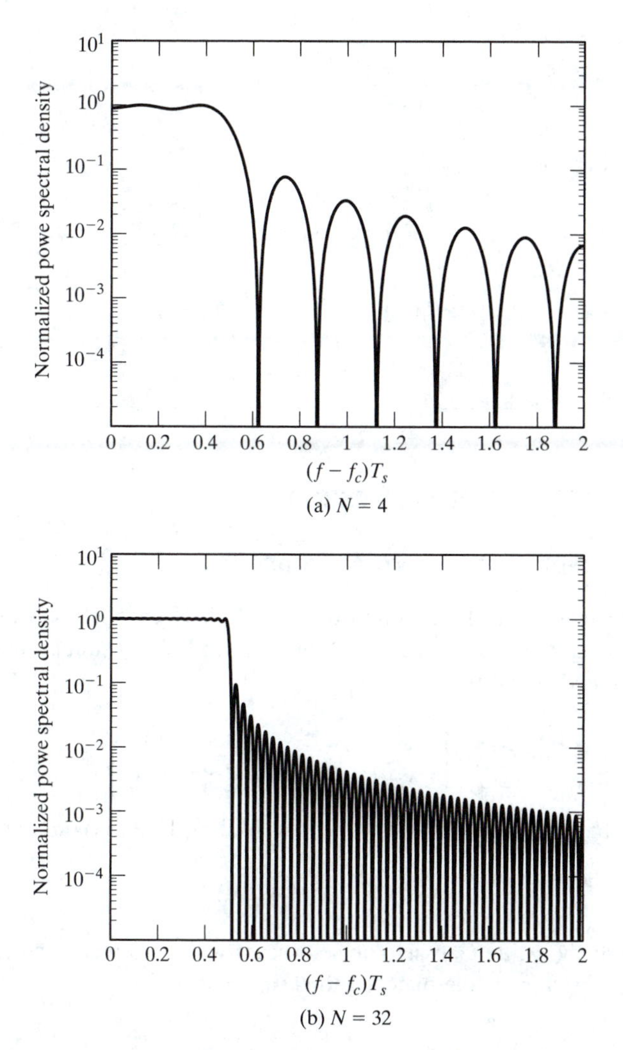

Figure 3.24 Normalized OFDM psd as a function of the normalized frequency difference.

since $E(X_n) = E(X_{n-k}) = 0$ and $E(X_n^2) = Var(X_n) = \sigma^2$. The autocorrelation function of $\{Y_n\}$ can be represented in terms of $R_X(k)$ as

$$\begin{aligned} R_Y(k) &= E[Y_n Y_{n-k}] \\ &= E[(X_n + 0.5X_{n-1})(X_{n-k} + 0.5X_{n-k-1})] \\ &= E[X_n X_{n-k} + 0.5X_n X_{n-k-1} + 0.5X_{n-1}X_{n-k} + 0.25X_{n-1}X_{n-k-1}] \\ &= R_X(k) + 0.5R_X(k+1) + 0.5R_X(k-1) + 0.25R_X(k) \end{aligned}$$

$$= \begin{cases} 1.25\sigma^2, & \text{if } k = 0 \\ 0.5\sigma^2, & \text{if } k = +1 \text{ or } k = -1. \\ 0, & \text{otherwise} \end{cases}$$

The psd of $Y(t)$ is therefore

$$S_Y(f) = \frac{1}{T}|G(f)|^2[1.25\sigma^2 + 0.5\sigma^2(e^{-j2\pi fT} + e^{j2\pi fT})]$$
$$= \frac{1}{T}|G(f)|^2[1.25\sigma^2 + \sigma^2 \cos(2\pi fT)].$$

3.5 PROBABILITY OF TRANSMISSION ERROR

3.5.1 Coherent Reception in an AWGN Channel

BPSK. Consider coherent BPSK reception over an AWGN channel, as shown in Figure 3.25. Without loss of generality, consider the detection of a symbol transmitted over the time interval $[0,\ T_b]$. The transmitted signal is

$$x(t) = \begin{cases} \sqrt{E_b}\varphi_1(t), & \text{for symbol "1"} \\ -\sqrt{E_b}\varphi_1(t), & \text{for symbol "0"} \end{cases}, \tag{3.5.1}$$

where $\varphi_1(t)$ is the basis function as defined in Eq. (3.2.6). In an AWGN channel, the received signal is

$$r(t) = x(t) + n(t),$$

where $n(t)$ represents the white Gaussian noise process with zero mean and two-sided psd $N_0/2$. The output of the correlator (or the matched filter at $t = T_b$)

$$r_1 = \int_0^{T_b} r(t)\varphi_1(t)\,dt$$

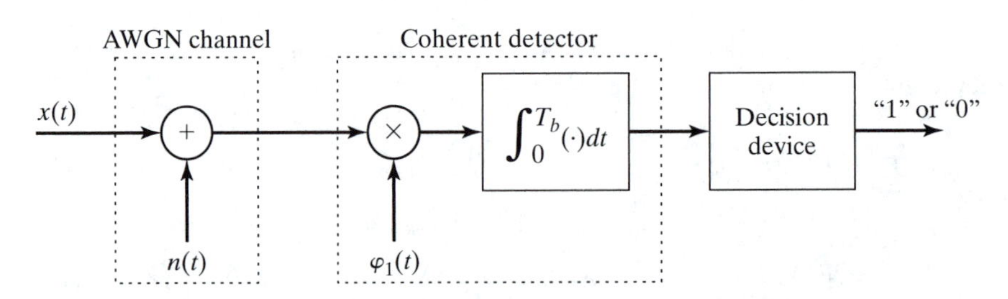

Figure 3.25 Coherent reception of BPSK in an AWGN channel.

is a Gaussian random variable with variance $\sigma_N^2 = N_0/2$ and conditional means $\mu_1 = \sqrt{E_b}$ given that "1" was sent and $\mu_0 = -\sqrt{E_b}$ given that "0" was sent. That is, the conditional probability density function (pdf) of the decision variable r_1 is given by

$$f_{r_1}(x) = \frac{1}{\sqrt{2\pi}\sigma_N} \exp\left[-\frac{(x-\mu_i)^2}{2\sigma_N^2}\right], \quad i = 0, 1; -\infty < x < \infty. \tag{3.5.2}$$

Corresponding to the decision regions shown in Figure 3.10, the decision rule is as follows:

$$\begin{cases} \text{if } r \geq 0, \text{ symbol "1" was sent;} \\ \text{if } r < 0, \text{ symbol "0" was sent.} \end{cases}$$

With equally likely symbols "1" and "0", the probability of symbol (bit) error, or bit error rate (BER), is

$$\begin{aligned} P_b &= P(r \geq 0 | \text{symbol "0" was sent}) P(\text{symbol "0" was sent}) \\ &\quad + P(r < 0 | \text{symbol "1" was sent}) P(\text{symbol "1" was sent}) \\ &= \frac{1}{2}\left\{\int_0^\infty \frac{1}{\sqrt{2\pi}\sigma_N} \exp\left[-\frac{(x-\mu_0)^2}{2\sigma_N^2}\right] dx + \int_{-\infty}^0 \frac{1}{\sqrt{2\pi}\sigma_N} \exp\left[-\frac{(x-\mu_1)^2}{2\sigma_N^2}\right] dx\right\} \\ &= Q\left(\sqrt{\frac{2E_b}{N_0}}\right), \end{aligned} \tag{3.5.3}$$

where $Q(t) \triangleq \frac{1}{\sqrt{2\pi}} \int_t^\infty \exp(-z^2/2)dz$ is called the Q-function, which defines the area under the standard Gaussian (with zero mean and unit variance) tail. The probability of correct reception is

$$P_c = 1 - P_e = 1 - Q\left(\sqrt{\frac{2E_b}{N_0}}\right). \tag{3.5.4}$$

In general, consider a memoryless binary modulation scheme with signals $x_1(t)$ and $x_2(t)$ having bit energy E_1 and E_2 respectively. The two signals have the correlation coefficient ρ and the squared Euclidean distance d^2. Given that symbols "1" and "0" are equally likely, using the matched filter receiver in the AWGN channel, the BER is given by

$$P_b = Q\left(\sqrt{\frac{(1-\rho)\bar{E}_b}{N_0}}\right) = Q\left(\sqrt{\frac{d^2}{2N_0}}\right), \tag{3.5.5}$$

where $\bar{E}_b = (E_1 + E_2)/2$ is the average bit energy. As the Q-function, $Q(\cdot)$, is monotonically decreasing as its argument increases, the BER decreases as the distance between the two signals in the signal space increases.

QPSK and $\pi/4$-DQPSK. Since QPSK is two BPSK at quadrature, the two BPSK signal components are orthogonal. Over an AWGN channel, the signal detections at the two branches are thus independent. The probability of detection error for the odd-numbered digits is the same as that for the even-numbered digits and is equal to that of BPSK. As a result, the probability of bit error for coherent QPSK is the same as that for coherent BPSK

$$P_b = Q\left(\sqrt{\frac{2E_b}{N_0}}\right). \tag{3.5.6}$$

A two-bit symbol is correctly detected only if both the odd-numbered and even-numbered digits are correctly detected. As the detection of the odd-numbered digit is independent of that of the even-numbered digit, the probability of correct symbol reception is

$$P_c = \left[1 - Q\left(\sqrt{\frac{2E_b}{N_0}}\right)\right]^2. \tag{3.5.7}$$

The probability of symbol error is therefore given by

$$P_s = 1 - P_c = 2Q\left(\sqrt{\frac{2E_b}{N_0}}\right) - \left[Q\left(\sqrt{\frac{2E_b}{N_0}}\right)\right]^2, \tag{3.5.8}$$

which can be approximated by $2Q\left(\sqrt{\frac{2E_b}{N_0}}\right)$ if $2E_b/N_0 \gg 1$.

In general, the probability of symbol error for MPSK with coherent demodulation is given by [123]

$$P_s \approx 2Q\left(\sqrt{2k\gamma_b}\sin\frac{\pi}{M}\right) \tag{3.5.9}$$

for $\gamma_b \gg 1$, where $k = \log_2 M$ and $\gamma_b = E_b/N_0$ is the SNR per bit. The BER of MPSK is well approximated by P_s/k.

For $\pi/4$-DQPSK with differential detection (e.g., using the receiver shown in Figure 3.15), the BER is given by [123]

$$P_b = Q(a, b) - \frac{1}{2}I_0(ab)\exp\left[-\frac{1}{2}(a^2 + b^2)\right], \tag{3.5.10}$$

where

$$a = \sqrt{2\left(1 - \frac{1}{\sqrt{2}}\right)\frac{E_b}{N_0}}, \quad b = \sqrt{2\left(1 + \frac{1}{\sqrt{2}}\right)\frac{E_b}{N_0}},$$

$I_0(x)$ is the zero-order modified Bessel function of the first kind, defined in Eq. (2.5.6), and $Q(a, b)$ is called the Marcum's Q function and is defined as

$$Q(a, b) = \int_b^\infty x\exp\left[-\frac{1}{2}(x^2 + a^2)\right] I_0(ax)\,dx. \tag{3.5.11}$$

MSK. The signal constellation of MSK shown in Figure 3.18 is the same as that of QPSK shown in Figure 3.11(a), except that in MSK both decision regions D_1 and D_3 are for symbol "1", and both D_2 and D_4 are for "0". Note that in MSK the total decision region for "1" or "0" is half of the signal space, similar to that for the odd- or even-numbered digits in QPSK. With the same locations of the signal points in the two signal spaces, it can be derived that the probability of bit error for MSK is the same as the probability of bit error for QPSK. That is, the probability of symbol (bit) error for MSK is

$$P_b = Q\left(\sqrt{\frac{2E_b}{N_0}}\right). \tag{3.5.12}$$

In summary, with coherent detection, BPSK, QPSK, and MSK have the same BER. Figure 3.26 shows plots of the BER performance of BPSK, QPSK, and MSK with coherent detection, and $\pi/4$-DQPSK with differential detection. It can be observed that the coherent detection has performance superior to that of differential detection.

Example 3.6 Bit-Error Rate for a Binary System

Consider a binary transmission system where, at the end of each symbol interval, the output of the demodulator is

$$r = \begin{cases} \alpha + n + Z, & \text{if "1" was sent} \\ -\alpha + n + Z, & \text{if "0" was sent} \end{cases},$$

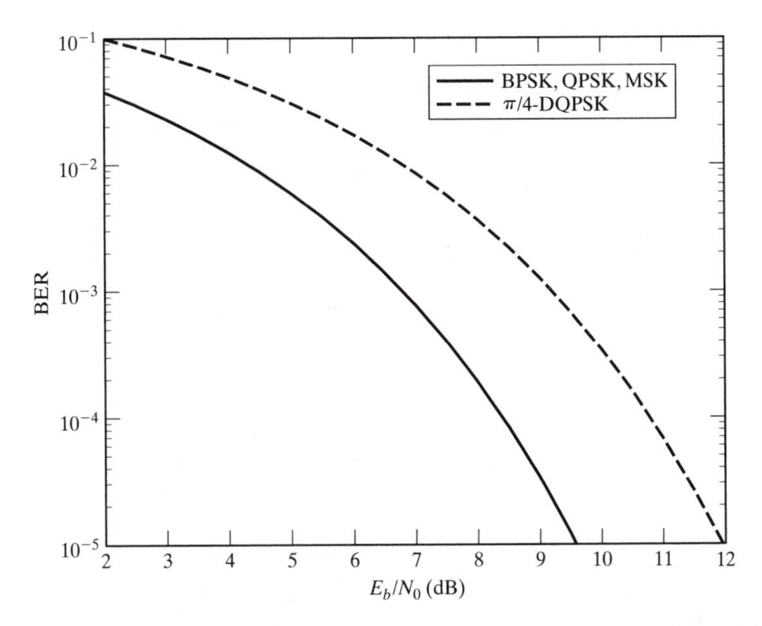

Figure 3.26 BER performance of coherently detected BPSK, QPSK and MSK, and differentially detected $\pi/4$-DQPSK in an AWGN channel.

where α (>0) is a constant, n is a Gaussian random variable with zero mean and variance σ^2, Z is a discrete random variable with probability distribution $P(Z = z) = 0.25$, $P(Z = 0) = 0.5$, and $P(Z = -z) = 0.25$. Assuming that symbols "1" and "0" are equally likely, derive an expression for the probability of bit error in terms of α, σ, and z.

Solution The mean of random variable Z is

$$E[Z] = z \times P(Z = z) + 0 \times P(Z = 0) + (-z) \times P(Z = -z) = 0.$$

As a result, we have $E(n + Z) = E(n) + E(Z) = 0$ since n has a zero mean. Furthermore, as symbols "1" and "0" have equal probability, the optimal decision threshold of the decision device following the demodulator is 0. The decision rule is as follows:

$$\begin{cases} \text{if } r \geq 0, \text{ then symbol "1" was sent;} \\ \text{if } r < 0, \text{ then symbol "0" was sent.} \end{cases}$$

Given $Z = 0$, the probability of bit error is

$$\begin{aligned} P_{b|Z=0} &= P(r < 0|\text{"1"}, Z = 0)P(\text{"1"}) + P(r \geq 0|\text{"0"}, Z = 0)P(\text{"0"}) \\ &= P(\alpha + n + Z < 0|Z = 0) \cdot \frac{1}{2} + P(-\alpha + n + Z \geq 0|Z = 0) \cdot \frac{1}{2} \\ &= \frac{1}{2}[P(n < -\alpha) + P(n \geq \alpha)] \\ &= Q\left(\frac{\alpha}{\sigma}\right). \end{aligned}$$

Similarly, given $Z = z$, we have

$$\begin{aligned} P_{b|Z=z} &= P(r < 0|\text{"1"}, Z = z)P(\text{"1"}) + P(r \geq 0|\text{"0"}, Z = z)P(\text{"0"}) \\ &= P(\alpha + n + Z < 0|Z = z) \cdot \frac{1}{2} + P(-\alpha + n + Z \geq 0|Z = z) \cdot \frac{1}{2} \\ &= \frac{1}{2}[P(n < -\alpha - z) + P(n \geq \alpha - z)] \\ &= \frac{1}{2}\left[Q\left(\frac{\alpha + z}{\sigma}\right) + Q\left(\frac{\alpha - z}{\sigma}\right)\right] \end{aligned}$$

and, given $Z = -z$, we have

$$P_{b|Z=-z} = \frac{1}{2}\left[Q\left(\frac{\alpha - z}{\sigma}\right) + Q\left(\frac{\alpha + z}{\sigma}\right)\right].$$

Hence, the probability of bit error is

$$\begin{aligned} P_b &= P_{b|Z=0}P(Z = 0) + P_{b|Z=z}P(Z = z) + P_{b|Z=-z}P(Z = -z) \\ &= \frac{1}{4}\left[2Q\left(\frac{\alpha}{\sigma}\right) + Q\left(\frac{\alpha + z}{\sigma}\right) + Q\left(\frac{\alpha - z}{\sigma}\right)\right]. \end{aligned}$$

Example 3.7 MSK with Noncoherent Demodulation

Consider the detection of MSK signals using a noncoherent receiver over an AWGN channel. Derive the probability of bit error.

Solution With noncoherent demodulation, the property of phase continuity in MSK is not exploited in the demodulation. As a result, the MSK signal is demodulated in the same way as a binary FSK signal without any phase constraint. Furthermore, the two MSK signals are no longer orthogonal. Without loss of generality, consider the transmission of the MSK signal over the symbol interval $[0, T_b]$. The MSK signal in $t \in [0, T_b]$ is

$$x(t) = \begin{cases} \sqrt{\dfrac{2E_b}{T_b}} \cos[2\pi f_1 t + \phi(0)], & \text{for symbol "0"} \\ \sqrt{\dfrac{2E_b}{T_b}} \cos[2\pi f_2 t + \phi(0)], & \text{for symbol "1"} \end{cases},$$

where $f_2 - f_1 = 1/(2T_b)$, and $\phi(0)$ is the carrier phase at $t = 0$. Note that (a) $f_1 \gg 1/T_b$ and $f_2 \gg 1/T_b$, and (b) $\phi(0)$ can take on the values 0, $\pi/2$, π, and $3\pi/2$ with equal likelihood. Figure 3.27 shows the structure of a noncoherent receiver for binary FSK, which consists of a frequency discriminator and a decision device. In the absence of additive noise, it can be easily

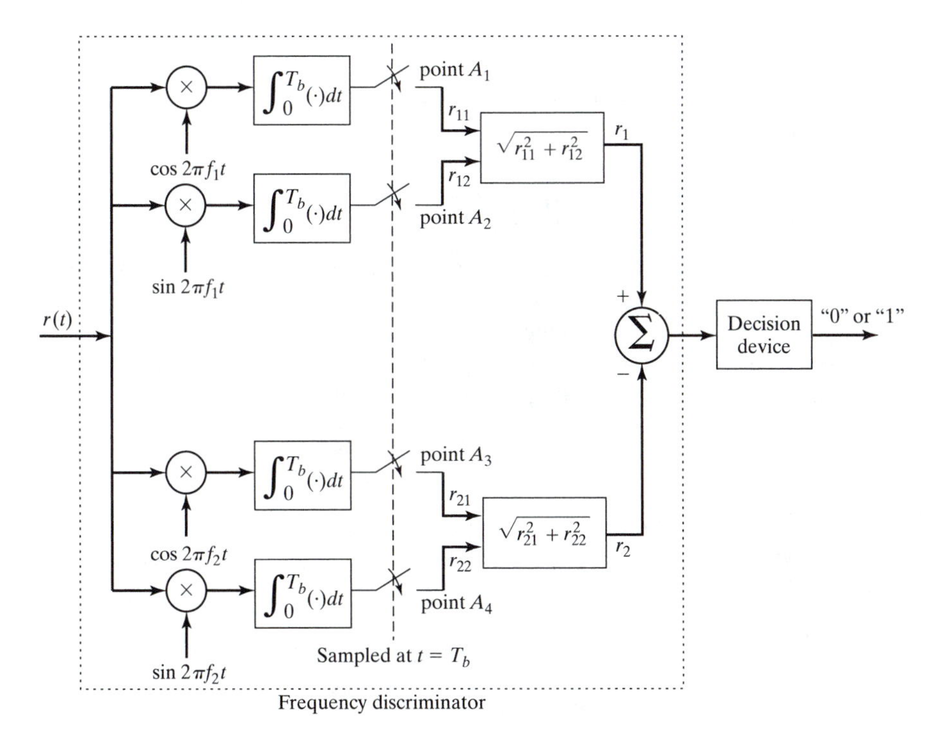

Figure 3.27 Noncoherent receiver structure for MSK.

derived that, given symbol "0" was sent,

$$r_{11} = \sqrt{E_b T_b/2}\cos[\phi(0)]$$
$$r_{12} = \sqrt{E_b T_b/2}\sin[\phi(0)]$$
$$r_{21} = |\rho|\sqrt{E_b T_b/2}\sin[\phi(0)]$$
$$r_{22} = |\rho|\sqrt{E_b T_b/2}\cos[\phi(0)]$$

and, given symbol "1" was sent,

$$\begin{array}{lcl} r_{11} & = & |\rho|\sqrt{E_b T_b/2}\sin[\phi(0)+\pi] \\ r_{12} & = & |\rho|\sqrt{E_b T_b/2}\cos[\phi(0)+\pi] \\ r_{21} & = & \sqrt{E_b T_b/2}\cos[\phi(0)] \\ r_{22} & = & \sqrt{E_b T_b/2}\sin[\phi(0)] \end{array},$$

where $|\rho| = 2/\pi$ is the absolute value of the correlation coefficient between the MSK signal and the carrier signal of the other frequency locally generated at the receiver. The outputs of the envelope detectors are then given by

$$r_1 = \sqrt{E_b T_b/2} \quad \text{and} \quad r_2 = |\rho|\sqrt{E_b T_b/2}$$

for symbol "0" and

$$r_1 = |\rho|\sqrt{E_b T_b/2} \quad \text{and} \quad r_2 = \sqrt{E_b T_b/2}$$

for symbol "1". Consider the input additive white Gaussian noise $n(t)$ with two-sided psd $N_0/2$. At point A_1, the noise component is

$$n_1 = \int_0^{T_b} n(t)\cos 2\pi f_1 t\, dt,$$

which is a Gaussian random variable with zero mean and variance $\sigma_N^2 = N_0 T_b/4$. Similarly, the noise component at point A_i, $i = 2, 3, 4$, is a Gaussian random variable with zero mean and variance $N_0 T_b/4$. Also, all the noise components, n_1, n_2, n_3, and n_4 are pair-wise independent. Based on the symmetry of the signal components and the same probabilistic behavior of the noise components, the optimum threshold is 0. Let $r = r_1 - r_2$. Then the decision rule is as follows:

$$\begin{cases} \text{if } r > 0, \text{ symbol "0" was sent;} \\ \text{if } r \leq 0, \text{ symbol "1" was sent.} \end{cases}$$

In the presence of the noise components, the pdf of the decision variable r, given symbol "0" was sent, is the same as that of $-r$, given symbol "1" was sent. Therefore, given that symbols "0" and "1" are equally likely, the BER is

$$\begin{aligned} P_b &= \frac{1}{2}P(r > 0|\text{"1"}) + \frac{1}{2}P(r \leq 0|\text{"0"}) \\ &= P(r > 0|\text{"1"}) \\ &= P(r_1^2 > r_2^2|\text{"1"}). \end{aligned}$$

Given that symbol "1" was sent, we have

$$r_1 = \sqrt{[|\rho|(\sqrt{E_bT_b/2})\sin(\phi(0)+\pi)+n_1]^2 + [|\rho|(\sqrt{E_bT_b/2})\cos(\phi(0)+\pi)+n_2]^2}$$

and

$$r_2 = \sqrt{[(\sqrt{E_bT_b/2})\cos\phi(0)+n_3]^2 + [(\sqrt{E_bT_b/2})\sin\phi(0)+n_4]^2}.$$

Since $\phi(0)$ can take on the values of $0, \pi/2, \pi, 3\pi/2$, each with a probability of $1/4$, r_1 and r_2 can be equivalently represented as

$$\sqrt{\left(|\rho|\sqrt{\frac{E_bT_b}{2}}+n_x\right)^2+n_y^2} \quad \text{and} \quad \sqrt{\left(\sqrt{\frac{E_bT_b}{2}}+n_x\right)^2+n_y^2},$$

respectively, where n_x and n_y are independent Gaussian random variables with zero mean and equal variance $\sigma_n^2 = N_0T_b/4$. As a result, from Appendix D, r_1 and r_2 follow a Rician distribution with pdfs given by

$$f_{r_1}(x|\text{“1”}) = \begin{cases} \dfrac{x}{\sigma_n^2}\exp\left(-\dfrac{x^2+\dfrac{|\rho|^2E_bT_b}{2}}{2\sigma_n^2}\right) I_0\left(\dfrac{x|\rho|}{\sigma_n^2}\sqrt{\dfrac{E_bT_b}{2}}\right), & x \ge 0 \\ 0, & x < 0 \end{cases}$$

and

$$f_{r_2}(x|\text{“1”}) = \begin{cases} \dfrac{x}{\sigma_n^2}\exp\left(-\dfrac{x^2+\dfrac{E_bT_b}{2}}{2\sigma_n^2}\right) I_0\left(\dfrac{x}{\sigma_n^2}\sqrt{\dfrac{E_bT_b}{2}}\right), & x \ge 0 \\ 0, & x < 0 \end{cases},$$

respectively. From the conditional pdfs, it can be derived that the BER is given by [123]

$$P_b = Q(a,b) - \frac{1}{2}I_0(ab)\exp\left[-\frac{1}{2}(a^2+b^2)\right],$$

where

$$a = \sqrt{\frac{E_b}{2N_0}\left(1-\sqrt{1-|\rho|^2}\right)} \quad \text{and} \quad b = \sqrt{\frac{E_b}{2N_0}\left(1+\sqrt{1-|\rho|^2}\right)},$$

$Q(a, b)$ is the Marcum's Q function defined in Eq. (3.5.11), and $I_0(x)$ is the zero-order modified Bessel function of the first kind, defined in Eq. (2.5.6). Figure 3.28 shows the BER performance. For comparison, the BER performance with coherent detection is also included. It is observed that the coherent demodulation achieves a gain of more than 7 dB (at a BER of 10^{-5}) in the input SNR/bit by exploiting the phase continuity property of MSK and the orthogonality between the two MSK signals. The strong correlation between the two MSK signals in the noncoherent detection increases the BER significantly.

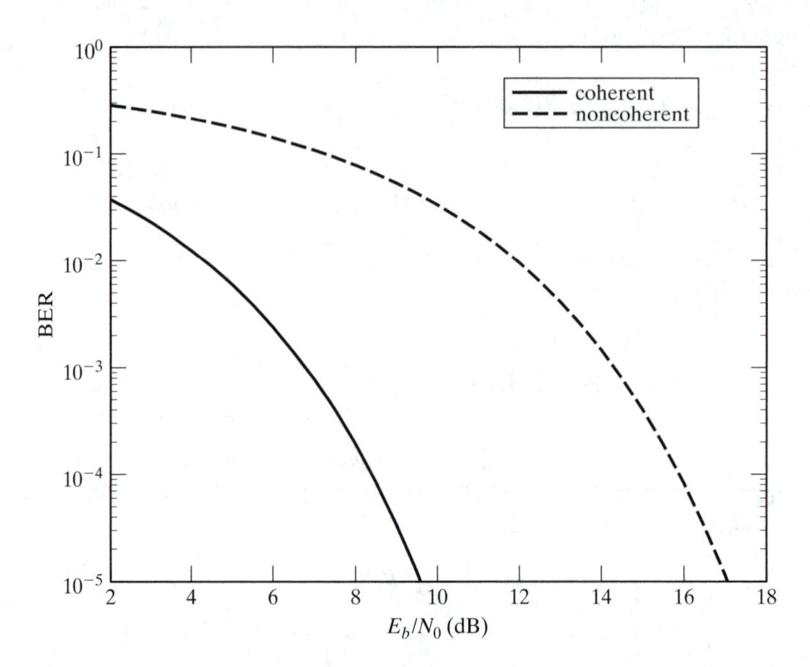

Figure 3.28 BER performance MSK with coherent and noncoherent demodulation.

3.5.2 Coherent Reception in a Flat Slow Rayleigh Fading Channel

We consider a stationary flat and slow fading channel where (*a*) the delay spread introduced by the multipath propagation environment is negligible compared with the symbol interval (hence, the channel does not introduce intersymbol interference) and (*b*) channel fading status does not change much over a number of symbol intervals. The first condition means that the effect of the channel can be represented by a complex gain $\alpha(t)\exp[j\theta(t)]$, where $\alpha(t)$ is the amplitude fading and $\theta(t)$ is the phase distortion. Given that a signal $x(t)$, with symbol interval T_s, is transmitted, the received signal is

$$r(t) = \alpha(t)\exp[j\theta(t)]x(t) + n(t), \tag{3.5.13}$$

where $n(t)$ is white Gaussian noise with zero mean and two-sided power spectral density $N_0/2$. The second condition means that it is possible for the receiver to estimate $\theta(t)$ and remove it. As a result, in the following BER performance analysis, we assume $\theta(t) = 0$ without loss of generality.

Consider a cellular system where the effect of propagation path loss and shadowing on the received signals is compensated for by power control and the received signals experience only multipath Rayleigh fading. In other words, we consider that the transmitted signal has a constant bit energy E_b and the received signal has instantaneous bit energy equal to $\alpha^2 E_b$, where α is the amplitude fading in the symbol interval. Based on the transmission performance analysis for an AWGN channel, for a given modulation scheme, its probability of bit error can be represented as a function, $P_b(\cdot)$, of the received bit energy to the one-sided noise power spectral density E_b/N_0

(or SNR/bit), denoted by γ_b. In the following, we will extend the analysis to a fading channel in two steps: (a) to find the conditional probability of bit error $P_{b|\alpha}(\gamma_b|\alpha)$, given the amplitude fading α; and (b) to average the conditional probability $P_{b|\alpha}(\gamma_b|x)$ with respect to the pdf of α at $\alpha = x$, in order to take into account the effect of all possible amplitude fading values on the transmission performance. Therefore,

$$P_b(\overline{\gamma}_b) = \int_{-\infty}^{\infty} P_{b|\alpha}(\gamma_b|x) f_\alpha(x)\, dx, \tag{3.5.14}$$

where $\overline{\gamma_b}$ is the average received SNR/bit with respect to α^2,

$$\begin{aligned}\overline{\gamma_b} &= \int_{-\infty}^{\infty} x^2 \gamma_b f_\alpha(x)\, dx \\ &= \gamma_b E(\alpha^2)\end{aligned}$$

and $f_\alpha(\cdot)$ is the pdf of the amplitude fading α. For a Rayleigh fading channel, α follows a Rayleigh distribution with pdf

$$f_\alpha(x) = \begin{cases} \dfrac{x}{\sigma_\alpha^2} \exp\left(-\dfrac{x^2}{2\sigma_\alpha^2}\right), & x \geq 0 \\ 0, & x < 0 \end{cases}. \tag{3.5.15}$$

It can be derived that $E(\alpha^2) = 2\sigma_\alpha^2$. In this case, we have $\overline{\gamma_b} = 2\sigma_\alpha^2 \gamma_b$. For coherent BPSK, QPSK, and MSK, from Eq. (3.5.3), we have

$$P_{b|\alpha}(\gamma_b|x) = Q(\sqrt{2x^2\gamma_b}). \tag{3.5.16}$$

Substituting Eqs. (3.5.16) and (3.5.15) into Eq. (3.5.14), we have

$$\begin{aligned}P_b(\bar{\gamma}_b) &= \int_0^{\infty} Q(\sqrt{2x^2\gamma_b}) \frac{x}{\sigma_\alpha^2} \exp\left(-\frac{x^2}{2\sigma_\alpha^2}\right) dx \\ &= \frac{1}{2\sigma_\alpha^2} \int_0^{\infty} Q(\sqrt{2y\gamma_b}) \exp\left(-\frac{y}{2\sigma_\alpha^2}\right) dy \quad (\text{where } y = x^2) \\ &= \frac{1}{2\sigma_\alpha^2} \int_0^{\infty} \left[\frac{1}{\sqrt{2\pi}} \int_{\sqrt{2y\gamma_b}}^{\infty} \exp(-u^2/2)\, du\right] \exp\left(-\frac{y}{2\sigma_\alpha^2}\right) dy \\ &= \frac{1}{2\sigma_\alpha^2} \frac{1}{\sqrt{2\pi}} \int_0^{\infty} \exp(-u^2/2) \left[\int_0^{u^2/2\gamma_b} \exp\left(-\frac{y}{2\sigma_\alpha^2}\right) dy\right] du \\ &= \frac{1}{2}\left[1 - \sqrt{\frac{\overline{\gamma}_b}{1+\overline{\gamma}_b}}\right].\end{aligned} \tag{3.5.17}$$

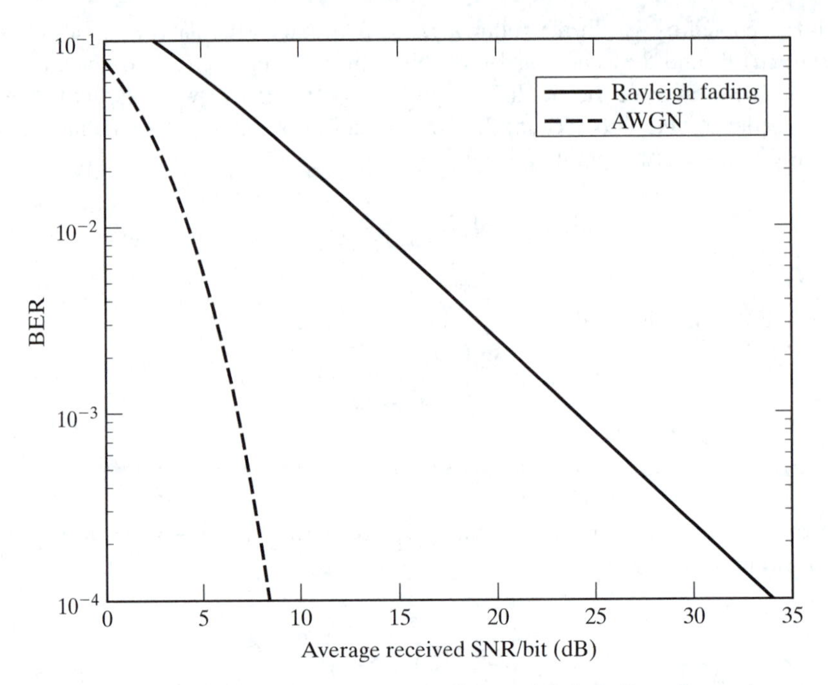

Figure 3.29 BER performance over flat Rayleigh fading channels.

By using $\frac{1}{\sqrt{1+x}} \approx 1 - \frac{x}{2}$ for $x \ll 1$, we have the following approximation for coherent BPSK, QPSK, and MSK for $\bar{\gamma}_b \gg 1$:

$$P_b \approx \frac{1}{4\bar{\gamma}_b}.$$

Figure 3.29 shows the BER transmission performance of coherent BPSK, QPSK, and MSK in a Rayleigh fading channel and an AWGN channel, respectively. It is observed that the amplitude fading severely degrades the transmission performance. In an AWGN channel, for a large value of γ_b, the probability of error decreases exponentially with respect to γ_b^2. However, in a Rayleigh fading channel, the probability of error decreases linearly with respect to $\bar{\gamma}_b$ for $\bar{\gamma}_b \gg 1$. The smaller the P_b value required, the worse the performance degradation. This is because, in the fading environment, the received signal amplitude changes with time. When the signal is in deep fading (i.e., $\alpha(t) \ll 1$), the instantaneous SNR/bit drops significantly, resulting in a very high chance of transmission error, even though most of the time the instantaneous γ_b value of the received signal is larger than the corresponding one in an AWGN channel.

Example 3.8 BPSK in a Fading Channel

Consider digital transmission via BPSK over a fading channel. At any instant, the channel provides a gain of 1.0 with a probability of 0.9 and a gain of 0.05 with a probability of 0.1. Derive the BER under the assumption of coherent detection.

Solution The BER performance of BPSK over the AWGN channel is

$$P_{b,AWGN} = Q\left(\sqrt{\frac{2E_b}{N_0}}\right).$$

Given that the channel gain is equal to 1.0, the conditional BER is $P_{b1} = P_{b,AWGN}$; Given that the channel gain is equal to 0.05, the conditional BER is $P_{b2} = Q\left(\sqrt{\frac{0.05^2 \times 2E_b}{N_0}}\right)$. The overall BER is then

$$\begin{aligned} P_b &= P(\text{bit error}|\text{gain} = 1.0)P(\text{gain} = 1.0) + P(\text{bit error}|\text{gain} = 0.05)P(\text{gain} = 0.05) \\ &= 0.9P_{b1} + 0.1P_{b2} \\ &= 0.9Q\left(\sqrt{\frac{2E_b}{N_0}}\right) + 0.1Q\left(\sqrt{\frac{0.05^2 \times 2E_b}{N_0}}\right). \end{aligned}$$

Given that the transmitted bit energy is E_b, the average received bit energy is $\bar{E}_b = (1.0^2 E_b) \times 0.9 + (0.05^2 E_b) \times 0.1 = 0.90025E_b$. The BER, in terms of the received SNR/bit, $\bar{\gamma}_b = \bar{E}_b/N_0$, is then

$$P_b = 0.9Q(\sqrt{2\bar{\gamma}_b/0.90025}) + 0.1Q(\sqrt{2\bar{\gamma}_b \times 0.0025/0.90025}).$$

Figure 3.30 shows the BER performance over the fading channel. For comparison, the performance for the AWGN channel is also included. The effect of the channel fading on increasing

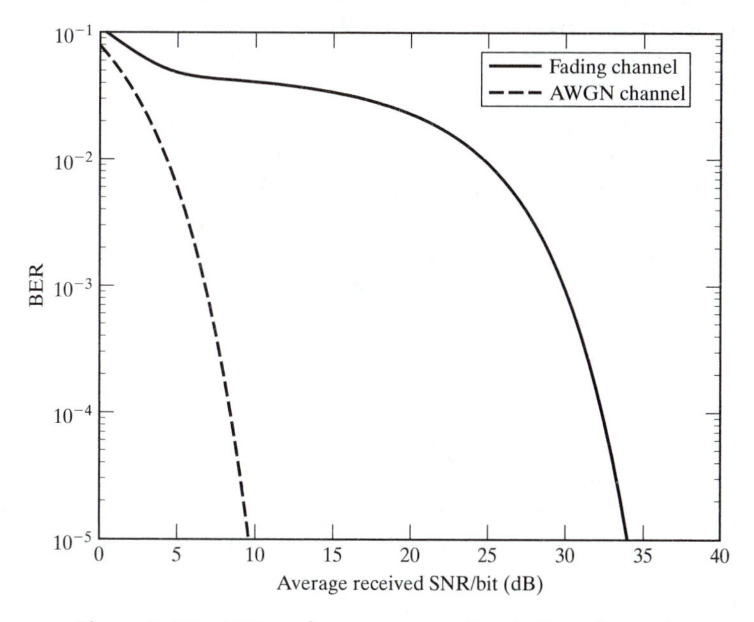

Figure 3.30 BER performance over the fading channel.

the BER is clearly observed. The performance degradation in the fading channel is completely due to the 10% chance of having the deep fading with channel gain equal to 0.05. The loss is approximately 24.5 dB at the BER of 10^{-5}.

SUMMARY

In this chapter, we have addressed bandpass signaling techniques normally used for transmission over a radio channel, and examined the psd of the modulated signals. We have also studied the transmission error performance in an AWGN channel and in a slow Rayleigh fading channel. It has been shown that the transmission accuracy is severely degraded by channel fading. In the next chapter, we will discuss two channel impairment mitigation techniques, namely: diversity reception to combat channel flat fading, and channel equalization to reduce intersymbol interference due to channel time dispersion.

ENDNOTES

1. MSK was invented by Doelz and Heald in 1961 [41]. For the development of MSK and how it compares with other modulation schemes such as QPSK, see the paper by Pasupathy [110].
2. GMSK was first proposed by Murota and Hirade in 1981 [102]. For the psd of GMSK signals, see the paper [102].
3. $\pi/4$-DQPSK was invented by Baker [9]. For application of the modulation to mobile cellular systems, see the paper by Akaiwa and Nagata [4].
4. OFDM was originally proposed by Cimini [76]. For more discussion, see papers by Casas and Leung [23, 24], and by Zou and Wu [166].
5. For detailed discussion of power spectral density, see [70, 132, 123, 157].
6. This chapter deals with modulation schemes used in the second generation cellular systems (i.e., $\pi/4$-DQPSK, GMSK, and OFDM) and the basic associated schemes such as QPSK and MSK. For a broad coverage of digital modulation, see textbooks by Haykin [62], Proakis and Salehi [124], and Sklar [142]. For more advanced studies on digital modulation, see the textbooks by Proakis [123], Stüber [147], and Wilson [157].

PROBLEMS

P3-1 Consider voice transmission using BPSK signaling at a transmission rate of 2400 bps and a bandwidth efficiency of 1 bps/Hz. It is desired to increase the data rate to 19.2 kbps. If the number of points in the signal constellation is increased until the data rate becomes 19.2 kbps while the symbol rate remains at 2400 symbols/s, determine

a. the number of points in the constellation, and
b. the bandwidth efficiency factor of the modulation scheme.

Note: Bandwidth efficiency is defined as the ratio R_b/W, where R_b is the bit rate and W is the system bandwidth.

P3-2 If we increase the symbol transmission rate to 19.2 kbps, as in Problem P3-1, and use the same BPSK modulation, determine the additional power requirement for the transmission to maintain the same transmission accuracy as a 2.4 kbps modulation technique.

P3-3 Consider the signal

$$x(t) = \begin{cases} \dfrac{At}{T_s} \sin(2\pi f_c t), & 0 \le t \le T_s \\ 0, & \text{otherwise} \end{cases}.$$

a. Determine the impulse response of the matched filter for the signal.
b. Determine the output of the matched filter at $t = T_s$.
c. Suppose the signal $x(t)$ is passed through a correlator that correlates the input $x(t)$ with $x(t)$. Determine the value of the correlator output at $t = T_s$. Compare your result with that in part (b).

P3-4 Draw block diagrams of both transmitter and receiver (with coherent detection) of a QPSK system for transmitting data at a rate of 2400 bps and a carrier frequency of $f_c = 1800$ kHz.

P3-5 Suppose the binary bit stream "**01001110**" is sent using $\pi/4$-DQPSK, with the left most bits applied to the transmitter first. Assuming that $\Phi_0 = 0$, determine the carrier phase Φ_n, the in-phase component I_n, and the quadrature component Q_n, for the input sequence.

P3-6 Determine the minimum frequency separation for orthogonality between the binary FSK signals where carrier phase continuity is not required.

P3-7 Verify that the impulse response $h(t)$ given in Eq. (3.3.9) and the transfer function $H(f)$ given in Eq. (3.3.8) for the filter in the GMSK transmitter are a Fourier transform pair.

P3-8 OFDM signaling may be used to cope with multipath delay spread.

a. Discuss why and how OFDM is a suitable signaling scheme for coping with delay spread introduced by the propagation channel.
b. Discuss how Doppler can affect the performance of OFDM, and what countermeasure(s) may be introduced to compensate for the Doppler effect.

P3-9 Derive the psd of $\pi/4$-DQPSK under the assumptions that (a) symbols "1" and "0" are equally likely, (b) all the symbols are independent of each other, and (c) a rectangular baseband pulse with duration T_s is used.

P3-10 Consider a QPSK signal that is represented by the equivalent low-pass signal

$$v(t) = \sum_{n=-\infty}^{\infty} a_n g(t - nT_s),$$

where a_n takes on four possible and equally likely values $\{1, j, -1, -j\}$. The sequence $\{a_n\}$ is statistically independent from symbol to symbol.

a. Determine and sketch the psd of $v(t)$ when

$$g(t) = \begin{cases} A, & 0 \le t \le T_s \\ 0, & \text{otherwise} \end{cases}.$$

b. Repeat part (a) when

$$g(t) = \begin{cases} A, & 0 \le t \le T_s/2 \\ -A, & T_s/2 < t \le T_s \\ 0, & \text{otherwise} \end{cases}.$$

c. Compare the spectra obtained in parts (a) and (b) in terms of the bandwidth to the first spectral null.

P3-11 Determine an expression of the psd for the signal $x(t) = v(t)\cos(2\pi f_c t)$, with the baseband pulse train

$$v(t) = \sum_{k=-\infty}^{\infty} a_n g(t - nT_b),$$

for the two cases, where $g(t)$ is a pulse shape defined on $[0, T_b]$, and the random variables $\{a_n\}$ are defined in terms of independent and identically distributed random variables $\{b_n\}$ as follows:

a. b_n is a binary random variable, with values -1 and $+1$ equally likely, and $a_n = b_n + 0.5b_{n-1}$, $n = \ldots, -2, -1, 0, 1, 2, \ldots$.
b. $b_n \sim N(0, \sigma^2)$ and $a_n = b_n + \frac{1}{3}a_{n-1}$.

P3-12 Information digits a_n which can take on values $+1$ or -1 with equal likelihood are processed as shown in Figure 3.31. The digits b_n are ternary, with values $+2$, 0 and -2. These digits are transmitted by using no pulse for 0 and a pulse $g(t)$ or $-g(t)$ for digits $+2$ and -2, respectively. $G(f) = \mathcal{F}[g(t)]$. Show that the resulting signal has the following psd

$$\Phi_Y(f) = \frac{4}{T}|G(f)|^2 \cos^2(\pi f T).$$

P3-13 Verify that, in Figure 3.25, the output of the coherent detector is a Gaussian random variable, which has variance $N_0/2$, mean $\sqrt{E_b}$, given that "1" was sent, and mean $-\sqrt{E_b}$, given that "0" was sent.

P3-14 Suppose that BPSK is used for transmitting information over an AWGN channel of zero mean and two-sided psd $N_0/2$ watts/Hz. The transmitted energy is $E_b = A^2 T_b/2$, where T_b is the bit interval and A is the signal amplitude. Determine the signal amplitude required to achieve an error probability of 10^{-6} if the data rate is: (a) 10 kbps, (b) 100 kbps.

P3-15 Consider the use of BPSK and QPSK signaling over an AWGN channel, where the noise has zero mean and two-sided psd $N_0/2$ watts/Hz.

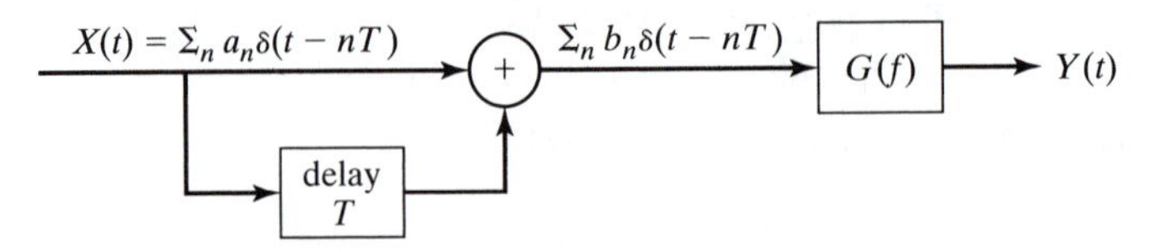

Figure 3.31 A transmitter structure.

a. Explain the similarities and the differences between these two signaling schemes in terms of transmission rate and error performance.
b. From the error performance point of view, if QPSK is slightly worse than BPSK, explain why.

P3-16 For BPSK signaling over an AWGN channel in which the two binary signals are transmitted with equal likelihood, the probability of bit error is given by

$$P_b = Q\left(\sqrt{\frac{2E_b}{N_0}}\right),$$

where E_b is the signal bit energy and $N_0/2$ is the two-sided noise psd. Suppose that the ratio E_b/N_0 is 10 dB.

a. Calculate the probability of bit error for BPSK signaling in the AWGN channel.
b. Suppose $N_0/2 = 0.5 \times 10^{-10}$ watts/Hz, $E_b/N_0 = 10$ dB, and the data rate is $R_b = 100$ kbps. Calculate the amplitude A of the low-pass rectangular envelope.

P3-17 Given a carrier phase synchronization error x, the BER for coherent BPSK is

$$P_e = Q\left(\sqrt{\frac{2E_b}{N_0}\cos^2(x)}\right).$$

If x is a random variable uniformly distributed over $[-a, +a]$, where $0 < a \ll \pi$, determine the average BER as a function of a.

P3-18 Verify the BER performance given in Eq. (3.5.5) for binary modulation schemes.

P3-19 Consider the transmission of a BPSK signal over a fading channel. The channel introduces an amplitude gain α, in addition to the additive white Gaussian noise with zero mean and two-sided psd of $N_0/2$. The channel gain α is a random variable, taking on values $\{0, 0.25, 0.5, 0.75\}$ with equal likelihood. The receiver employs a filter matched to the transmitted waveform for detection. Determine the BER as a function of E_b/N_0, where E_b is the bit energy of the transmitted signal. What value does the BER approach as E_b/N_0 approaches infinity?

P3-20 Consider a flat fading channel that exhibits Rayleigh fading characteristics. Let $x(t)$ be the transmitted signal. The received signal is given by

$$r(t) = \alpha(t) \cdot x(t) + n(t)$$

where $\alpha(t)$ is the amplitude gain of the Rayleigh fading channel and $n(t)$ is AWGN of zero-mean and two-sided psd $N_0/2$ watts/Hz.

a. Determine the distribution of $\alpha^2(t)$ at any t.
b. If $\gamma_b = E_b/N_0$ is the received SNR/bit in the absence of channel fading, what is the received SNR/bit for the Rayleigh fading channel?
c. What is the distribution of the received SNR/bit in the presence of Rayleigh fading?

P3-21 Suppose that the transmitted signal is an equally likely binary antipodal signal set, i.e., $-x_0(t) = x_1(t)$ and $\int_0^T x_m^2(t)\,dt = E_b,\ \ m = 0, 1$. The channel consists of a direct path and a large number of undesired paths, the macroscopic effect of which results in a totally random signal-dependent noise. Only the signal emerging from the direct path is accurately recoverable. The received signal is $r(t) = x_m(t) + z(t) + n(t), \quad 0 \le t \le T$, where $x_m(t)$ is the signal from the direct path, $z(t) = \alpha x_m(t)$ is the signal-dependent noise, and $n(t)$ is the signal-independent white Gaussian noise with zero mean and two-sided psd $N_0/2$. Assuming that (a) α is a Gaussian random variable with zero mean and variance σ_α^2 and is independent from symbol to symbol; (b) $z(t)$ and $n(t)$ are statistically independent random processes; and (c) a filter which is matched to the transmitted signal $x_1(t)$ is employed to receive $r(t)$, derive an expression for the probability of bit error. Explain the asymptotic behavior of the probability as a function of E_b/N_0.

P3-22 a. Verify that the probability of binary FSK with noncoherent detection in an AWGN channel is (see Example 3.7)

$$P_b = \frac{1}{2}\exp(-0.5E_b/N_0),$$

where E_b is the signal bit energy, N_0 is the one-sided psd of the additive white Gaussian noise, and the correlation between the two FSK signals $\rho = 0$;

b. Find the average probability of error for a flat slow Rayleigh fading channel.

P3-23 We want to evaluate the BER performance of $\pi/4$-DQPSK using differential detection over a slow Rayleigh fading channel via computer simulation. The receiver structure is shown in Figure 3.15. For simplicity, we can simulate the transmission system equivalently at baseband, as long as the channel complex gain remains approximately constant over each symbol interval.

a. Evaluate the BER performance over an AWGN channel by simulation and verify the simulation result using the analytical value given in Eq. (3.5.10);
b. Evaluate the BER performance over a flat Rayleigh fading channel with normalized fading rate $\nu_m T_s = 0.001,\ 0.01,\ 0.1$, respectively, where ν_m is the maximum Doppler shift and T_s is the symbol interval, for the average received SNR/bit from 0 dB to 40 dB;
c. Comment on the effect of (1) channel fading and (2) channel fading rate on the BER performance. Explain.

4

Receiver Techniques for Fading Dispersive Channels

In addition to multiplicative and additive noise, the propagation channels may exhibit both frequency and time dispersion that lead to signal distortion known as fading and intersymbol interference (ISI). Diversity combining and channel equalization, as methods to mitigate channel impairments, are described in this chapter.

4.1 OVERVIEW OF CHANNEL IMPAIRMENT MITIGATION TECHNIQUES

As explained in Chapter 2, in a wireless mobile communications system, due to user mobility and multipath propagation, transmitted signals often experience channel fading and time dispersion. Channel gain fluctuations can be decomposed into long-term fading and short-term fading. Long-term fading is mainly due to shadowing and variations in the distance between the mobile and base station. It changes with time at a relatively slow rate. Power control can be used to compensate for the effect of long-term fading on digital transmission quality and on cellular system capacity. Short-term fading is mainly due to multipath propagation. It changes with time at a much faster rate compared with that of long-term fading. Power control in general is not effective to combat multipath fading; instead, diversity and error-correction coding are used. To overcome the effect of propagation delay dispersion due to multipath, channel equalizers for FDMA and TDMA (narrowband) systems and Rake receivers for CDMA (wideband) systems can be used.

Power Control. Power control is used to maintain the minimum necessary transmitted power for reliable communication. As mentioned in Chapter 1, and will be discussed in Chapter 5, in cellular systems, the service area is divided into radio cells. Mobile users in each cell are served by a single base station. Different cells can use the same frequency spectrum subject to the interference constraint. Power control is essential in cellular systems for high system capacity and satisfactory transmission quality by (*a*) limiting interference among users in different cells using the same frequency channel (i.e., cochannel interference), and (*b*) alleviating the near–far effect which reduces interference among users in the same or different cells using adjacent frequency channels. In addition, it prolongs the battery power of mobile terminals and thus allows for lighter and lower cost mobile terminals. In CDMA systems, power control is required for system operation, as CDMA system capacity is interference limited (to be discussed in Chapter 6).

Near–far effect is a condition in which a nearby transmitter captures the receiver of the mobile or base station so that the latter is unable to detect the signal of a second transmitter located farther away. The near–far effect can be equalized through power control.

There are basically two types of power control: open-loop power control, and closed loop power control. In open-loop power control, the transmitter estimates the channel condition (gain) and adjusts its transmission power accordingly. The channel estimate is based on the transmission on the other link, under the assumption that the uplink and downlink channel conditions are closely correlated. In FDD systems, the uplink and downlink use different carrier frequencies, which leads to different Doppler shifts and different channel fading conditions. However, the effect of the carrier frequency difference on the long-term fading due to shadowing and distance dependent path loss is not as significant as that on the short-term multipath fading. In TDD systems, both the uplink and downlink share the same frequency channel at different time periods. If the long-term fading does not change much from the time period for one link to the next time period for the other link, the channel estimation should be reasonably accurate and the open-loop power control is effective. With no need for feedback information from the receiver, open-loop power control does not require a resource overhead for signaling and is relatively fast in controlling the power, especially when there is a sudden change in the channel condition (such as when the mobile user moves into the shadow of a large building). However, it suffers from poor channel estimation accuracy from time to time. In closed-loop power control, the receiver estimates (*a*) the received power level, (*b*) the received signal to interference ratio, or (*c*) the transmission error rate, compares the measurement with the target desired value, and sends the information for power control adjustment to the transmitter. The closed-loop power control is more accurate as it is based on the condition of the actual transmission link; however, it suffers from feedback delay, especially when the channel condition changes fast, and requires a resource overhead for the feedback signaling. The two types of power control are combined in practice. For example, at the beginning of a new transmission (such as after a silent period in packetized voice transmission) when the feedback information is not available, open-loop power control is used to determine the initial proper transmitted power; after a transmission period, the closed-loop power control replaces the open-loop power control.

Diversity. As shown in Subsection 3.5.2, the short-term multipath fading can severely reduce transmission accuracy. For example, to achieve a BER of 10^{-4} for BPSK with coherent reception over a flat Rayleigh fading channel, the required average received SNR per bit is approximately

26 dB higher than that required for an AWGN channel, as shown in Figure 3.29. Diversity is an effective way to combat channel fading.

Diversity improves transmission performance by making use of more than one independently faded version of the transmitted signal. If several replicas of the signal, carrying the same information, are received over multiple channels that exhibit independent fading with comparable strengths, the chances that all the independently faded signal components experience deep fading simultaneously are greatly reduced. This will significantly improve transmission accuracy as transmission errors are most likely to happen when the instantaneous SNR is low during a deep fading period. Diversity is a commonly used technique in wireless systems to combat channel fading, due to the following facts:

a. The degradation of transmission quality due to channel fading cannot be simply overcome by increasing the transmitted signal power. This is because, even with high transmitted power, when the channel is in deep fading, the instantaneously received SNR per bit can still be very low, resulting in a high probability of transmission error during the deep fading period;
b. In wireless communications, the power available on the reverse link is severely limited by the battery capacity in hand-held subscriber units. With diversity, the required transmitted power can be greatly reduced;
c. Cellular communications systems are mostly interference limited and, once again, mitigation of channel fading by diversity reception can translate into improved interference tolerance which, in turn, means greater ability to support additional users and therefore higher system capacity.

Detailed discussion of diversity, including diversity schemes, combining techniques, and performance improvement, is given in Section 4.2.

Channel Equalization. From Chapter 2 we also know that, in high rate transmission, when the signal bandwidth is larger than the channel coherence bandwidth, the channel exhibits frequency-selective fading. In addition to amplitude fluctuation and carrier phase distortion, the channel introduces propagation delay dispersion, which results in intersymbol interference (ISI). ISI can significantly increase transmission error rate. Different from additive background noise, ISI increases as the transmitted signal power increases and, therefore, its effect cannot be reduced simply by increasing the transmitted signal power. In FDMA and TDMA systems, equalization is an effective way to combat ISI due to the channel time dispersion. The device which equalizes the dispersive effect of a channel with memory is referred to as an equalizer. For channel equalization at baseband, we can design an equalizer and place it between the demodulator and decision device such that the output of the equalizer (i.e., the input of the decision device) is ISI free. If a linear equalizer is used, then ideally the transfer function of the equalizer should be the inverse of the channel transfer function, so that the effect of the channel at the decision device input is completely compensated for by the equalizer. For a time-variant wireless channel, the channel transfer function changes with time. As a result, the equalizer should be able to adapt to the channel variation for ISI-free transmission. In general, an adaptive channel equalizer should be used to improve transmission accuracy over a time-dispersive wireless channel. Channel equalization techniques are studied in detail in Section 4.3.

Spread Spectrum and Rake Receiver. In direct sequence CDMA systems, an information carrying signal is spread in the frequency domain by a pseudorandom noise (PN) sequence having a rate (called chip rate) much higher than the information symbol rate. Each symbol of the PN sequence is called a chip. After the spread spectrum modulation, the transmitted signal normally has a bandwidth much larger than the coherence bandwidth of a wireless channel. That is, the channel exhibits frequency-selective fading at the chip level (and probably at the data symbol level too). In order to recover the transmitted information sequence, the receiver needs to generate the same PN sequence and synchronize the locally generated sequence with the incoming one to despread the received signal. By making use of the autocorrelation property of the PN sequence (see Appendix E), the receiver can use multiple correlators to separate the received signal components of different propagation delays to a chip interval. Then, by estimating the channel gain experienced by each signal component, the receiver can combine all the signal components coherently by properly compensating for the propagation delay and carrier phase distortion. In fact, after aligning all the received signal components in time by delay compensation, the components are combined in the same way as in diversity combining (to be discussed in Subsection 4.2.2). With the ability to separate and then constructively combine the signal components with different delays, the receiver can collect the transmitted signal energy from different propagation paths and, therefore, overcome the transmission performance degradation due to delay dispersion. The operation of the receiver is somewhat analogous to that of a garden rake, therefore the name Rake receiver. Rake receiver is an example of implicit diversity systems.

Channel Coding and Interleaving. Channel coding is an elaborate cross-checking technique to overcome transmission errors over a noisy channel [86, 115]. It introduces redundancy in a deterministic manner, at the transmitter, to the information sequence and exploits the controlled redundancy at the receiver demodulator output for error correction. The redundancy may seem to be a waste of system resources such as frequency bandwidth. However, if an error-correction code is designed properly, using the same resources (i.e., frequency bandwidth and transmission power), we should be able to transmit the coded sequence at a faster rate such that, when the transmission rate for the information bits remains the same as in the uncoded system, the transmission accuracy for the coded system is higher. As a result, we can achieve a coding gain, which translates to higher transmission accuracy, plus high spectral and/or power efficiency. Channel coding is effective in combating independent random symbol errors. Correlative channel fading usually results in bursty errors during deep fading periods, which reduces the forward error correction capability of channel coding. Interleaving is a popular approach to address this issue. The coded symbols are first interleaved before being mapped to modulated waveforms in the transmitter. At the receiver, the demodulator output symbols are then de-interleaved before being applied to the decoder. If the interleaver length is sufficient, the negative effect of the channel fading correlation on the coding gain can be eliminated. The cost for the interleaving is the extra processing delay and hardware memory units, especially when the normalized channel fading rate is low; also, the extra delay may pose a problem for real-time communications.

Further improvement is possible by combining channel coding with modulation, a scheme called trellis-coded modulation (TCM) [153]. TCM can improve the reliability of a digital transmission system without increasing transmit power or the required bandwidth. This is achieved by combining the choice of a higher-order modulation scheme with that of a convolutional code at the transmitter and by combining the demodulation and decoding in one step at the receiver.

Channel coding can also be combined with diversity (via multiple antennas), a technique called space-time coded modulation (STCM) [150, 151]. The space-time coding is a bandwidth and power efficient method for wireless communications. With multiple transmitter and/or receiver antennas, and based on the channel state information, the spatial properties of the space-time codes can guarantee that the diversity burden is paced at the transmitter while maintaining optional receiver diversity (which is particularly important for battery powered handsets); at the same time, the temporal properties guarantee that the diversity advantage is achieved without any sacrifice in the transmission rate.

In the following section, we will study diversity reception and adaptive channel equalization for accurate transmission over multipath fading dispersive channels.

4.2 DIVERSITY

4.2.1 Diversity Mechanisms

The following sections describe the various ways of obtaining independently faded signals:

Frequency Diversity. The desired message is transmitted simultaneously over several frequency slots. The separation between adjacent frequency slots should be larger than the channel coherence bandwidth such that channel fading over each slot is independent of that in any other slot. By using redundant signal transmission, this diversity improves link transmission quality at the cost of extra frequency bandwidth.

Time Diversity. The desired message is transmitted repeatedly over several time periods. The time separation between adjacent transmissions should be larger than the channel coherence time such that the channel fading experienced by each transmission is independent of the channel fading experienced by all of the other transmissions. In addition to extra system capacity (in terms of transmission time) due to the redundant transmission, this diversity introduces a significant signal processing delay, especially when the channel coherence time is large. In practice, time diversity is more frequently exploited through interleaving, forward-error correction, and automatic retransmission request (ARQ).

Space Diversity. The desired message is transmitted by using multiple transmitting antennas and/or receiving antennas. The space separation between adjacent antennas should be large enough to ensure that the signals from different antennas are independently faded. In a Rayleigh fading environment, it can be shown that, if two antennas are separated by half of the carrier wavelength, the corresponding two signals experience independent fading. Taking into account the shadowing effect, usually a separation of at least 10 carrier wavelengths is required between two adjacent antennas. This diversity does not require extra system capacity; however, the cost is the extra antennas needed.

Angle Diversity. The desired message is received simultaneously by several directive antennas pointing in widely different directions. The received signal consists of scattering waves coming from all directions. It has been observed that the scattered signals associated with the different (nonoverlapping) directions are uncorrelated. Angle diversity can be viewed as a special case of space diversity since it also requires multiple antennas.

Path Diversity. In CDMA cellular networks (to be discussed in Chapter 6), the use of direct sequence spread spectrum modulation techniques permits the desired signal to be transmitted over a frequency bandwidth much larger than the channel coherence bandwidth. The spread spectrum signal can resolve multipath signal components as long as the path delays are separated by at least one chip period. A Rake receiver can separate the received signal components from different propagation paths by using code correlation and can then combine the signal components constructively. In CDMA, exploiting the path diversity reduces the transmitted power needed and increases the system capacity.

As diversity is used to overcome flat fading and equalization is used to overcome intersymbol interference due to channel time dispersion, in this section, for simplicity, we assume that all the channels exhibit flat fading with propagation delay equal to zero. Figure 4.1 shows an equivalent model of a system with coherent demodulation and L-th order diversity. In the figure, channel $l(l = 1, 2, \ldots, L)$ provides a channel gain $\alpha_l(t)\exp[j\theta_l(t)]$ to the transmitted signal $x(t)$ and introduces an additive white Gaussian noise $w_l(t)$ with zero mean and two-sided psd $N_0/2$.

4.2.2 Linear Combining

After obtaining independently faded signal components at the output of the demodulators, the next step is to combine these signal components for transmitted symbol detection. Various methods have been proposed for combining independently faded signal components, and the tradeoff among these methods is the receiver complexity versus transmission performance improvement. Linear combining techniques include those discussed in the following sections.

Maximal Ratio Combining. This combining technique assumes that the receiver is able to accurately estimate the amplitude fading $\alpha_l(t)$ and carrier phase distortion $\theta_l(t)$ for each diversity channel. With knowledge of the complex channel gains, $\alpha_l(t)\exp[j\theta_l(t)]$, $l = 1, 2, \ldots, L$,

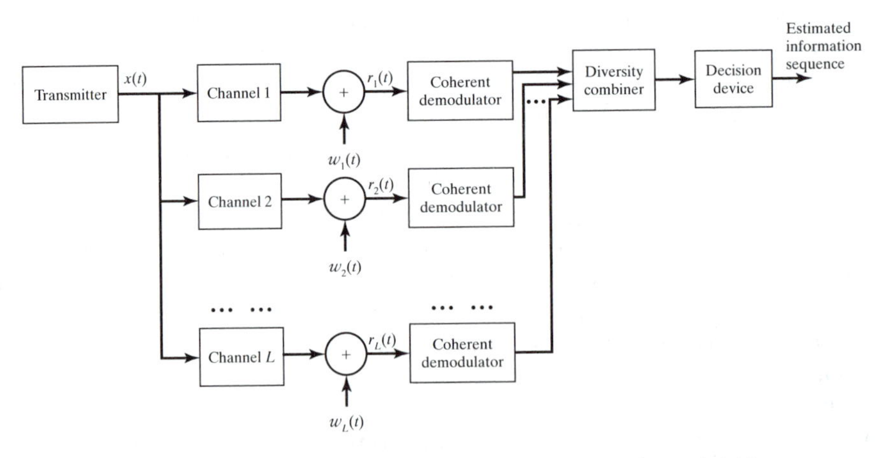

Figure 4.1 Illustration of diversity with coherent demodulation.

the receiver coherently demodulates the received signal from each branch. The phase distortion $\theta_l(t)$ of the received signal is removed from the lth branch by multiplying the signal component by $\exp[-j\theta_l(t)]$. The coherently detected signal is then weighted by the corresponding amplitude gain $\alpha_l(t)$. The weighted received signals from all the L branches are then summed together and applied to the decision device. Maximal ratio combining achieves the best performance [16].

Equal-Gain Combining. The maximal ratio combining approach requires an accurate estimate of the channel amplitude gain $\alpha_l(t)$, which increases the receiver complexity. An alternative approach is to weight all the signals equally after coherent detection (which removes the phase distortion $\theta_l(t)$). The coherently detected signals from all the L branches are simply added and applied to the decision device. As the receiver does not need to estimate the amplitude fading $\alpha_l(t)$, its complexity is reduced as compared with that of maximal ratio combining.

Selective Combining. In this scheme, the receiver monitors the SNR value of each diversity channel and chooses the one with the maximum SNR value for signal detection. Compared with the preceding two schemes, selective diversity is much easier to implement without much performance degradation, especially for the reverse link transmission where the diversity branches can be physically located in different base stations, which would make it difficult to use maximal ratio combining or equal-gain combining. Figure 4.2 illustrates

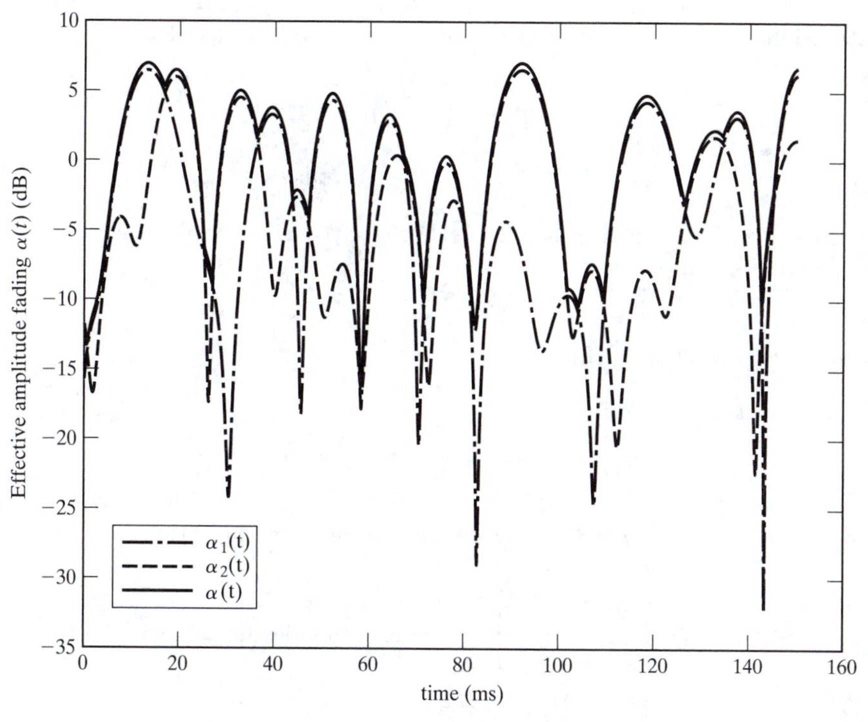

Figure 4.2 Illustration of amplitude fading with $L = 2$ selective diversity.

how selective diversity with $L = 2$ can reduce the chances of deep fading, where the effective fading of the received signal amplitude is $\alpha(t) = \max\{\alpha_1(t), \alpha_2(t)\}$. (For clarity, $\alpha(t) = \max\{\alpha_1(t), \alpha_2(t)\} + 0.5$ in the figure.)

Among the three combining schemes, maximal ratio combining achieves the best performance, followed by equal-gain combining. However, with a low diversity order, the performance differences among the three combining techniques are not very significant.

Example 4.1 SNR Improvement by Selective Diversity over Rayleigh Fading

Consider Lth-order diversity with selective combining in a Rayleigh fading propagation environment. Each diversity channel exhibits independent and identically distributed (iid) fading. Find the pdf of the SNR in the diversity reception.

Solution At any instant, the amplitude fading of the lth channel, α_l, $l \in \{1, 2, \ldots, L\}$, has a pdf of

$$f_\alpha(x) = \begin{cases} \dfrac{x}{\sigma_\alpha^2} \exp\left(-\dfrac{x^2}{2\sigma_\alpha^2}\right), & \text{if } x \geq 0 \\ 0, & \text{otherwise} \end{cases}, \tag{4.2.1}$$

where $\sigma_\alpha^2 = \frac{1}{2}E(\alpha^2)$. Let γ_l be the received SNR per bit of the lth channel at any instant, with mean $E(\gamma_l) = \Gamma_c$ for $l \in \{1, 2, \ldots, L\}$. As γ_l is proportional to α_l^2, from Eq. (4.2.1), it can be easily derived that γ_l follows an exponential distribution with parameter Γ_c,

$$f_{\gamma_l}(x) = \begin{cases} \dfrac{1}{\Gamma_c} \exp\left(-\dfrac{x}{\Gamma_c}\right), & \text{if } x \geq 0 \\ 0, & \text{otherwise} \end{cases}. \tag{4.2.2}$$

With selective diversity, at any instant, the effective received SNR per bit, γ, is the one from the strongest received signal,

$$\gamma = \max\{\gamma_1, \gamma_2, \ldots, \gamma_L\}.$$

For $x \geq 0$, γ has the following cdf

$$\begin{aligned} F_\gamma(x) &= P(\gamma \leq x) \\ &= P(\gamma_1 \leq x \cap \gamma_2 \leq x \cap \ldots \cap \gamma_L \leq x) \\ &= \prod_{l=1}^{L} P(\gamma_l \leq x) \text{ with independently faded channels} \\ &= \left[\int_0^x f_{\gamma_l}(z)dz\right]^L \text{ with identically distributed fading} \\ &= \left[1 - \exp\left(-\frac{x}{\Gamma_c}\right)\right]^L, \end{aligned}$$

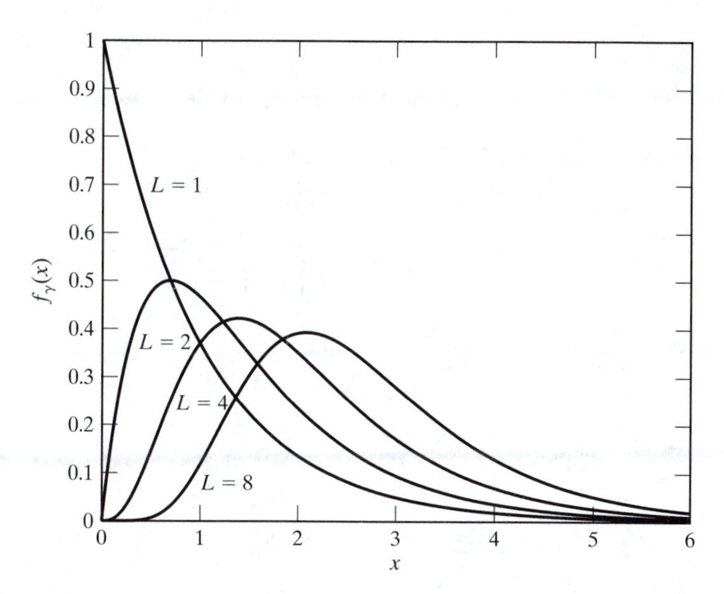

Figure 4.3 Probability density function of the received SNR per bit with selective diversity.

and for $x < 0$, the cdf of γ is $F_\gamma(x) = 0$. The pdf of γ is then

$$f_\gamma(x) = \frac{dF_\gamma(x)}{dx} = \frac{L}{\Gamma_c} \exp\left(-\frac{x}{\Gamma_c}\right)\left[1 - \exp\left(-\frac{x}{\Gamma_c}\right)\right]^{L-1}, \quad x \geq 0.$$

Figure 4.3 plots the received SNR per bit for unit mean SNR per bit, $\Gamma_c = 1$. It can be observed that, as L is increased from 1 to 2, the chances that $\gamma \ll 1$ are significantly reduced. As L is further increased from 2 to 4 and to 8, the pdf curve shifts from left to right, indicating less chance for the instantaneous SNR to have a very small value, which results in better transmission performance.

Example 4.2 Independent Diversity Channels with Unequal SNRs

Consider Lth-order diversity with selective combining. The L diversity channels exhibit independent Rayleigh fading. Let γ_l be the received SNR per bit of the lth channel at any instant, $l = 1, 2, \ldots, L$. The first $L - 1$ channels have the same average received SNR per bit (i.e., $E(\gamma_1) = E(\gamma_2) = \ldots = E(\gamma_{L-1}) = \Gamma_c$). The Lth channel has a value of mean SNR per bit much smaller than that of the other channels (i.e., $E(\gamma_L) \ll \Gamma_c$). Study the effect of the unequal SNRs on the diversity reception.

Solution From Example 4.1, we know that the effective SNR per bit with the diversity, $\gamma = \max\{\gamma_1, \gamma_2, \ldots, \gamma_L\}$, has the following cdf

$$F_\gamma(x) = \prod_{l=1}^{L}\left[1 - \exp\left(-\frac{x}{\Gamma_l}\right)\right], \tag{4.2.3}$$

where $\Gamma_l = E(\gamma_l)$. As diversity is used to combat channel deep fading, we are interested in the SNR improvement achieved by the diversity during deep fading periods. In the following, we consider two cases: $x \ll \Gamma_L$ and $\Gamma_L < x \ll \Gamma_c$ in Eq. (4.2.3).

a. For $x \ll \Gamma_L$, we have $x \ll \Gamma_l$, $l = 1, 2, \ldots, L$ as $\Gamma_L \ll \Gamma_c$. Using the approximation $\exp(-x) \approx 1 - x$ for $x \ll 1$, we can approximate the cdf of γ by

$$F_\gamma(x) \approx \prod_{l=1}^{L} \left(\frac{x}{\Gamma_l}\right) = \left(\frac{x}{\Gamma_L}\right)\left(\frac{x}{\Gamma_c}\right)^{L-1},$$

where $(x/\Gamma_L) \gg (x/\Gamma_c)$.

b. For $\Gamma_L < x \ll \Gamma_c$, from Eq. (4.2.3), we have the following approximation for the cdf

$$F_\gamma(x) \approx \left[1 - \exp\left(-\frac{x}{\Gamma_L}\right)\right]\left(\frac{x}{\Gamma_c}\right)^{L-1}.$$

If $\Gamma_L \ll x \ll \Gamma_c$, $[1 - \exp(-x/\Gamma_L)] \approx 1$, we have

$$F_\gamma(x) \approx \left(\frac{x}{\Gamma_c}\right)^{L-1},$$

which corresponds to the diversity reception with order $L - 1$.

Together with the results given in Example 4.1, we see that the chances of deep fading (both when $x \ll \Gamma_L$ and when $\Gamma_L < x \ll \Gamma_c$) using diversity reception increases when Γ_L is reduced from Γ_c to a value much smaller than Γ_c. In particular, the contribution of the Lth channel disappears when $\Gamma_L \ll x \ll \Gamma_c$, which is intuitively easy to understand.

4.2.3 Performance Improvement

Consider the transmission of a digital modulated signal $x(t)$ over flat slow Rayleigh fading channels using coherent demodulation with Lth order diversity. As shown in Figure 4.1, the received signal component from the lth diversity channel is

$$r_l(t) = \alpha_l(t) \exp[j\theta_l(t)] x(t) + w_l(t), \quad l = 1, 2, \ldots, L. \tag{4.2.4}$$

It is assumed that (a) the channel fading processes are mutually statistically independent, (b) the additive white Gaussian noise processes are mutually statistically independent, and (c) the channel fading processes and additive noise processes are independent of each other. For a slow fading channel, the complex channel gain can be assumed to be a complex constant over each symbol interval. The demodulator in each channel is optimum for an AWGN channel (e.g., using filters matched to the orthonormal functions $\varphi_n(t)$, $n = 1, 2, \ldots, N$, which define the signal space of the transmitted signal $x(t)$, as discussed in Section 3.2). Therefore, the output of the demodulator of the lth branch at the end of the kth symbol interval is $\alpha_{lk} \exp(j\theta_{lk})\vec{x}_k + \vec{n}_{lk}$, where $\alpha_{lk} \exp(j\theta_{lk})$ is the complex channel gain of the lth channel over the kth symbol interval, $\vec{x}_k$ is the vector

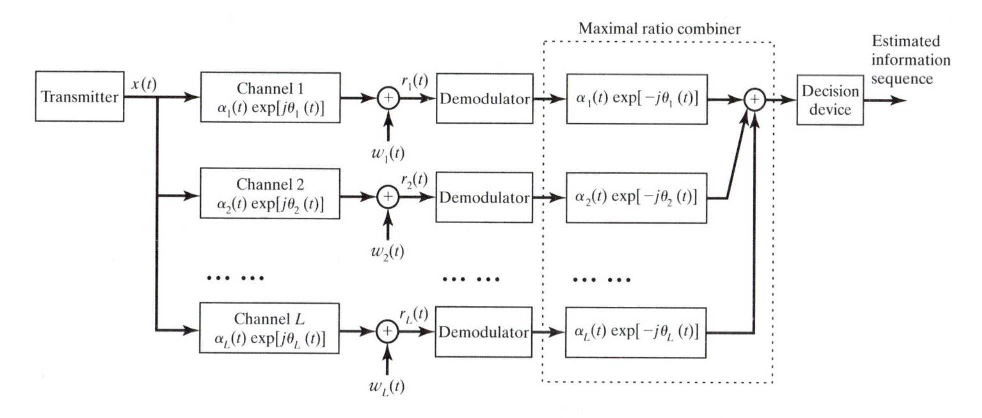

Figure 4.4 Diversity reception with maximal ratio combining.

representation of the transmitted signal over the kth symbol interval in the N-dimensional signal space and is also the demodulator output for an AWGN channel, and $\vec{n}_{lk} = (n_{lk,1}, n_{lk,2}, \ldots, n_{lk,N})$ is the corresponding vector representation of the noise component at the demodulator output due to $w_l(t)$. As discussed in Chapter 3, it can be easily shown that $||\vec{x}_k||^2 = E_s$ (E_s is the symbol energy), and that each component in $\vec{n}_{lk}$ is a Gaussian random variable with zero mean and variance $N_0/2$ that is independent of any other noise component in the same diversity channel or in a different diversity channel.

Consider maximal ratio combining, as shown in Figure 4.4. The decision variable for the kth transmitted symbol can be represented by

$$\begin{aligned}\vec{r}_k &= \sum_{l=1}^{L}[\alpha_{lk}\exp(-j\theta_{lk})][\alpha_{lk}\exp(j\theta_{lk})\vec{x}_k + \vec{n}_{lk}]\\ &= \left[\sum_{l=1}^{L}\alpha_{lk}^2\right]\vec{x}_k + \left[\sum_{l=1}^{L}\alpha_{lk}\exp(-j\theta_{lk})\vec{n}_{lk}\right]\\ &= g_k\vec{x}_k + \vec{n}_k,\end{aligned}$$

where

$$g_k = \sum_{l=1}^{L}\alpha_{lk}^2$$

and

$$\vec{n}_k = \sum_{l=1}^{L}\alpha_{lk}\exp(-j\theta_{lk})\vec{n}_{lk}.$$

For the noise vector, $\vec{n}_k = (n_{k,1}, n_{k,2}, \ldots, n_{k,N})$, each component is

$$n_{k,n} = \sum_{l=1}^{L}\alpha_{lk}\exp(-j\theta_{lk})n_{lk,n},\ n = 1, 2, \ldots, N.$$

Therefore, given the weighting gain $\alpha_{lk}\exp(-j\theta_{lk})$, $l = 1, 2, \ldots, L$, each noise component $n_{k,n}$ is a Gaussian random variable with zero mean and variance

$$\sigma_{k,n}^2 = \frac{N_0}{2}\sum_{l=1}^{L}\alpha_{lk}^2.$$

For BPSK, $N = 1$. The vectors $\vec{r}_k$, $\vec{x}_k$, and $\vec{n}_k$ can be represented by the corresponding scalar variables r_k, x_k, and n_k, respectively. The decision variable for the kth transmitted symbol is

$$r_k = g_k x_k + n_k,$$

where $x_k = \sqrt{E_b}$ for symbol "1" and $x_k = -\sqrt{E_b}$ for symbol "0" as in Example 3.3. The SNR per bit at the output of the combiner for the kth symbol is then

$$\gamma_k = \frac{[g_k x_k]^2}{2\sigma_{k,n}^2} = \frac{\left[\sum_{l=1}^{L}\alpha_{lk}^2 x_k\right]^2}{N_0\sum_{l=1}^{L}\alpha_{lk}^2} = \frac{E_b}{N_0}\sum_{l=1}^{L}\alpha_{lk}^2, \tag{4.2.5}$$

where $\frac{E_b}{N_0}$ is the SNR value for the AWGN channel with $\alpha_{lk} = 1$ and $L = 1$. In a Rayleigh fading environment, the α_{lk}'s are iid Rayleigh random variables with parameter σ_α^2. Therefore, γ_k follows a chi-square distribution with $2L$ degrees of freedom. Its pdf is given by

$$f_\gamma(x) = \frac{x^{L-1}\exp(-x/\Gamma_c)}{(L-1)!\Gamma_c^L}, \quad x \ge 0, \tag{4.2.6}$$

where $\Gamma_c = 2\sigma_\alpha^2 E_b/N_0$ is the average SNR per bit in each diversity channel. From Eq. (4.2.5), the mean SNR per bit after the combining is

$$\Gamma_b = E[\gamma_k] = L\Gamma_c,$$

which increases linearly with L. Figure 4.5 shows the pdf of the SNR per bit for maximal ratio combining with $\Gamma_c = 1$. It can be observed that, as L increases, the pdf curve shifts from left to right, meaning that the chance for a small instantaneous SNR is greatly reduced. Comparing the curves in Figure 4.5 with those in Figure 4.3, we see that maximal ratio combining indeed gives better performance as the chances for a small instantaneous SNR are further reduced from those with selective diversity, especially with a large L. This is because maximal ratio combining makes use of the signal components in all the diversity channels and the mean SNR per bit increases linearly with L. However, in selective diversity, at any time, only the signal from the best channel is used for detection. On the other hand, if we compare the two curves with $L = 2$, the difference between them is not significant. From this, we can conclude that, with a low order of diversity reception, selective diversity, equal-gain diversity, and maximal ratio combining diversity

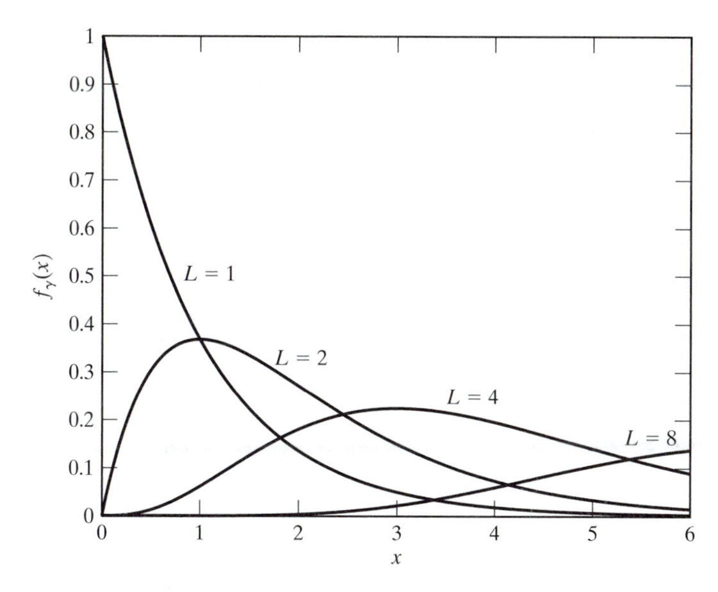

Figure 4.5 Probability density function of the received SNR per bit with maximal ratio combining.

achieve similar transmission performance, with the best performance exhibited by maximal ratio combining diversity and the worst performance exhibited by selective diversity.

The analysis of the probability of bit error over a Rayleigh fading channel can be extended to the case with diversity, so that

$$P_b = \int_0^\infty P_{e|\gamma}(x) f_\gamma(x)\, dx, \tag{4.2.7}$$

where $P_{e|\gamma}(x)$ is the conditional probability of bit error given that the received SNR per bit is $\gamma = x$. For coherent BPSK, from Eq. (3.5.3), we have

$$P_{e|\gamma}(x) = Q(\sqrt{2x}). \tag{4.2.8}$$

Substituting Eqs. (4.2.8) and (4.2.6) into Eq. (4.2.7), it can be derived that

$$P_b = [0.5(1-\mu)]^L \sum_{l=0}^{L-1} \binom{L-1+l}{l} [0.5(1+\mu)]^l, \tag{4.2.9}$$

where

$$\mu = \sqrt{\frac{\Gamma_c}{1+\Gamma_c}}.$$

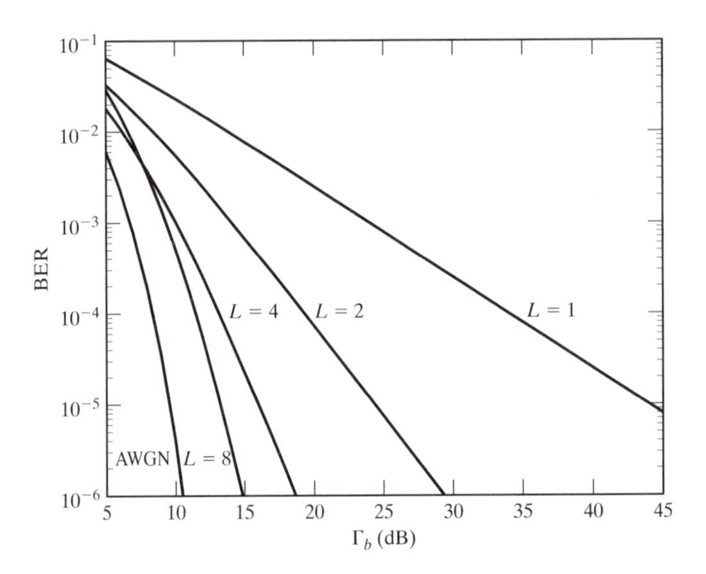

Figure 4.6 BER performance improvement using diversity with maximal ratio combining.

For $\Gamma_c \gg 1$, we have $0.5(1+\mu) \approx 1$ and $0.5(1-\mu) \approx 1/4\Gamma_c$. Furthermore,

$$\sum_{l=0}^{L-1} \binom{L-1+l}{l} = \binom{2L-1}{L}.$$

As a result, for $\Gamma_c \gg 1$, the probability of error can be approximated by

$$P_b \approx \left(\frac{1}{4\Gamma_c}\right)^L \binom{2L-1}{L}. \tag{4.2.10}$$

Figure 4.6 shows the BER transmission performance improvement achieved by diversity with maximal ratio combining for coherent BPSK, QPSK, and MSK over Rayleigh fading channels. The following observations can be made: (*a*) Diversity can dramatically improve the transmission performance; (*b*) The improvement is most significant when L increases from 1 to 2. As L increases, the BER curve moves towards that for an AWGN channel. In fact, it can be proved that when L approaches infinity, the BER curve with diversity will converge to the AWGN curve; (*c*) At a low Γ_b value, diversity with $L = 8$ may perform worse than diversity with $L = 4$. This is because, for the same Γ_b value, Γ_c for $L = 8$ is 3dB less than Γ_c for $L = 4$. With a low Γ_b, the effect of the much larger channel noise in the case of $L = 8$ on BER is stronger than the effect of diversity gain achieved by increasing L from 4 to 8.

Example 4.3 BPSK with Maximal-Ratio Diversity Combining

Consider digital transmission via BPSK with second-order diversity. The channel gain in each branch takes on a value of 1.0, with probability 0.9, and a value of 0.05, with probability 0.1, over each symbol interval. The additive noise in each branch is Gaussian and white with two-sided psd equal to $N_0/2$. The two branches have independent channel gain and noise component. Derive the BER performance for maximal ratio combining.

Solution Let g_1 and g_2 denote the channel gains of the first and second branches, respectively, over each symbol interval. With maximal ratio combining, the effective channel gain over the symbol interval is

$$g = g_1^2 + g_2^2,$$

and the additive noise component in the decision device input is a Gaussian random variable with zero mean and variance

$$\sigma_N^2 = \frac{N_0}{2}(g_1^2 + g_2^2) = \frac{gN_0}{2}.$$

Let E_b denote the transmitted bit energy. From Eq. (3.5.3), the conditional probability of bit error is

$$P(e|g_1, g_2) = Q\left(\sqrt{\frac{2g^2E_b}{gN_0}}\right) = Q\left(\sqrt{\frac{2gE_b}{N_0}}\right).$$

The probability of bit error is then

$$\begin{aligned} P_b &= \sum_{x_i}\sum_{y_j} P(e|g_1 = x_i, g_2 = y_j)P(g_1 = x_i, g_2 = y_j) \\ &= 0.9^2 P(e|g_1 = 1.0, g_2 = 1.0) + 0.1^2 P(e|g_1 = 0.05, g_2 = 0.05) \\ &\quad + 0.9 \times 0.1 P(e|g_1 = 1.0, g_2 = 0.05) + 0.1 \times 0.9 P(e|g_1 = 0.05, g_2 = 1.0) \\ &= 0.81 Q\left(\sqrt{\frac{4E_b}{N_0}}\right) + 0.01 Q\left(\sqrt{\frac{0.01E_b}{N_0}}\right) + 0.18 Q\left(\sqrt{\frac{2.005E_b}{N_0}}\right). \end{aligned}$$

Figure 4.7 shows the BER performance. For comparison, the BER curves for the AWGN channel and the fading channel without diversity (see Example 3.8) are also included. The performance improvement by second-order diversity is obvious, which is 12 dB at a BER of 10^{-4}.

Example 4.4 Differential BPSK (DBPSK) with Selective Diversity

It is known that the BER of DBPSK (BPSK with differential encoding) in an AWGN channel with zero mean and two-sided noise psd $N_0/2$ is given by [123]

$$P_b(\gamma_b) = \frac{1}{2}\exp(-\gamma_b), \tag{4.2.11}$$

where $\gamma_b = E_b/N_0$ is the SNR per bit of the received signal and E_b is the signal energy per bit. Derive the BER for DBPSK with Lth-order selective diversity in a Rayleigh fading environment.

Solution From Example 4.1, the pdf of the instantaneous SNR per bit with selective diversity over L iid Rayleigh fading channels is

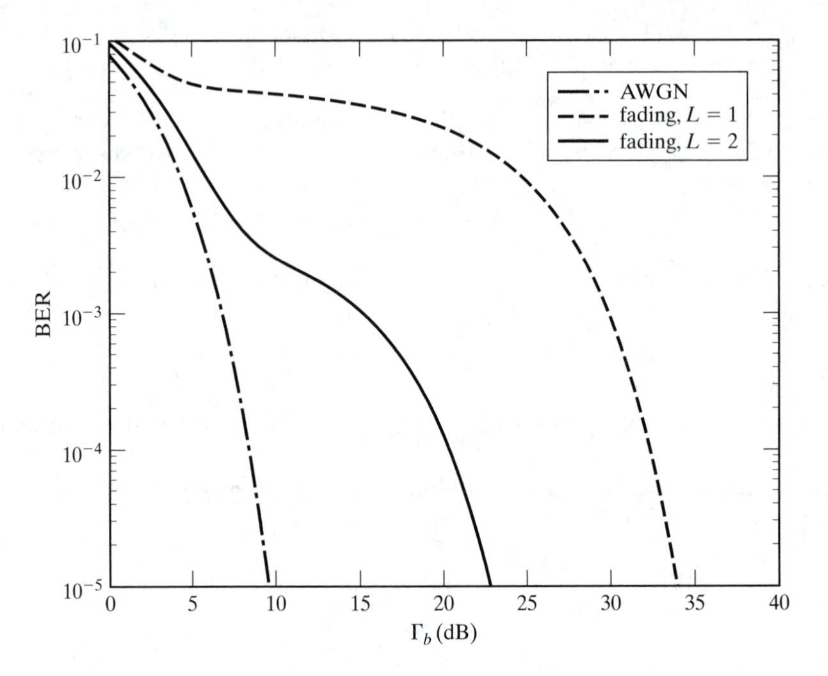

Figure 4.7 BER performance improvement via diversity.

$$f_\gamma(x) = \frac{L}{\Gamma_c} \exp\left(-\frac{x}{\Gamma_c}\right)\left[1 - \exp\left(-\frac{x}{\Gamma_c}\right)\right]^{L-1}, \quad x \geq 0, \tag{4.2.12}$$

where $\Gamma_c = \gamma_b E[\alpha_l^2]$ is the mean SNR per bit of each diversity channel. With Eq. (4.2.11), substituting Eq. (4.2.12) into Eq. (4.2.7), we obtain the BER of interest as

$$P_b = \int_0^\infty \frac{L}{2\Gamma_c} \exp\left[-\left(1 + \frac{1}{\Gamma_c}\right)x\right]\left[1 - \exp\left(-\frac{x}{\Gamma_c}\right)\right]^{L-1} dx.$$

Without diversity ($L = 1$), we have

$$P_b = \frac{1}{2(1 + \Gamma_c)}.$$

With diversity, $L = 2, 3, \ldots$, using the binomial expansion

$$(1 - x)^{L-1} = \sum_{l=0}^{L-1} (-1)^l \binom{L-1}{l} x^l,$$

it can be easily derived that P_b is

$$P_b = \frac{L}{2\Gamma_c} \sum_{l=0}^{L-1} (-1)^l \binom{L-1}{l} \int_0^{\infty} \exp\left[-\left(1+\frac{1+l}{\Gamma_c}\right)x\right] dx$$

$$= \frac{L}{2} \sum_{l=0}^{L-1} \frac{(-1)^l \binom{L-1}{l}}{1+l+\Gamma_c}.$$

Figure 4.8 shows the BER of DBPSK with selective diversity, where $\Gamma_b = L\Gamma_c$. The following observation can be made: (*a*) Comparing the BER curve for an AWGN channel with that for the fading channel without diversity ($L = 1$), the transmission performance is severely degraded by the Rayleigh fading; (*b*) Selective diversity can greatly improve the BER performance; (*c*) The performance improvement is more significant when L is increased from 1 to 2 than it is when L is further increased from 2 to 4, and to 8. If we compare Figure 4.6 with Figure 4.8, the relatively better BER performance shown in Figure 4.6 is due to the fact that (*a*) maximal ratio combining performs better than selective diversity, and (*b*) coherent BPSK performs better than differential BPSK in an AWGN channel. Note that the superior performance of BPSK with maximal ratio combining is obtained under the assumption that the receiver can have accurate estimates of the diversity channel gains $\alpha_{lk} \exp(j\theta_{lk})$ and can then weight the received signals accordingly. If the channel gain (especially the phase distortion) estimates are not accurate, the performance of BPSK with maximal ratio combining will not be as good as that shown in Figure 4.6. On the other hand, in addition to the much simpler receiver structure (achieved by selective diversity), DBPSK does not require accurate carrier phase synchronization and is robust to carrier phase distortion

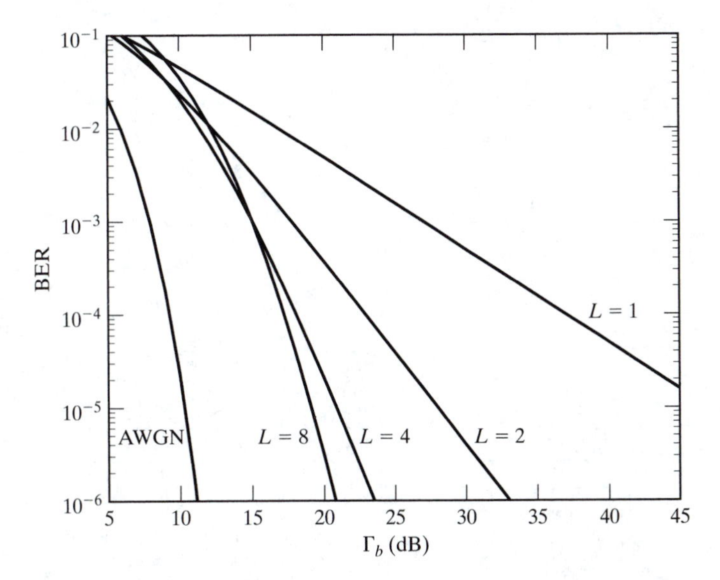

Figure 4.8 BER performance of DBPSK with selective diversity.

introduced by the fading channel as long as the phase distortion does not change much over a time duration of two symbol intervals (similar to $\pi/4$-DQPSK discussed in Section 3.3). Therefore, in practice, the performance difference between DBPSK with selective diversity and coherent BPSK with maximal ratio combining may not be as significant as that shown in Figures 4.6 and 4.8.

4.3 CHANNEL EQUALIZATION

Consider the equivalent representation of a wireless communications system at baseband, without channel encoding/decoding for simplicity. At the transmitter, the modulator maps the information sequence to the data sequence $\{\vec{x}_n\}$ in the N-dimensional signal space (as discussed in Section 3.2), depending on the modulation scheme used. For convenience, we can model the transmitter by a transmitter filter with an input signal $\vec{x}(t) = \sum_{n=-\infty}^{\infty} \vec{x}_n \delta(t - nT)$ such that the filter output is the complex envelope of the signal actually transmitted, where T is the symbol interval of the transmitted signal. As discussed in Chapter 2, a typical wireless channel introduces both fading and propagation delay dispersion to the transmitted signal, in addition to additive white Gaussian noise. From Chapter 3, we know that the optimum receiver for an AWGN channel is a matched filter receiver, consisting of a demodulator (a bank of matched filters) and a decision device. The decision device makes the decision at the end of each symbol interval according to the maximum-likelihood decision rule.

The transmitter filter, channel, and receiver filter together can be viewed as an effective channel for the input signal $\vec{x}(t)$, as shown in Figure 4.9. Let $c(t)$ denote the impulse response of the effective channel. Then the output $\vec{r}(t)$ of the effective channel, sampled at the end of the nth symbol interval, $t = nT$, can be represented as

$$
\begin{aligned}
\vec{r}(t)|_{t=nT} &= [\vec{x}(t) \star c(t) + \vec{n}(t)]|_{t=nT} \\
&= \left\{ \left[\sum_{l=-\infty}^{\infty} \vec{x}_l \delta(t - lT) \right] \star c(t) + \vec{n}(t) \right\} \Bigg|_{t=nT} \\
&= \left\{ \sum_{l=-\infty}^{\infty} \vec{x}_l c(t - lT) + \vec{n}(t) \right\} \Bigg|_{t=nT} \\
&= \sum_{l=-\infty}^{\infty} \vec{x}_l c(nT - lT) + \vec{n}(nT),
\end{aligned}
$$

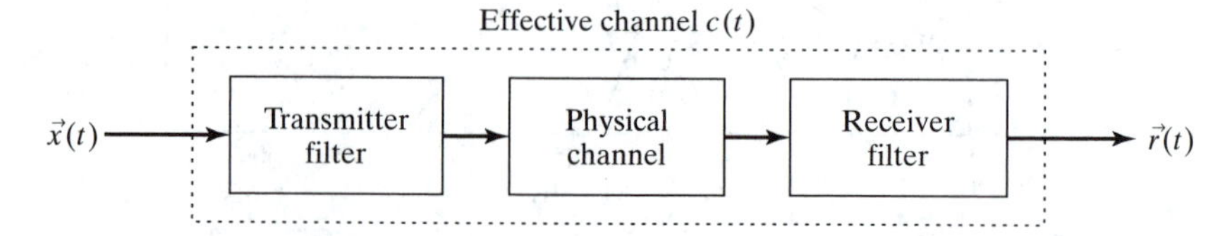

Figure 4.9 The effective channel with impulse response $c(t)$.

where $\vec{n}(t)$ is the filtered Gaussian noise vector due to the additive white Gaussian noise components introduced by the physical channel. In discrete-time sequence representation with T as the time unit, we have

$$\vec{r}_n = \sum_{l=-\infty}^{\infty} \vec{x}_l c_{n-l} + \vec{n}_n = \vec{x}_n \star c_n + \vec{n}_n, \quad (4.3.1)$$

where $\vec{r}_n = \vec{r}(nT)$, $c_n = c(nT)$, and $\vec{n}_n = \vec{n}(nT)$. Eq. (4.3.1) can be rewritten as

$$\vec{r}_n = c_0 \vec{x}_n + \sum_{l=-\infty, l \neq n}^{\infty} \vec{x}_l c_{n-l} + \vec{n}_n. \quad (4.3.2)$$

The first term is the desired signal component (modified by the channel gain c_0) which contains information of the nth transmitted symbol. As each entity in the effective channel can introduce time dispersion to the input signal, c_{n-l} at $n \neq l$ may not be zero. As a result, the second term represents the ISI component due to other transmitted symbols, and the last term is the noise component. The ISI component makes it much more likely for the decision device in the receiver to have a decision error, as compared to the case without ISI.

It can be shown that, to minimize the probability of transmission error, the optimum receiver (in a mean-square error sense) consists of a matched filter, an equalizer, and a maximum likelihood decision device. The matched filter is matched to the transmitter filter and the physical channel in tandem. In other words, it is matched to the received signal waveforms and, therefore, is able to collect all the received signal energy. The equalizer is a transversal filter, and is needed to compensate for intersymbol interference as the matched filter introduces further time dispersion to the received signal.

For notational simplicity, in the following, we assume that the transmitted signal can be represented by a one-dimensional signal space. As a result, the vector representations such as $\vec{x}_n$, $\vec{r}_n$, and $\vec{n}_n$ can be simplified to the corresponding real-valued scalar representations.

Example 4.5 Effect of ISI on Transmission Accuracy

In a digital transmission system using BPSK, the matched filter receiver designed for an AWGN channel is used. The physical channel introduces an additive Gaussian noise of zero mean and two-sided psd $N_0/2$. In addition, it introduces ISI. At the end of the nth symbol interval, the output of the demodulator is

$$r_n = x_n + 0.5 x_{n-1} + n_n,$$

where the desired signal component $x_n = \sqrt{E_b}$ if symbol "1" was sent and $x_n = -\sqrt{E_b}$ if symbol "0" was sent, E_b is the transmitted signal symbol energy, and the noise component n_n is a Gaussian random variable with zero mean and variance $N_0/2$. Determine the probability of transmission error.

Solution Without ISI, from Subsection 3.5.1, the probability of transmission error is

$$P_b = Q\left(\sqrt{\frac{2E_b}{N_0}}\right).$$

Table 4.1 Demodulator Output at the End of Current Symbol in the Absence of the Noise Component

Previous symbol	Current symbol	Demodulator output without ISI	Demodulator output with ISI
"0"	"0"	$-\sqrt{E_b}$	$-1.5\sqrt{E_b}$
"0"	"1"	$\sqrt{E_b}$	$0.5\sqrt{E_b}$
"1"	"0"	$-\sqrt{E_b}$	$-0.5\sqrt{E_b}$
"1"	"1"	$\sqrt{E_b}$	$1.5\sqrt{E_b}$

With ISI, the effect of ISI depends on whether or not the previous and the current transmitted symbols are the same. Table 4.1 lists the non-noise components in the demodulator output versus the transmitted symbols.

With the decision threshold setting at 0 for an AWGN channel and equally likely symbols "1" and "0", the probability of error is

$$\begin{aligned} P_b &= \sum P(\text{error}|\text{previous symbol, current symbol})P(\text{previous symbol, current symbol}) \\ &= \frac{1}{4}[P(\text{error}|\text{"0","0"}) + P(\text{error}|\text{"0","1"}) + P(\text{error}|\text{"1","0"}) + P(\text{error}|\text{"1","1"})] \\ &= \frac{1}{4}[P(-1.5\sqrt{E_b} + n_n > 0) + P(0.5\sqrt{E_b} + n_n < 0) \\ &\quad + P(-0.5\sqrt{E_b} + n_n > 0) + P(1.5\sqrt{E_b} + n_n < 0)] \\ &= \frac{1}{2}\left[Q\left(1.5\sqrt{\frac{2E_b}{N_0}}\right) + Q\left(0.5\sqrt{\frac{2E_b}{N_0}}\right)\right]. \end{aligned}$$

Figure 4.10 plots the probabilities of transmission error as a function of E_b/N_0 with and without ISI. It can be clearly observed that ISI severely degrades the transmission performance. At a BER of 10^{-3}, it requires an additional 5.5 dB in the transmitted power to overcome the ISI. The required increase of the transmitted power increases as the required BER decreases.

4.3.1 Linear Equalization

Figure 4.11 shows the structure of a linear equalizer, which is a tapped-delay-line filter with $(2N + 1)$ taps. The coefficient of the kth tap is denoted by b_k. The impulse response of the equalizer is

$$b(t) = \sum_{k=-N}^{N} b_k\delta(t - kT). \tag{4.3.3}$$

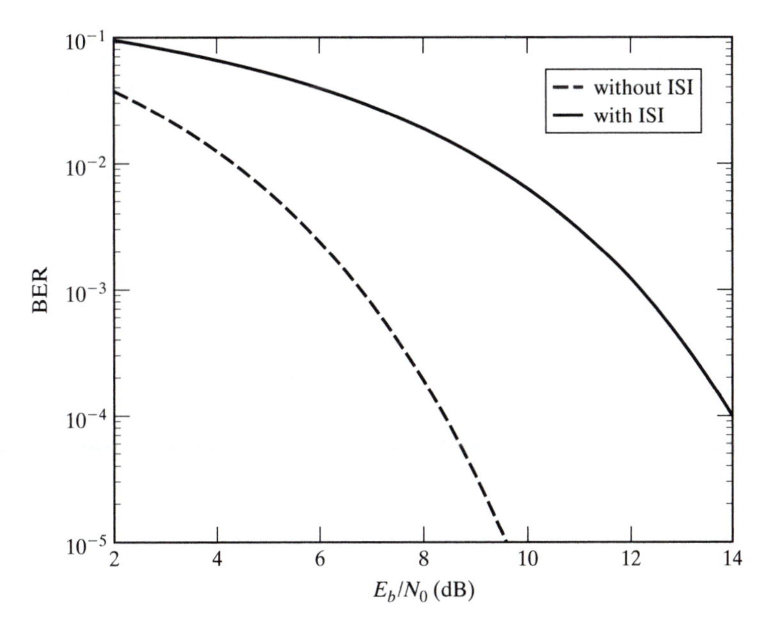

Figure 4.10 Comparison of the transmission accuracy with and without ISI.

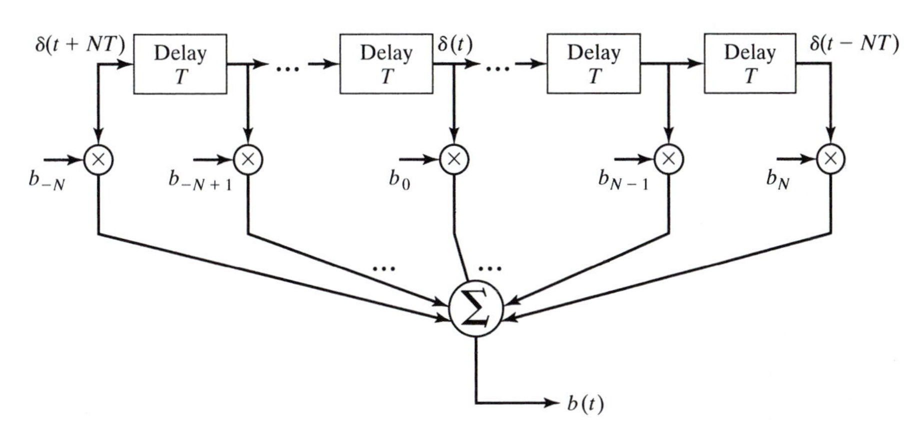

Figure 4.11 The tapped-delay-line linear equalizer.

The transfer function of the equalizer in the z domain is given by

$$B(z) = \sum_{k=-N}^{N} b_k z^{-k}.$$

Figure 4.12 shows the equalized system, where the effective channel and the equalizer are connected in tandem. Let $R(z) = \sum_{n=-\infty}^{\infty} r_n z^{-n}$ denote the z transform of the discrete time

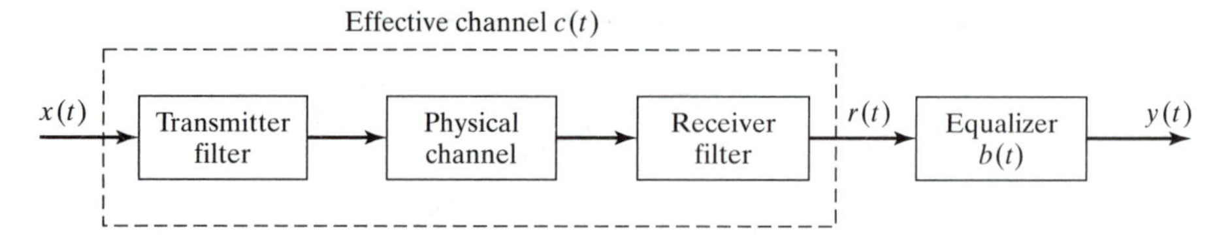

Figure 4.12 The equalized system.

sequence $\{r_n\}$. The output of the equalizer in the z domain is then

$$
\begin{aligned}
Y(z) &= R(z)B(z) \\
&= \sum_{n=-\infty}^{\infty} r_n z^{-n} \sum_{k=-N}^{N} b_k z^{-k} \\
&= \sum_{n=-\infty}^{\infty} \sum_{k=-N}^{N} b_k r_n z^{-(k+n)}.
\end{aligned}
$$

Letting $l = k + n$, the above can be rearranged as

$$
\begin{aligned}
Y(z) &= \sum_{l=-\infty}^{\infty} \sum_{k=-N}^{N} b_k r_{l-k} z^{-l} \\
&= \sum_{l=-\infty}^{\infty} y_l z^{-l},
\end{aligned}
$$

where

$$y_l = \sum_{k=-N}^{N} b_k r_{l-k}.$$

Zero-Forcing (ZF) Linear Equalizer. In the absence of the additive noise introduced by the physical channel, the output of the effective channel at $t = nT$ is

$$r_n = x_n \star c_n. \tag{4.3.4}$$

In the z domain, we have

$$R(z) = X(z)C(z), \tag{4.3.5}$$

where $X(z) = \sum_{n=-\infty}^{\infty} x_n z^{-n}$ and $C(z) = \sum_{n=-\infty}^{\infty} c_n z^{-n}$ are the z transforms of the discrete time sequence $\{x_n\}$ and $\{c_n\}$, respectively. As a result, the output signal of the equalizer in the z domain is

$$Y(z) = R(z)B(z) = X(z)C(z)B(z). \tag{4.3.6}$$

Figure 4.13 illustrates the equivalent discrete-time representation of the equalized system.

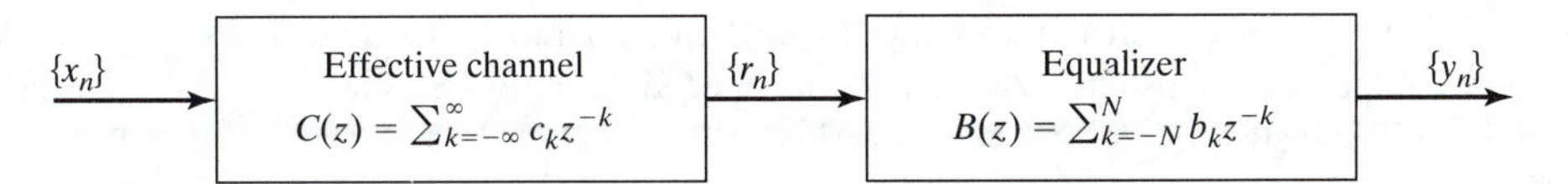

Figure 4.13 The equivalent discrete-time representation of the equalized system.

Let $H(z)$ denote the transfer function of the equalized system. From Figure 4.13, we have

$$H(z) = \frac{Y(z)}{X(z)} = C(z)B(z),$$

which, in the time domain, corresponds to

$$h_n = b_n \star c_n = \sum_{k=-N}^{N} b_k c_{n-k}.$$

The convolution is illustrated in Figure 4.14 for $N = 2$.

For ISI-free transmission, the desired output $Y(z)$ of the equalizer should be $X(z)$, which requires that

$$H(z) = 1. \tag{4.3.7}$$

In the time domain, condition Eq. (4.3.7) means that

$$h_n = \begin{cases} 1, & n = 0 \\ 0, & n \neq 0 \end{cases}. \tag{4.3.8}$$

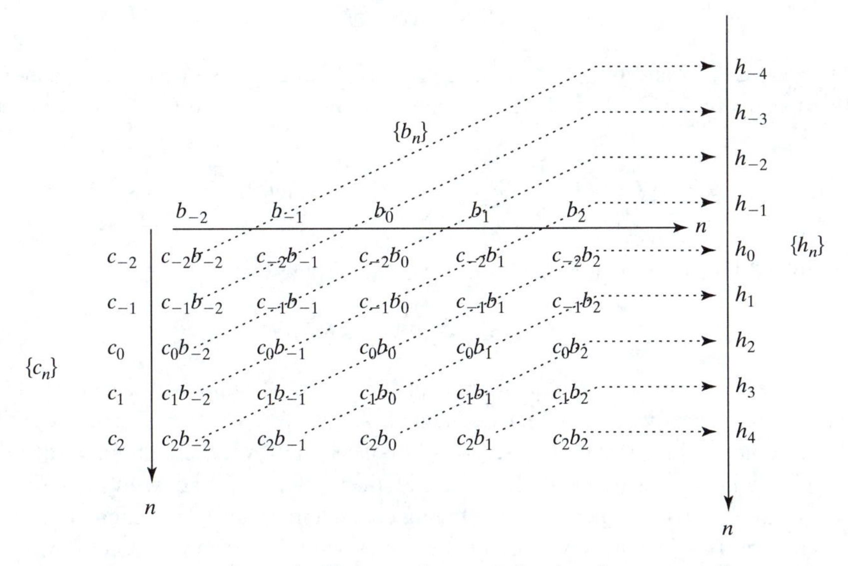

Figure 4.14 Convolution map for zero-forcing linear equalizer.

That is, ISI-free transmission requires that $h_n = 0$ for all integers n except for $n = 0$. When the channel impulse response has such uniformly spaced zero crossings at T-spaced intervals, the channel is said to satisfy *Nyquist's first criterion*.

Given the effective channel $C(z)$, we want to design the equalizer $B(z)$ such that the conditions of Eq. (4.3.7) or Eq. (4.3.8) for ISI-free transmission can be satisfied. Here the equalizer coefficients are determined in such a way as to force the impulse response of the equalized system, $h(t)$, to be zero at $t = \pm T, \pm 2T, \ldots$. As a result, the equalizer is called a zero-forcing linear equalizer. However, in reality, with a finite number $2N + 1$ of taps in the equalizer, we can only guarantee that

$$\hat{h}(n) = \begin{cases} 1, & n = 0 \\ 0, & n = \pm 1, \pm 2, \ldots, \pm N \end{cases} \tag{4.3.9}$$

by choosing the tap coefficients to satisfy the following equation

$$\begin{bmatrix} c_0 & \cdots & c_{-N+1} & c_{-N} & c_{-N-1} & \cdots & c_{-2N} \\ \vdots & & \vdots & \vdots & \vdots & & \vdots \\ c_{N-1} & \cdots & c_0 & c_{-1} & c_{-2} & \cdots & c_{-N-1} \\ c_N & \cdots & c_1 & c_0 & c_{-1} & \cdots & c_{-N} \\ c_{N+1} & \cdots & c_2 & c_1 & c_0 & \cdots & c_{-N+1} \\ \vdots & & \vdots & \vdots & \vdots & & \vdots \\ c_{2N} & \cdots & c_{N+1} & c_N & c_{N-1} & \cdots & c_0 \end{bmatrix} \begin{bmatrix} b_{-N} \\ \vdots \\ b_{-1} \\ b_0 \\ b_1 \\ \vdots \\ b_N \end{bmatrix} = \begin{bmatrix} 0 \\ \vdots \\ 0 \\ 1 \\ 0 \\ \vdots \\ 0 \end{bmatrix}. \tag{4.3.10}$$

Note that, as N approaches ∞, $\hat{h}(n)$ approaches $h(n)$ in Eq. (4.3.7).

The equalization process can also be explained in the frequency domain. The condition of Eq. (4.3.7) means that

$$B(z) = 1/C(z).$$

At $z = \exp(j\omega T)$, $B(z)$ and $C(z)$ are the discrete-time Fourier transforms of the sequences $\{b_n\}$ and $\{c_n\}$, respectively. Let $C(\omega) = \mathcal{F}\,[c(t)]$ denote the Fourier transform of $c(t)$, and let

$$\tilde{C}(\omega) = \frac{1}{T} \sum_{l=-\infty}^{\infty} C\left(\omega - \frac{2\pi l}{T}\right), \; |\omega| \le \frac{\pi}{T}$$

denote the folded transfer function of $C(\omega)$. It can be derived that

$$C(z)|_{z=\exp(j\omega T)} = \tilde{C}(\omega). \tag{4.3.11}$$

As a result, a zero-forcing linear equalizer with $N \to \infty$ is simply an inverse filter, which inverts the transfer function of the effective channel. A zero-forcing linear equalizer with a finite integer N value approximates this inverse. This is conceptually very straightforward. In general, an all-pass channel does not introduce ISI. On the other hand, a channel with memory such as a bandlimited channel or a frequency selective fading channel introduces ISI which, in the frequency domain, is characterized by a nonconstant transfer function in the spectral region of the transmitted signal. The equalizer is to compensate for the distortion on the transmitted signal introduced by the (effective) channel such that the equalized system has a constant transfer function in the frequency region.

Example 4.6 Zero-Forcing Linear Equalizer with Infinite Taps

For the communication system in Example 4.5, design a linear equalizer to combat ISI, and determine the probability of transmission error with the channel equalization.

Solution The effective channel has an impulse response (in the absence of additive noise) given by

$$c(t) = \delta(t) + 0.5\delta(t - T).$$

The channel transfer function is then

$$C(z) = \sum_{k=-\infty}^{\infty} c_k z^{-k} = 1 + 0.5z^{-1}.$$

To completely equalize the channel, the transfer function of the equalizer should be

$$\begin{aligned} B(z) &= \frac{1}{C(z)} \\ &= \frac{1}{1 + 0.5z^{-1}} \\ &= 1 - 0.5z^{-1} + 0.5^2 z^{-2} - \ldots + (-0.5z^{-1})^k + \ldots, \; |0.5z^{-1}| < 1. \end{aligned}$$

The equalization function can be implemented by a tapped-delay-line linear filter with a large number of taps (to approximate the infinite number of taps).

In the presence of additive noise n_n at the end of the nth symbol interval at the effective channel output, the noise component at the equalizer output at the same instant is

$$v_n = n_n \star b_n = \sum_{k=0}^{\infty} (-0.5)^k n_{n-k}.$$

Since n_n is a zero-mean Gaussian random variable with variance $N_0/2$ and is independent from sample to sample, v_n is also Gaussian with zero mean and variance equal to

$$\sigma_v^2 = \sum_{k=0}^{\infty} [(-0.5)^k]^2 (N_0/2) = 2N_0/3.$$

As a result, the probability of bit error with equalization is

$$P_b = Q\left(\sqrt{\frac{E_b}{2N_0/3}}\right) = Q\left(\sqrt{\frac{3E_b}{2N_0}}\right).$$

Figure 4.15 shows the BER performance with and without the equalizer. The BER curve for the AWGN channel without ISI is also plotted for comparison. It is clear that equalization improves

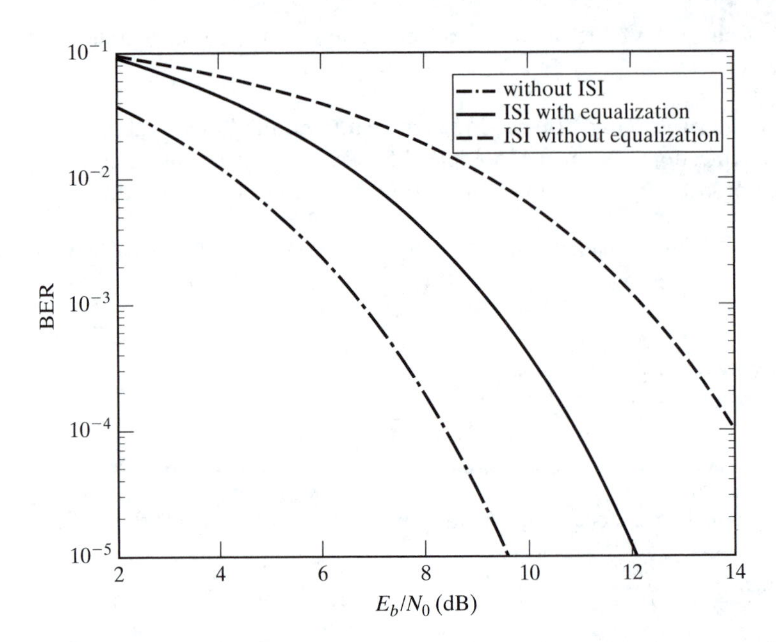

Figure 4.15 Comparison of the transmission accuracy with and without equalization.

the transmission accuracy. Also, because the equalizer increases the noise variance (power), the equalized system does not have a transmission accuracy as good as the system without ISI.

Example 4.7 Zero-Forcing Linear Equalizer with Finite Taps

The impulse response of a dispersive effective channel is

$$c(t) = \exp\left(-\frac{|t|}{3T}\right), -\infty < t < \infty,$$

where T is the transmitted symbol interval. Design a 3-tap zero-forcing linear equalizer for the channel.

Solution The discrete-time representation of the effective channel impulse response is $\{c_n\}$, where

$$c_n = c(t)|_{t=nT} = \exp\left(-\frac{|n|T}{3T}\right) = \exp\left(-\frac{|n|}{3}\right).$$

Let $\mathbf{b} = (b_{-1}, b_0, b_1)^T$ denote the tap coefficient vector of the 3-tap linear equalizer. The discrete-time representation of the equalized system impulse response is

$$h_n = c_n \star b_n = \sum_{k=-1}^{1} b_k c_{n-k}.$$

We want to determine **b** such that

$$h_n = \begin{cases} 1, & n = 0 \\ 0, & n = \pm 1 \end{cases}.$$

From Eq. (4.3.10), **b** can be computed by

$$\begin{bmatrix} 1 & \exp(-1/3) & \exp(-2/3) \\ \exp(-1/3) & 1 & \exp(-1/3) \\ \exp(-2/3) & \exp(-1/3) & 1 \end{bmatrix} \begin{bmatrix} b_{-1} \\ b_0 \\ b_1 \end{bmatrix} = \begin{bmatrix} 0 \\ 1 \\ 0 \end{bmatrix}.$$

This leads to

$$\begin{bmatrix} b_{-1} \\ b_0 \\ b_1 \end{bmatrix} = \begin{bmatrix} 1 & \exp(-1/3) & \exp(-2/3) \\ \exp(-1/3) & 1 & \exp(-1/3) \\ \exp(-2/3) & \exp(-1/3) & 1 \end{bmatrix}^{-1} \begin{bmatrix} 0 \\ 1 \\ 0 \end{bmatrix} = \begin{bmatrix} -1.4726 \\ 3.1103 \\ -1.4726 \end{bmatrix}.$$

In fact, with the 3-tap equalizer, we have

$$h_n = \begin{cases} 1, & n = 0 \\ 0, & n = \pm 1, \pm 2, \pm 3, \pm 4, \pm 5, \pm 6, \pm 7, \pm 8 \end{cases}.$$

Therefore, the 3-tap linear equalizer is very effective in combating the ISI introduced by the effective channel.

Minimum Mean-Square Error (MMSE) Linear Equalizer. In reality, the noise component due to the physical channel cannot be ignored. Example 4.6 shows that the zero-forcing equalizer increases the noise power and, therefore, reduces the transmission accuracy as compared with that of no ISI. If the folded transfer function of the effective channel has high attenuation in some frequency regions, the equalizer (i.e., the inverse filter) can significantly enhance noise in the frequency regions. This is undesirable, especially for a frequency selective fading channel which is very likely to have high attenuation at certain frequencies. As a result, in designing the equalizer, we should consider the transmission performance degradation due to both ISI and channel noise. One way to do this is to choose the equalizer coefficients such that the mean squared error (MSE), which is the expected sum of the squares of all the ISI terms plus the noise power at the output of the equalizer, is minimized. Such an equalizer is called an MMSE linear equalizer. The equalizer maximizes the signal-to-distortion ratio at its output with the constraint of a finite number of equalizer taps.

In the presence of additive Gaussian noise at the receiver input, the output signal $r(t)$ of the effective channel is given by Eq. (4.3.1). The output of the equalizer, sampled at $t = nT$, is

$$y_n = \sum_{k=-N}^{N} b_k r_{n-k}. \tag{4.3.12}$$

The desired output of the equalizer at $t = nT$ is x_n. The equalization error at $t = nT$ is defined as the difference between the desired output x_n and the actual output y_n

$$\epsilon_n = x_n - y_n. \tag{4.3.13}$$

The problem of choosing the equalizer tap coefficients is to solve the following minimization problem:

$$\min_{\mathbf{b}} E[\epsilon_n^2],$$

where $\mathbf{b} = (b_{-N}, \dots, b_{-1}, b_0, b_1, \dots, b_N)^T$ represents the linear equalizer of $2N + 1$ taps, and the superscript T denotes matrix transposition. The MSE of the equalization, denoted by $\mathcal{E}$, can be derived as

$$\begin{aligned}\mathcal{E} &= E[\epsilon_n^2] \\ &= E\left[\left(x_n - \sum_{k=-N}^{N} b_k r_{n-k}\right)^2\right] \\ &= \sum_{l=-N}^{N}\sum_{k=-N}^{N} b_k b_l R_r(k-l) - 2\sum_{l=-N}^{N} b_l R_{xr}(k) + E(x_n^2),\end{aligned} \tag{4.3.14}$$

where

$$R_r(k-l) = E[r_{n-l} r_{n-k}] \tag{4.3.15}$$

$$R_{xr}(k) = E[x_n r_{n-k}] \tag{4.3.16}$$

and the expectation is taken with respect to the random information sequence corresponding to $\{x_n\}$ and the additive noise. The minimum MSE solution is obtained by first differentiating $\mathcal{E}$ with respect to the equalizer coefficients $\{b_k\}$ $(k = 0, \pm 1, \pm 2, \dots, \pm N)$ and then setting the derivatives to zero. Solving for

$$\frac{\partial \mathcal{E}}{\partial b_k} = 0$$

leads to the necessary conditions for the minimum MSE as

$$\sum_{l=-N}^{N} b_l R_r(k-l) = R_{xr}(k), k = 0, \pm 1, \pm 2, \dots, \pm N. \tag{4.3.17}$$

In matrix form, it is given by

$$\mathbf{R}_r \mathbf{b} = \mathbf{R}_{xr} \tag{4.3.18}$$

or

$$\mathbf{b} = \mathbf{R}_r^{-1} \mathbf{R}_{xr} \tag{4.3.19}$$

where $\mathbf{R}_r$ and $\mathbf{R}_{xr}$ are correlation matrix and vector, respectively, defined as

$$\mathbf{R}_r = \begin{bmatrix} R_r(0) & \dots & R_r(N) & \dots & R_r(2N) \\ R_r(-1) & \dots & R_r(N-1) & \dots & R_r(2N-1) \\ \dots & \dots & \dots & \dots & \dots \\ R_r(-2N+1) & \dots & R_r(-N+1) & \dots & R_r(1) \\ R_r(-2N) & \dots & R_r(-N) & \dots & R_r(0) \end{bmatrix}$$

and

$$\mathbf{R}_{xr} = (R_{xr}(-N), \dots, R_{xr}(-1), R_{xr}(0), R_{xr}(1), \dots, R_{xr}(N))^T.$$

Equation (4.3.18) is known as the Wiener-Hopf equation. From the equation we see that, unlike the case of the zero-forcing equalizer, in the MMSE equalizer the tap coefficients depend on the statistical properties (the correlation matrix and vector) of the noise as well as the ISI. In practice, the autocorrelation matrix $\mathbf{R_r}$ and the cross-correlation vector $\mathbf{R_{xr}}$ are unknown *a priori.* For a time-invariant channel, however, the correlation matrix and vector can be estimated by transmitting a test signal over the channel and using the time-average estimates

$$\hat{R}_r(k-l) = \frac{1}{\tilde{N}} \sum_{n=1}^{\tilde{N}} r_{n-l} r_{n-k} \tag{4.3.20}$$

and

$$\hat{R}_{xr}(k) = \frac{1}{\tilde{N}} \sum_{n=1}^{\tilde{N}} x_n r_{n-k} \tag{4.3.21}$$

in place of the ensemble averages to solve for the equalizer coefficients given by Eq. (4.3.17), where $\tilde{N}$ is the number of data samples in the test signal.

Adaptive Linear Equalizer. In a mobile cellular environment, the characteristics of the wireless dispersive fading channel change randomly with time. In order for an equalizer to effectively combat ISI, the equalizer coefficients should change according to the channel status so as to track the channel variations. Such an equalizer is called an adaptive equalizer since it adapts to the channel variations. For a linear adaptive equalizer, the tap coefficients should be chosen in such a way that the MSE, $\mathcal{E} = E[\epsilon_n^2]$, is minimized.

From Eqs. (4.3.12)–(4.3.14), we have

$$\begin{aligned} \frac{\partial \mathcal{E}}{\partial b_k} &= E\left[2\epsilon_n \frac{\partial \epsilon_n}{\partial b_k}\right] \\ &= -2R_{\epsilon r}(k), \end{aligned}$$

where

$$R_{\epsilon r}(k) = E[\epsilon_n r_{n-k}].$$

As the channel is time varying, the optimal tap coefficient vector changes with time. If the channel variation rate is much smaller than the symbol rate of the transmitted signal, we can determine the optimal tap coefficient vector $\mathbf{b}_{opt}$ iteratively under the assumption that $\mathbf{b}_{opt}$ does not change much over the period of the iterations. The iteration relation for the tap coefficient vector can be represented as

$$\mathbf{b}(n+1) = \mathbf{b}(n) + \Delta\mathbf{b}(n),$$

where $\mathbf{b}(n)$ is the estimated $\mathbf{b}$ value at $t = nT$. The term $\mathbf{b}(n)$ is the old value of the tap coefficient vector, and $\Delta\mathbf{b}(n)$ is the correction applied to it to compute the new (updated) value $\mathbf{b}(n+1)$ of the tap coefficient vector. When the kth element of $\mathbf{b}(n)$, $b_k(n)$, $k = -N, -N+1, \ldots, N$, is not optimum, $\partial\mathcal{E}/\partial b_k$ is nonzero. On the other hand, $b_k(n)$ approaches the optimum value of b_k when the gradient $\partial\epsilon/\partial b_k$, and hence $R_{\epsilon r}(k)$, approaches zero. Iterative computation is a hill-climbing approach. The goal is to reach the extremal point as quickly as possible. Hill-climbing requires two parameters: direction and stepsize. When the gradient $\partial\epsilon/\partial b_k$ is nonzero, it can be used as the direction parameter. Therefore, one way to choose $\Delta b_k(n)$ is to let it be proportional to $\partial\epsilon/\partial b_k$ or $R_{\epsilon r}(k)$. Let α be the proportional factor (stepsize). The iterative equation can be written as

$$b_k(n+1) = b_k(n) + \alpha E[\epsilon_n r_{n-k}]$$

for $k = -N, \cdots, -1, 0, 1, \cdots, N$. The ensemble average, $E[\cdot]$, is difficult to implement. It can be shown that $\epsilon_n r_{n-k}$ is an unbiased estimate of $E[\epsilon_n r_{n-k}]$. For simplicity of implementation, we can use $\epsilon_n r_{n-k}$ in place of $E[\epsilon_n r_{n-k}]$ and write

$$b_k(n+1) = b_k(n) + \alpha(\epsilon_n r_{n-k}), k = -N, \cdots, -1, 0, 1, \cdots, N.$$

For $\mathbf{b}(n)$ to converge to $\mathbf{b}_{opt}$, given by the Wiener-Hopf equation, α should be in the range [123]

$$0 < \alpha < \frac{2}{(2N+1) \times (\text{signal power plus noise power at the equalizer input})}.$$

The iterative algorithm is called the least-mean-square (LMS) algorithm for the tap coefficients. Figure 4.16 shows the structure of an adaptive linear equalizer that uses the LMS algorithm and the previously detected symbols to update the tap coefficients.

In practice, very often a training sequence is transmitted before the information data sequence to compute the initial optimum tap coefficients of the adaptive equalizer. The training sequence normally is a deterministic sequence with noise-like characteristics—the so-called pseudorandom sequence. The sequence is known to the receiver, and is used to adjust its tap coefficients to the optimum values. After the training sequence, the adaptive equalizer then uses the previously detected symbols to estimate the equalization error, and the LMS algorithm described above to update the tap coefficients.

4.3.2 Decision Feedback Equalization

The linear equalizers are very effective in equalizing channels where ISI is not severe. The severity of the ISI is directly related to the spectral characteristics. In the case that there are spectral nulls in the transfer function of the effective channel (which is often encountered in

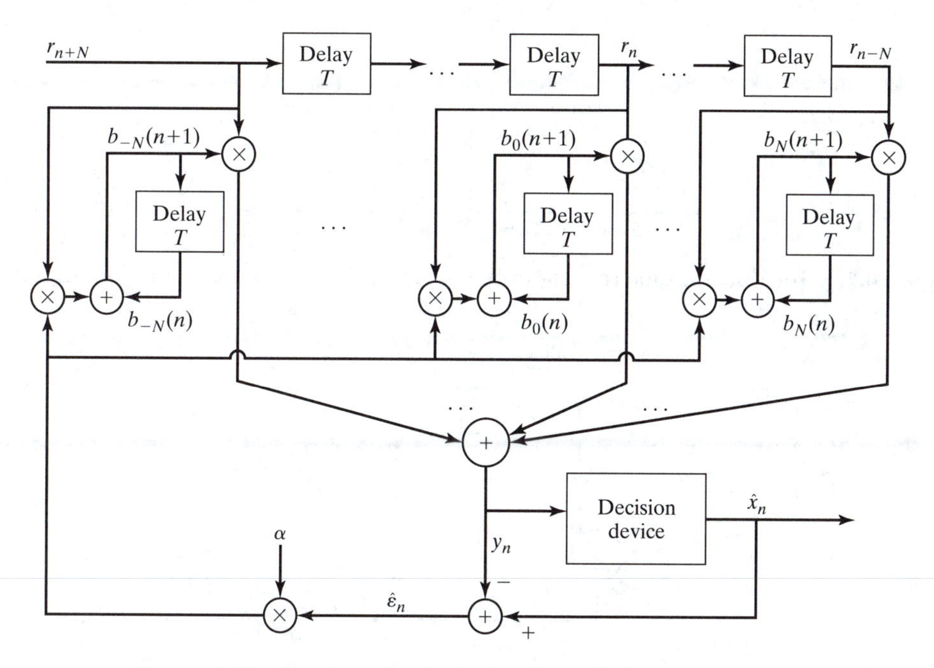

Figure 4.16 Decision-directed adaptive MSE linear equalizer.

wireless communications), the additive noise at the receiver input will be dramatically enhanced by the linear equalizers. To overcome this problem, nonlinear equalizers can be used. The decision feedback equalizer (DFE) is particularly useful for channels with severe amplitude distortions and has been widely used in wireless communications.

In the presence of channel noise, from Eq. (4.3.1), the effective channel output signal at $t = nT$ is

$$r_n = c_0 x_n + \sum_{l=-\infty}^{-1} c_l x_{n-l} + \sum_{l=1}^{\infty} c_l x_{n-l} + n_n. \tag{4.3.22}$$

In Eq. (4.3.22), the first term represents the desired signal component with the channel gain c_0. The second term is due to the precursors of the channel impulse response that occur before the main sample c_0 associated with the desired data symbol. The third term is due to the postcursors of the channel impulse response that occur after the main sample c_0. Figure 4.17 illustrates the precursors and postcursors for the impulse response of an effective channel. From Eq. (4.3.22), we see that the ISI term associated with the postcursors is related to the previously transmitted symbols. Assuming no detection errors, the information of the previous symbols is available to the receiver. As a result, the equalizer can make use of this information to combat the ISI component and, therefore, to provide a transmission performance improvement over that of the linear equalizers.

Figure 4.18 shows the structure of a typical DFE implemented at baseband. The DFE has a $(N + 1)$-tap feedforward filter and an M-tap feedback filter, where the input to the feedforward

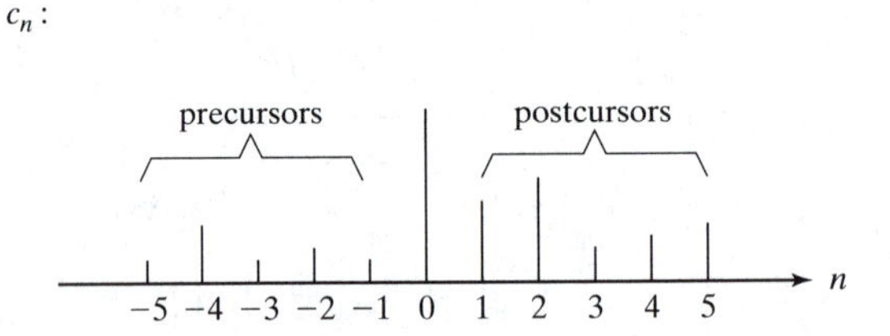

Figure 4.17 The discrete-time representation of the dispersive channel impulse response.

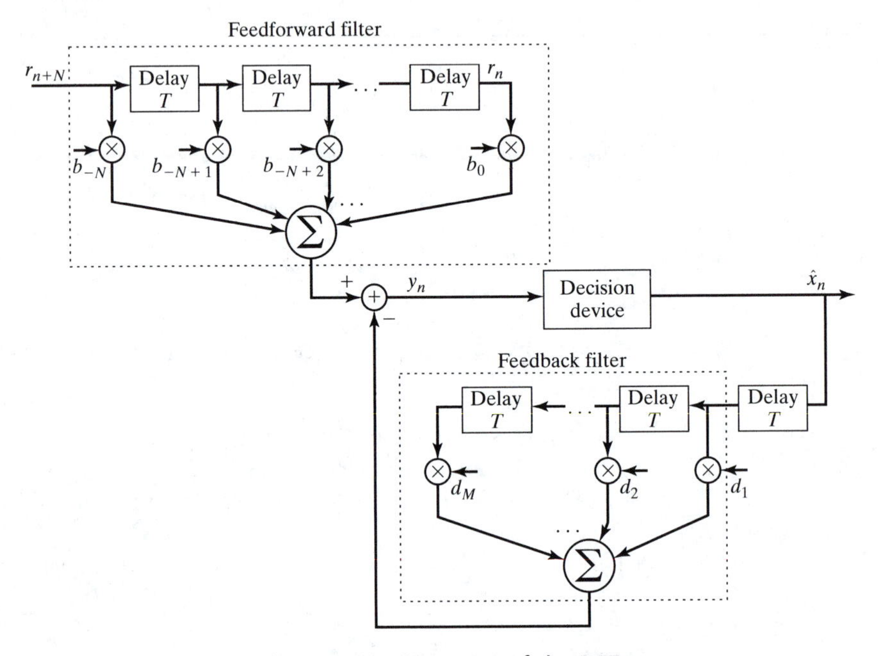

Figure 4.18 Structure of the DFE.

filter is the output $\{r_{n+N}\}$ of the effective channel sampled at $t = nT$, and the input to the feedback filter is the decision on previous symbols $\{\hat{x}_{n-1}\}$ which is an estimate of $\{x_{n-1}\}$. Both the feedforward filter and the feedback filter are implemented as tapped-delay-line filters with symbol-spaced taps. The feedforward filter is like a linear equalizer. The decisions made on the previously transmitted symbols are used via the feedback filter. The basic idea is that, if the value of the symbols already detected are known (past decisions are assumed to be correct), then the ISI contributed by these symbols can be canceled exactly, by subtracting the past symbol values with appropriate weighting from the equalizer output. The tap coefficients of both the feedforward filter and the feedback filter can be determined based on different design criteria, as discussed in the following.

Zero-Forcing (ZF) DFE. Here we study a particular version of the DFE, known as a zero-forcing DFE. It operates as follows. Consider the dispersive channel impulse response as shown in Figure 4.17, where c_n achieves the maximum value at $n = 0$ and normally the number of taps in the feedback filter, M, is chosen to be the same as the number of the postcursors (i.e., $c_n = 0$ for $n > M$). The transfer function of the channel in the z domain is

$$C(z) = \sum_{k=-\infty}^{\infty} c_k z^{-k}. \tag{4.3.23}$$

The transfer function of the feedforward linear filter in the DFE is

$$B(z) = \sum_{k=-N}^{0} b_k z^{-k}. \tag{4.3.24}$$

The transfer function of the effective channel consisting of the dispersive channel (to be equalized) and the feedforward filter is then

$$C(z)B(z) = \sum_{n=-\infty}^{\infty} h_n z^{-n},$$

where

$$h_n = \sum_{k=-N}^{0} b_k c_{n-k}.$$

Ideally, the tap coefficients of the feedforward filter should be chosen in such a way that $h_n = 0$ for $n < 0$ without changing the relative values of the remaining components for $n > 0$ (i.e., to force the ISI due to the precursors in the channel impulse response to be zero at the sampling moment for detecting the current symbol). Under the constraint of the finite tap number $N + 1 (> M)$, we can only force h_n to be zero for $n = -N, -N+1, \ldots, -2, -1$, by choosing the values of the tap coefficients according to the following equation

$$\begin{bmatrix} c_0 & c_{-1} & \cdots & c_{-N} \\ c_1 & c_0 & \cdots & c_{-N+1} \\ \vdots & \vdots & & \vdots \\ c_N & c_{N-1} & \cdots & c_0 \end{bmatrix} \begin{bmatrix} d_{-N} \\ d_{-N+1} \\ \vdots \\ d_0 \end{bmatrix} = \begin{bmatrix} 0 \\ 0 \\ \vdots \\ 1 \end{bmatrix}. \tag{4.3.25}$$

The tap coefficients of the feedback filter, b_j, are chosen to be h_j, $j = 1, 2, \ldots, M$, so that the effect of the postcursors can be removed by the feedback filter. As a result, the output of the feedback filter is subtracted from the output of the feedforward filter to form the input to the decision device, given by

$$\begin{aligned} y_n &= \sum_{k=-N}^{0} b_k r_{n-k} - \sum_{j=1}^{M} d_j \hat{x}_{n-j} \\ &= \sum_{k=-N}^{0} b_k r_{n-k} - \sum_{j=1}^{M} h_j \hat{x}_{n-j}. \end{aligned}$$

Under the assumption that the residual precursors in the system impulse response after the feedforward filter are negligible, the output of the feedforward filter can be approximately given by

$$\sum_{k=-N}^{0} b_k r_{n-k} \approx x_n + \sum_{j=1}^{M} h_j x_{n-j} + v_n,$$

where $v_n = \sum_{k=-N}^{0} b_k n_{n-k}$ is due to the received noise components. If we further assume that the previous symbols are correctly detected ($\hat{x}_{n-j} = x_{n-j},\ j = 1, 2, \ldots, M$), then the DFE output is

$$\hat{x}_n = D[y_n] \approx D\left[\left(x_n + \sum_{j=1}^{M} h_j x_{n-j} + v_n\right) - \left(\sum_{j=1}^{M} h_j \hat{x}_{n-j}\right)\right] \tag{4.3.26}$$

$$= D[x_n + v_n]. \tag{4.3.27}$$

where $D[\cdot]$ denotes the decision function. $D[\cdot]$ is a non-linear device which has a weak signal suppression property. If x_n is stronger than v_n, the latter is suppressed and $\hat{x}_n = x_n$.

Under the two assumptions, the equalization error at the input to the decision device is

$$\epsilon_n = x_n - y_n = v_n,$$

which is due to the channel noise only. That is, the ISI due to the dispersive channel is completely eliminated.

Given that the received noise component n_n is an iid Gaussian random variable with zero mean and variance σ_n^2, the equalization error, v_n, is also a Gaussian random variable with zero mean and variance

$$\sigma_v^2 = \sigma_n^2 \sum_{k=-N}^{0} b_k^2.$$

Compared with the ZF linear equalizer, ZF-DFE in general reduces the noise amplification effect. However, the ZF-DFE is not optimum in any sense.

Minimum Mean-Square Error (MMSE) DFE. In general, feedback systems are nonlinear. However, if the previously detected symbols are correct, the DFE behaves like a linear system. The DFE is characterized by the feedforward tap coefficients $\{b_{-N}, b_{-N+1}, \ldots, b_0\}$ and the feedback tap coefficients $\{d_1, d_2, \ldots, d_M\}$. If the overall equalizer is linear, the feedforward and feedback sections can be partitioned and the tap coefficients can be chosen independently to achieve a performance objective, for example, to minimize the probability of transmission error. However, an analytical expression for the error probability in terms of the tap coefficients is not known. A practical, but suboptimum, approach is to minimize the mean-square error at the input to the decision device. Under the assumption of correct previously detected symbols, the equalized signal, sampled at $t = nT$, is

$$y_n = \sum_{k=-N}^{0} b_k r_{n-k} - \sum_{j=1}^{M} d_j x_{n-j}. \tag{4.3.28}$$

The desired equalizer output at the nth sampling instant x_n. As a result, the mean-square error is

$$\begin{aligned} \mathcal{E} &= E[\epsilon_n^2] \\ &= E[(x_n - y_n)^2] \\ &= E\left[\left(x_n - \sum_{k=-N}^{0} b_k r_{n-k} + \sum_{j=1}^{M} d_j x_{n-j}\right)^2\right]. \end{aligned}$$

The mean-square error $\mathcal{E}$ is minimum when the tap coefficient d_k of the feedforward filter, $k = -N, -N+1, \ldots, 0$, is chosen in such a way that

$$\begin{aligned} \frac{\partial \mathcal{E}}{\partial b_k} &= 2E\left[-r_{n-k}\left(x_n - \sum_{k=-N}^{0} b_k r_{n-k} - \sum_{j=1}^{M} d_j x_{n-j}\right)\right] \\ &= -2E[r_{n-k}\epsilon_n] \\ &= 0, \end{aligned} \tag{4.3.29}$$

and when the tap coefficient b_{N+j} of the feedback filter, $j = 1, 2, \ldots, M$, is chosen in such a way that

$$\begin{aligned} \frac{\partial \mathcal{E}}{\partial d_j} &= 2E\left[x_{n-j}\left(x_n - \sum_{k=-N}^{0} b_k r_{n-k} - \sum_{j=1}^{M} d_j x_{n-j}\right)\right] \\ &= 2E[x_{n-j}\epsilon_n] \\ &= 0. \end{aligned} \tag{4.3.30}$$

Under the assumption that $E(x_n x_{n-j}) = \sigma_x^2 \delta(j)$, where σ_x^2 is the transmitted symbol power and $\delta(j)$ is the Kronecker delta function defined in Eq. (3.2.1), using Eq. (4.3.22) to present the received sample r_n, Eqs. (4.3.29) and (4.3.30) can be written as a set of $(N + M + 1)$ linear equations of the form

$$\sum_{k=-N}^{0} b_k R_c(k-l) + R_n(k-l)/\sigma_x^2 = \sum_{j=1}^{M} d_j c_{l-j} + c_l \tag{4.3.31}$$

for $l = -N, -N+1, \ldots, 0$ and

$$d_j = \sum_{k=-N}^{0} b_k c_{j-k} \tag{4.3.32}$$

for $j = 1, 2, \ldots, M$, where

$$R_n(k-l) = E[n_{n-l} n_{n-k}]$$

is the correlation function of the channel noise $\{n_n\}$ and

$$R_c(k-l) = \sum_{m=-\infty}^{\infty} c_{m-l} c_{m-k}$$

is the correlation function of the channel impulse response sample sequence $\{c_n\}$. Given the correlation functions, $R_n(k-l)$ and $R_c(k-l)$, and the channel impulse response sequence, $\{c_n\}$, the tap coefficients of the feedforward and feedback filters that minimize the mean-square error can be determined from Eqs. (4.3.31) and (4.3.32).

Adaptive DFE. For a time-varying wireless channel, the DFE tap coefficients need to be adjusted constantly according to the instantaneous channel impulse response. In this case, we can also use the MMSE criterion to determine the tap coefficients for both the feedforward filter and the feedback filter. Letting

$$\begin{aligned}
\mathbf{b} &= (b_{-N}, b_{-N+1}, \ldots, b_{-1}, b_0)^T, \\
\mathbf{d} &= (d_1, d_2, \ldots, d_M)^T, \\
\mathbf{r}_n &= (r_{n+N}, r_{n+N-1}, \ldots, r_n)^T, \text{ and} \\
\hat{\mathbf{x}}_{n-1} &= (\hat{x}_{n-1}, \hat{x}_{n-2}, \ldots, \hat{x}_{n-M})^T,
\end{aligned}$$

the equalization error at $t = nT$ can be represented as

$$\epsilon_n = x_n - (\mathbf{r}_n^T \mathbf{b} - \hat{\mathbf{x}}_{n-1}^T \mathbf{d}). \tag{4.3.33}$$

We want to choose the tap coefficient vector $(\mathbf{b}, \mathbf{d})^T$ such that the mean-square equalization error, $\mathcal{E} = E[\epsilon_n^2]$, is minimized. On the basis that the feedforward and feedback sections are decomposable, the tap coefficients of the adaptive DFE can be obtained iteratively (similar to that in the adaptive linear equalizer) by

$$\mathbf{b}(n+1) = \mathbf{b}(n) + \alpha_1 \epsilon_n \mathbf{r}_n \tag{4.3.34}$$

and

$$\mathbf{d}(n+1) = \mathbf{d}(n) - \alpha_2 \epsilon_n \hat{\mathbf{x}}_{n-1}, \tag{4.3.35}$$

where n is the discrete time index, $\mathbf{b}(n)$ and $\mathbf{d}(n)$ are the estimated $\mathbf{b}$ and $\mathbf{d}$, respectively, at $t = nT$, and α_1 and α_2 are the stepsize parameters for the feedforward and feedback filters, respectively.

Practical implementation requires the replacement of ϵ_n by $\hat{\epsilon}_n \triangleq \hat{x}_n - [\mathbf{r}_n^T\mathbf{b}(n) - \hat{\mathbf{x}}_{n-1}^T\mathbf{d}(n)]$. In this way, all the quantities in Eqs. (4.3.34) and (4.3.35) are available at the receiver.

Given the same number of taps, whether or not a DFE can perform better than a linear equalizer in combating ISI depends on the impulse response of the effective channel. It has been shown that a DFE yields good performance in the presence of moderate to severe ISI as experienced in a wireless channel. A DFE can compensate for amplitude distortion without as much noise enhancement as a linear equalizer. This is because, in a DFE, using the feedback filter to cancel the ISI based on previously detected symbols allows more freedom in the choice of the tap coefficients of the feedforward filter. The combined impulse response of the effective channel and the feedforward filter may have nonzero samples following the main pulse. As the feedforward filter of a DFE is not required to act as the inverse filter for the effective channel, excessive noise enhancement can be avoided. On the other hand, a DFE suffers from the inherent error propagation problem due to the fact that the feedback filter uses the detected symbols. When a particular incorrect decision is fed back, the DFE output reflects this error during the next few symbols as the incorrect decision traverses the feedback delay line. Thus, there is a large likelihood of more incorrect decisions following the first one, a phenomenon called *error propagation*. The error propagation increases with the number of taps in the feedback filter. Fortunately, the error

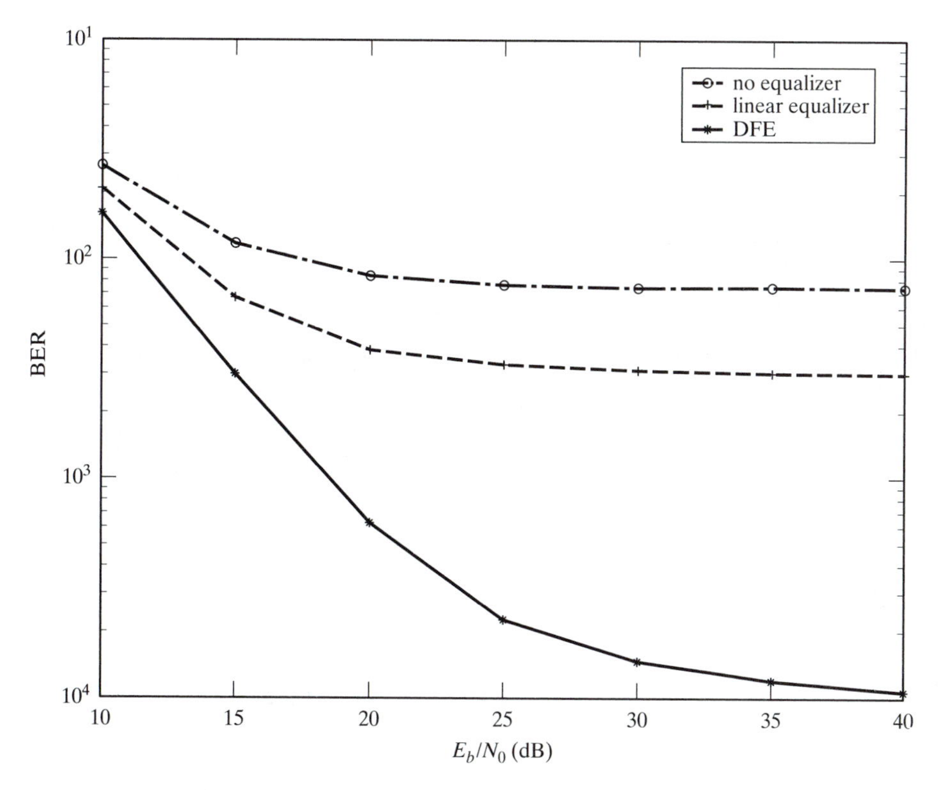

Figure 4.19 BER performance using accurate equalization error in the adaptive algorithms.

propagation in a DFE is not catastrophic. For typical channels with a low BER, errors occur in short bursts that degrade performance only slightly. The performance of the adaptive linear equalizer and the adaptive DFE for a dispersive fading channel can be evaluated by computer simulation.

Example 4.8 Performance of Adaptive Linear Equalizer and Adaptive DFE

Consider a 2-path wireless propagation channel, where the first path experiences Rician fading with a k-factor of 7.5 dB, and the second path experiences Rayleigh fading. The two paths have the same power of the diffusive signal component. Consider BPSK with coherent demodulation, and assume perfect carrier phase synchronization. The normalized fading rate is $\nu_m T = 0.0005$, where ν_m is the maximum Doppler shift and T is the BPSK symbol interval. The propagation delay difference between the two paths is equal to the symbol interval T. It is desired to evaluate the BER transmission performance when using (a) no equalizer, (b) a 7-tap adaptive linear equalizer, and (c) an adaptive DFE with a 3-tap feedforward filter and a 2-tap feedback filter. In the simulation, one sample is generated for each transmitted/received symbol.

Simulation results

Figure 4.19 shows the BER performance as a function of E_b/N_0 under the assumption that exact equalization error is available to the receiver, where E_b is the average received signal energy per

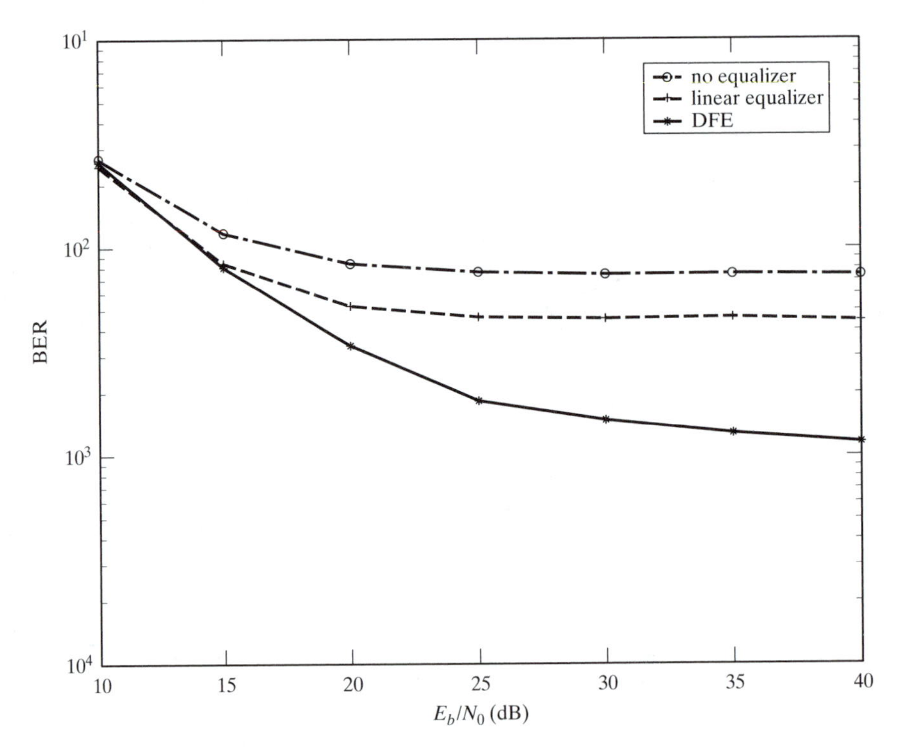

Figure 4.20 BER performance using estimated equalization error in the adaptive algorithms.

bit from both propagation paths, and N_0 is the one-sided psd of additive white Gaussian noise. In this case, there is no error propagation in the DFE. It is clear that both the linear equalizer and the DFE improve the transmission performance over that without equalization, as the ISI can be severe from time to time when the first path is in deep fading while the second path is not. It is also clear that the DFE performs much better than the linear equalizer, due to (*a*) the limited number of taps in the linear equalizer and (*b*) the noise enhancement effect, especially at low E_b/N_0 values.

Figure 4.20 shows the BER performance where the adaptive algorithms use the estimated equalization error to adjust the tap coefficients. In this case, the DFE suffers from inherent error propagation. Both the linear equalizer and the DFE improve transmission performance and the DFE outperforms the linear equalizer. However, the performance improvements are not as significant as in the case when exact equalization error is available. The perturbation in estimating the equalization errors degrades the performance of the equalizers and introduces error propagation to the DFE.

SUMMARY

In this chapter, we have addressed the reception of signals emerging from a fading time-dispersive channel. Techniques for mitigating wireless channel impairments include power control for long-term fading, diversity for short-term multipath fading, adaptive channel equalizers for intersymbol interference in TDMA and FDMA systems, and Rake receiver for short-term multipath dispersive fading in CDMA systems. In particular, we have studied diversity and equalization in detail. Diversity reception makes use of independently faded received signal components to reduce the chances of overall deep fading and therefore increase the transmission accuracy. An adaptive channel equalizer follows channel variations over time. The zero-forcing linear equalizers reduce intersymbol interference to zero at the decision device input, and the decision feedback equalizer makes use of the previously detected information symbols to cancel the interference component due to the previously transmitted symbols.

ENDNOTES

1. For an overview on channel impairment mitigation techniques, see the paper by Sklar [141].
2. Using space diversity to overcome short-term fading was first discovered in the 1920s [14, 113].
3. For more advanced studies on diversity, see the paper by Parson *et al.* [46], the book edited by Jakes [42], and books by Schwartz, Bennett, and Stein [136], Lee [82], Stüber [147] and Proakis [123].
4. For more detailed discussion on channel equalization, see the book by Clark [29].
5. For more discussion on the zero-forcing adaptive linear equalizer, see the papers by Lucky [90, 91]; for more discussion on adaptive DFE, see the papers by George, Bowen and Storey [52] and by Monsen [99].
6. For a comprehensive tutorial on adaptive channel equalization, see the paper by Qureshi [125]. For details of adaptive filtering algorithms for the tap coefficient vector of the linear equalizer and DFE, see the book by Haykin [61].

7. Neither the linear equalizer nor the DFE is an optimum equalizer in terms of minimizing the probability of transmission error in detecting the information sequence. The optimum detector is called the maximum likelihood sequence detector and is discussed in the paper by Forney [51].

PROBLEMS

P4-1 Describe how diversity can combat the effect of deep fading and explain why maximal ratio combining gives the best performance.

P4-2 Two coherent receivers with diversity in a Rayleigh fading environment are studied. One receiver uses maximum ratio combining and the other uses selective diversity. It is found that the transmission performance difference between the two receivers increases with the order of diversity. Why?

P4-3 In diversity reception, it is desired that the diversity channels experience independent fading. Why? Consider diversity with two channels and explain how transmission performance degrades as the correlation between the channel gains increases.

P4-4 Consider a 2-branch diversity system using BPSK with coherent detection. The channel gain in each branch can take on the values of 1.0 and 0.05. The channel gains in the 2 branches are correlated with the joint probability distribution given by $P(1.0, 1.0) = 0.4$, $P(0.05, 0.05) = 0.4$, and $P(1.0, 0.05) = 0.2$.

a. For maximal ratio combining, calculate the probability of bit error as a function of the received SNR/bit and compare the result with that of Example 4.3, where the channel gains are independent.
b. For selective combining, calculate the probability of bit error as a function of the received SNR/bit.
c. Compare the probabilities of bit error in part (*a*) and part (*b*).

P4-5 Consider a 2-branch diversity system using BPSK with coherent detection. The channel gain in each branch is 1.0 with a probability of 0.9 and 0.05 with a probability of 0.1. The additive noise in each branch is white Gaussian with two-sided psd of $N_0/2$. The two branches have independent channel gains. Calculate the probability of bit error as a function of the received SNR/bit for (*a*) selective diversity and (*b*) equal-gain combining. Draw the BER curves and compare them.

P4-6 Consider digital transmission in a Rayleigh fading environment. The transmitted signal $x(t)$ has bit energy E_b and bit interval T_b. Diversity reception with equal-gain combining is used. The received signal from each diversity channel is corrupted by additive white Gaussian noise with zero mean and two-sided psd $N_0/2$. Filters matched to the transmitted waveforms are used for demodulation. Express the mean received SNR per bit of the decision variable in terms of the diversity order L and the mean SNR/bit per channel.

P4-7 Consider diversity reception with 2 independent Rayleigh-fading channels and selective combining.

a. Derive the pdf of the received SNR/bit at the combiner output.
b. If BPSK is used for transmission, find the probability of bit error with diversity reception.

P4-8 A multipath fading channel has a delay spread of $T_m = 0.1$ s and a Doppler spread of $B_d = 0.05$ Hz. The total channel bandwidth at bandpass available for signal transmission is $W = 30$ Hz.

a. In order to reduce the effects of intersymbol interference, what value for the transmitted symbol interval T_s would be appropriate?
b. Determine the coherence bandwidth and coherence time of this channel.
c. The channel is to be used to transmit data via BPSK in a frequency diversity mode. Explain how you would use the available channel bandwidth to obtain frequency diversity and determine the maximum diversity order.

P4-9 An optimum receiver for a time-dispersive channel consists of a matched filter followed by an equalizer. The matched filter is designed to minimize the effect of random noise and the equalizer is designed to minimize the effect of ISI. Using mathematical arguments, demonstrate that

a. the matched filter tends to accentuate the effect of ISI, and
b. the equalizer tends to accentuate the effect of additive channel noise.

P4-10 Verify the relation between the z transform $C(z)$ of $\{c_n\}$ and the Fourier transform $C(\omega)$ of $c(t)$ given in Eq. (4.3.11).

P4-11 For the adaptive linear equalizer, show that $\epsilon_n r_{n-k}$ is an unbiased estimate of $E[\epsilon_n r_{n-k}]$, where ϵ_n is the equalization error at $t = nT$, r_{n-k} is the output of the effective channel at $t = (n-k)T$, and T is the basic time unit.

P4-12 For the adaptive linear equalizer, show that if the stepsize α satisfies the following condition

$$0 < \alpha < \frac{2}{(2N+1) \times (\text{signal power plus noise power at the equalizer input})},$$

then the tap coefficients of the equalizer will converge to the optimum values given by the Wiener-Hopf equation.

P4-13 In a digital transmission system using BPSK, the matched filter receiver designed for an AWGN channel is used. The physical channel introduces an additive Gaussian noise of zero mean and two-sided psd $N_0/2$. In addition, it introduces ISI. At the end of the nth symbol interval, the output of the demodulator is

$$r_n = x_n + 0.9x_{n-1} + n_n,$$

where the desired signal component $x_n = \sqrt{E_b}$ if symbol "1" was sent and $x_n = -\sqrt{E_b}$ if symbol "0" was sent, E_b is the transmitted signal symbol energy; the noise component n_n is a Gaussian random variable with zero mean and variance $N_0/2$.

a. Determine the probability of transmission error (1) without equalization and (2) with an ideal zero-forcing linear equalizer of 7 taps.
b. Design an ideal zero-forcing DFE with a 5-tap feedforward filter and a 2-tap feedback filter for the channel and determine the probability of transmission error.
c. Plot the BER curves for the above three cases and compare the performance of the two equalizers.

P4-14 Consider an effective channel with impulse response

$$c(t) = 0.9\delta(t + 4T) - 0.4\delta(t + 3T) + 1.2\delta(t) + 0.5\delta(t - T) - 0.7\delta(t - 4T),$$

where T is the transmission symbol interval.

a. Design a 5-tap zero-forcing linear equalizer for the effective channel.
b. Compute the impulse response of the equalized system and determine whether the equalizer is effective in combating the ISI.
c. Does the equalizer significantly enhance the input white noise component?
d. Is it necessary to increase the number of taps in the equalizer for better transmission performance? Why?

P4-15 Verify that Eqs. (4.3.29) and (4.3.30) can be equivalently represented by Eqs. (4.3.31) and (4.3.32), given that $E(x_n x_{n-j}) = \sigma_x^2 \delta(j)$.

P4-16 An effective channel has the following transfer function:

$$C(z) = (1 + 0.5z^{-1})(1 + 2z^{-1}).$$

a. Determine the ideal zero-forcing linear equalizer for the channel.
b. Design a zero-forcing DFE for the channel, where the feedforward filter is to equalize the $(1 + 0.5z^{-1})$ component and the feedback filter is to equalize the $(1 + 2z^{-1})$ component.
c. Design a zero-forcing DFE for the channel, where the feedforward filter is to equalize the $(1 + 2z^{-1})$ component and the feedback filter is to equalize the $(1 + 0.5z^{-1})$ component.
d. Compare the effects of the above three equalizers on the additive Gaussian noise component with zero mean and variance σ^2 at the channel output.

P4-17 Consider a 2-path wireless propagation channel with a maximum Doppler shift ν_m, where the first path experiences Rician fading with the k-factor equal to 5 dB, and the second path experiences Rayleigh fading. The two paths have the same power for the diffusive signal component. Consider BPSK with coherent demodulation, assuming perfect carrier phase synchronization. The propagation delay difference between the two paths is equal to the symbol interval T. Using computer simulation, obtain the BER performance of the following cases:

a. Using a 5-tap adaptive linear equalizer under the assumption that the desired signal is available to the equalizer and the normalized fading rate $\nu_m T$ is 0.0005, where T is the BPSK symbol interval;
b. Using a 9-tap adaptive linear equalizer with $\nu_m T = 0.0005$ and under the assumption that the desired signal is available to the equalizer;
c. Using a 9-tap adaptive linear equalizer with $\nu_m T = 0.0005$ and without the assumption that the desired signal is available to the equalizer;
d. Using a 9-tap adaptive linear equalizer with $\nu_m T = 0.0001$ and without the assumption that the desired signal is available to the equalizer;

e. Using a 9-tap adaptive linear equalizer with $\nu_m T = 0.0005$ and under the assumption that the desired signal is available to the equalizer, with the k-factor equal to 7 dB.

Comment on the effects of (1) tap number, (2) whether the desired signal is available to the equalizer, (3) the normalized fading rate, and (4) the k-factor on the BER performance.

P4-18 Consider a 2-path wireless propagation channel, where the first path experiences Rician fading with the k-factor equal to 5 dB, and the second path experiences Rayleigh fading. The two paths have the same power for the diffusive signal component. Consider BPSK with coherent demodulation, assuming perfect carrier phase synchronization. The normalized fading rate is $\nu_m T = 0.0005$, where ν_m is the maximum Doppler shift and T is the BPSK symbol interval. The propagation delay difference between the two paths is equal to the symbol interval T. Using computer simulation, obtain the BER performance in the following cases:

a. Using a DFE with a 3-tap feedforward filter and a 2-tap feedback filter, under the assumption that the desired signal is available to the equalizer;
b. Using a DFE with a 3-tap feedforward filter and a 2-tap feedback filter, without the assumption that the desired signal is available to the equalizer;
c. Using a DFE with only a 2-tap feedback filter, under the assumption that the desired signal is available to the equalizer;
d. Using a DFE with only a 1-tap feedback filter under the assumption that the desired signal is available to the equalizer.

Compare the four curves of BER versus SNR/bit and comment on the effects of the error propagation and the DFE tap numbers on the performance.

5

Fundamentals of Cellular Communications

With the limitation in spectral width, the maximum number of users (capacity) that can be supported in a wireless system is an important performance measure. If the system is supported by a single base station, a high power transmitter is needed to support a large number of users. The system capacity (number of users) can be enlarged by arranging small cells, each requiring only a low power transmitter, in a cellular array. In this chapter, the rationale behind cellular operation as a means of increasing system capacity is explained. The increase in system capacity comes from the use of smaller cells, reuse of frequencies, and antenna sectoring. The ramifications of frequency reuse and the gain obtainable from cell splitting and antenna sectoring are described and discussed in detail.

5.1 INTRODUCTION

As discussed in Chapter 2, the wireless propagation channel exhibits impairments far more severe than those inherent in a guided wire. Severe impairments mean that the wireless channel yields a poorer signal-to-interference ratio (SIR) and, hence, higher transmission bit error rate at the output of the receiver compared with those in a wired channel. For example, the BER in an optical fiber is in the neighborhood of 10^{-10} while that in a wireless channel with multipath fading may be in the range of 10^{-2} to 10^{-3}. This high error rate has the effect of limiting the amount of throughput and rendering signal reception unreliable. Impairments in the propagation channel will result in a reduction of usable spectral width. Mitigation of interference due to channel impairments has the effect of enlarging the spectral width. A wider spectrum, in turn, means more bandwidth to support more users. Certain sophisticated modulation and coding techniques have built-in properties that the receiver can exploit to compensate for channel impairments. Some of the modulation and detection methods that are relevant to wireless communications are discussed in Chapters 3 and 4.

A wider spectrum will allow for greater sharing of the spectral width to support more users and increase the throughput. System capacity can be interpreted as the largest number of users that can be supported during any one use of the channel. A larger population will mean a larger geographical coverage. If a single transmitter were used to cover a large geographical area, a very high power transmitter and very high antenna would be required. With a single high power transmitter, all users will share the same set of frequencies, or radio channels. The system capacity, in terms of the maximum number of users that can be supported, offered by a single high power transmitter will hit a limit. The reason for this limitation is that the radio resources (i.e., frequencies) are not efficiently utilized. If the same set of radio resources were assigned to serve a smaller geographical area and then reused to serve another small geographical area, it would be possible to expand the system capacity. However, care must be taken to ensure that use of the same set of frequencies to serve more than one geographical area does not introduce reception interference between users in the two areas. Or, if there is some interference, it must be at an acceptable level. This means that the geographical regions that use the same set of radio frequencies must be physically separated from each other so that the power level of the signal that spills out from one region to a neighboring region does not produce unacceptable interference. This way of replicating identically structured and operated geographical regions gives rise to the concept of *cellular communications*.

A geographical region can be a single small area called a *cell* that is served by a single base station, or a cluster of cells. In cellular systems, each of the base stations in the cluster is connected to a mobile switching center (MSC) by wirelines. An MSC has more computing power and can perform many more functions than an individual base station. Therefore, most of the communications operations are handled by the MSC.

5.2 FREQUENCY REUSE AND MOBILITY MANAGEMENT

5.2.1 Cellular Communications and Frequency Reuse

If a given set of frequencies, or radio channels, can be reused without increasing the interference, then the large geographical area covered by a single high power transmitter can be divided into a number of small areas, each allocated a subset of frequencies. With a smaller geographical coverage, lower power transmitters with lower antennas can be used. Provided that the physical separation of two cells is sufficiently wide, the same subset of frequencies can be used in both cells. This is the concept of *frequency reuse* [92]. The ability to reuse the frequencies offers a means to expand the total system capacity without the need to employ high power transmitters. This plan of dividing the total large geographical coverage area into many small contiguous areas and using a low power transmitter with low antenna in each small area is referred to as *cellular communications*.

Each small area is served by a transmit/receive unit called a *base station* (BS). A base station is the common resource shared by multiple users. User-to-user communication passes through the base station. Users transmit their signals to the base station, and the base station then relays the information-bearing signals to their destinations. We say that the users access the base station so that the base station is also called an *access point* (AP). The propagation channel that handles the transmissions from the base station to the users is called the *forward channel* or *downlink* while the channel used for transmissions from the users to the base station is called the *reverse channel* or *uplink*. Techniques for multiaccessing the base station will be discussed in Chapter 6.

The radio coverage by one base station is referred to as a *cell*, which is also called a *footprint*. Cells arranged in a two-dimensional array form a cellular structure. From a conceptual point of view, it is highly desirable to construct the cellular system such that the cells do not overlap, and are tightly packed without dead spots. This form of layout requires the use of regular topologies (e.g., a square, triangular or hexagonal topology). Because of the differences in terrain and population densities, real footprints are amorphous in nature. On the other hand, cellular layouts using irregular structures are inefficient and limit growth. For this reason, cellular layouts and performance studies are based on regular topologies, even though they may just be conceptual. Also, regular topologies allow for systematic growth. In this chapter, all our discussions will be based on a hexagonal topology. The design and performance obtained using regular topologies may not correspond to real environments, but these topologies should provide valuable information and guidelines for structuring practical layouts.

While the cellular approach allows the use of low power transmitters and frequency reuse to expand system capacity, these advantages do not come without a price. Since a salient feature of wireless communications is the flexibility to support user roaming, smaller geographical areas will mean that mobile users can move out of one cell and into another cell quite frequently. To maintain continuous operation of an ongoing session, when the mobile migrates out of its current serving base station into the footprint of another, the connection must be handed off from the serving base station to the new base station.

An effective and efficient handoff mechanism must be in place to maintain service continuity and preserve end-to-end QoS (quality of service) requirements. The procedure to perform and to manage handoff is referred to as *handoff management*.

The principle of cellular communications is that a mobile host (MH)[1] is assigned a *home network*. The MH is identified by an *address*, called its home address. An agent in the home network, referred to as the *home agent*, keeps track of the MH's current location to facilitate delivery of messages destined for the MH. As the MH migrates away from its home network, the association between the MH and its home agent must be maintained so that the home agent can keep track of the MH's current location for message delivery purposes. In cellular communications, the procedure to keep track of the user's current location so as to maintain the association between the MH and its home agent is referred to as *location management*. Both handoff management and location management are necessitated by user mobility. These management functions are considered as two components of *mobility management*.

[1]For mobility and resource management at the networking layer, the term mobile host (MH) is often used instead of mobile station (MS).

5.2.2 Mobility Management

Mobility management consists of handoff management and location management. While mobility management issues are dealt with in Chapters 7 and 8, the salient aspects of handoff and location operations are described in the following paragraphs.

Handoff Management. When a mobile moves into a different cell while a session is in progress, the session has to be transferred to a new channel belonging to the new cell. This operation is referred to as handoff. The handoff operation involves the identification of a new base station and the allocation of channels to support both data and control signals in the new base station. As alluded to above, the MSC has the computing power to perform many different functions. Therefore, the handoff operation is normally handled by the MSC. The MSC keeps track of the resource usage of all cells under its jurisdiction. When a mobile moves into a different cell while a conversation is in progress, the MSC determines the availability of the unused channels in the new cell and makes transfer decisions. If the new base station has channels that can be allocated to handle both information carrying and control signals to accommodate the handoff connection, the handoff takes place. Otherwise handoff cannot occur.

Location Management. As previously mentioned, an MH is associated with a home network and its home address resides with its home agent. When the MH moves away from its home network, it enters a foreign territory. The network where the MH resides when it is away from home is called the foreign network. The MH has to register with its home agent, through the foreign agent, to let the home agent know its current location to facilitate message delivery. When an MH powers on, it registers with its home agent. When it moves to a foreign network, it has to register with its home agent, via the foreign agent. This means that there is an association between the home agent and the foreign agent. When the home agent has messages destined for the MH, it forwards them to the MH via the foreign agent. During the registration process, the home agent needs to ensure, from the identification conveyed through the foreign agent, that the mobile submitting a registration is the correct MH. The process of verifying that the identity provided during the registration process does belong to the correct MH is referred to as an *authentication* process.

5.3 CELL CLUSTER CONCEPT

In a wireless communications system, the channels used in the forward and reverse directions are separated in time or in frequency to permit duplexing. The total number of channels available in a cellular system is finite. The capacity of the cellular system is defined by the total number of channels available. The system capacity as a function of the total number of available channels depends on how the available channels are deployed. Specifically, if the available channels are reused for transmission of additional traffic, it is possible to expand the system capacity (i.e., support more users). Two or more different cells can use the same set of frequencies or radio channels if the nearest cells are separated such that the interference between cells at any given frequency is at an acceptable level. Cells which use the same set of frequencies are referred to as *cochannel cells*, and the interference between cochannel cells is referred to as *cochannel interference*. Frequencies or channels represent radio resources. In this text, we will use the terms *frequencies* and *channels* interchangeably.

The space between adjacent cochannel cells can be filled with other cells that use different frequencies to provide frequency isolation. A group of cells that use a different set of frequencies in each cell is called a *cell cluster*. Let N be the cluster size in terms of the number of cells within it and K be the total number of available channels without frequency reuse. The N cells in the cluster would then utilize all K available channels. In this way, each cell in the cluster contains one-Nth of the total number of available channels. In this sense, N is also referred to as the *frequency reuse factor* of the cellular system.

5.3.1 Capacity Expansion by Frequency Reuse

Suppose each cell is allocated J channels ($J \leq K$). If the K channels are divided among the N cells into unique and disjoint channel groups, each with J channels, then

$$K = JN. \tag{5.3.1}$$

Collectively the N cells in a cluster use the complete set of available frequencies. Since K is the total number of available channels, from Eq. (5.3.1), it can be seen that a decrease in the cluster size N is accompanied by an increase in the number of channels J allocated per cell. Thus, by decreasing the cluster size, it is possible to increase the capacity per cell.

The cluster can be replicated many times to form the entire cellular communications system. Let M be the number of times the cluster is replicated and C be the total number of channels used in the entire cellular system with frequency reuse. C is then the system capacity and is given by

$$C = MJN. \tag{5.3.2}$$

If N is decreased and J is proportionally increased so that Eq. (5.3.1) is satisfied, it is necessary to replicate the smaller cluster more times in order to cover the same geographical area. This means that the value of M has to be increased. Since $JN(= K)$ remains constant and M is increased, Eq. (5.3.2) shows that the system capacity C is increased. That is, when N is minimized, C is maximized. We will see shortly that minimizing N will increase cochannel interference.

Example 5.1 Cellular System Capacity

Consider a cellular system in which there are a total of 1001 radio channels available for handling traffic. Suppose the area of a cell is 6 km^2 and the area of the entire system is 2100 km^2.

a. Calculate the system capacity if the cluster size is 7.
b. How many times would the cluster of size 4 have to be replicated in order to approximately cover the entire cellular area?
c. Calculate the system capacity if the cluster size is 4.
d. Does decreasing the cluster size increase the system capacity? Explain.

Solution Given:

The total number of available channels $K = 1001$
Cluster size $N = 7$

Area of cell $A_{cell} = 6$ km^2
Area of cellular system $A_{sys} = 2100$ km^2

a. Since the number of channels per cell is $J = K/N$, then $J = 1001/7 = 143$ channels/cell. The coverage area of a cluster is

$$A_{cluster} = N \times A_{cell} = 7 \times 6 = 42 \text{ km}^2.$$

The number of times that the cluster has to be replicated to cover the entire cellular system is $M = A_{sys}/A_{cluster} = 2100/42 = 50$. Therefore,

$$C = MJN = 50 \times 143 \times 7 = 50{,}050 \text{ channels}.$$

b. For $N = 4$, $A_{cluster} = 4 \times 6 = 24$ km^2. Therefore,

$$M = A_{sys}/A_{cluster} = 2100/24 = 87.5 \simeq 87.$$

c. With $N = 4$, $J = 1001/4 \approx 250$ channels/cell. The system capacity is then

$$C = 87 \times 250 \times 4 = 87{,}000 \text{ channels}.$$

d. From (a) and (c), it is seen that a decrease in N from 7 to 4 is accompanied by an increase in M from 50 to 87, and the system capacity is increased from 50,050 channels to 87,000 channels. Therefore, decreasing the cluster size does increase the system capacity.

5.3.2 Cellular Layout for Frequency Reuse

As mentioned earlier, the discussion of cellular communications in this text is based on a two-dimensional chaining of hexagonal cells. The rule to find the nearest cochannel neighbor of a particular cell is as follows.

Rule for Determining the Nearest Cochannel Neighbors. The following two-step rule can be used to determine the location of the nearest cochannel cell:

Step 1: Move i cells along any chain of hexagons;
Step 2: Turn 60 degrees counterclockwise and move j cells.

The method of locating cochannel cells in a cellular system using the preceding rule is shown in Figure 5.1 for $i = 3$ and $j = 2$, where the cochannel cells are the shaded cells.

The cluster concept and frequency reuse in a cellular network are illustrated in Figure 5.2, where cells with the same number use the same set of frequencies. These are cochannel cells that must be separated by a distance such that the cochannel interference is below a prescribed QoS threshold. The parameters i and j measure the number of nearest neighbors between cochannel cells; the cluster size, N, is related to i and j by the equation

$$N = i^2 + ij + j^2. \tag{5.3.3}$$

For example, in Figure 5.2(b), we have $i = 1$ and $j = 2$, so that $N = 7$. With a cluster size $N = 7$, the frequency reuse factor is seven since each cell contains one-seventh of the total number of available channels.

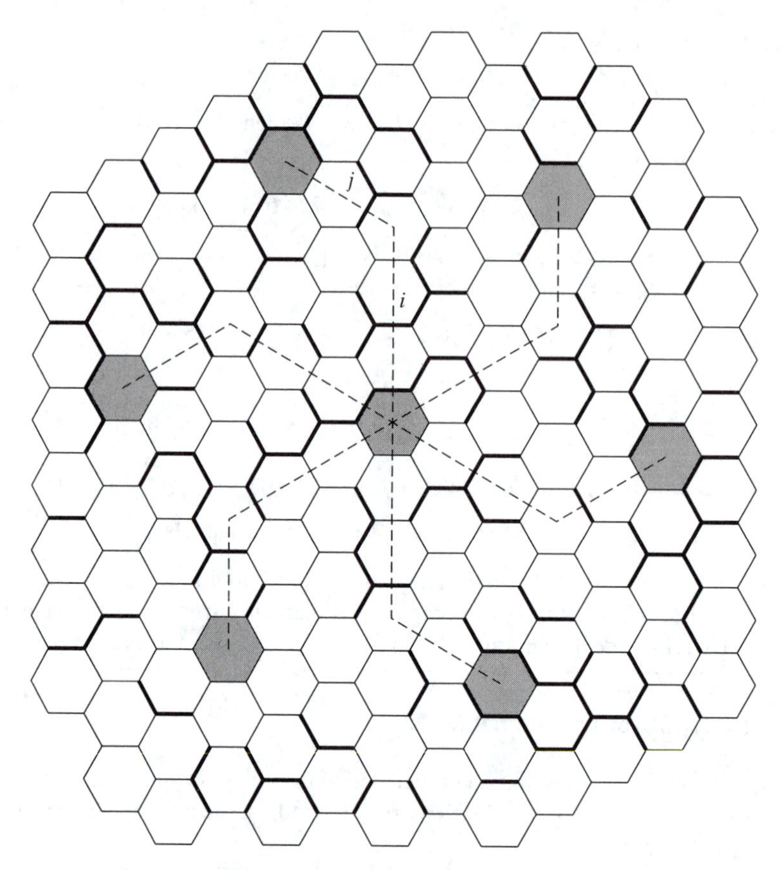

Figure 5.1 Locating cochannel cells in a cellular system.

Advantages of Cellular Systems. The advantages of operating in a cellular arrangement include:

a. the use of low power transmitter and,
b. an allowance for frequency reuse.

Frequency reuse needs to be structured so that cochannel interference is kept at an acceptable level. As the distance between cochannel cells increases, cochannel interference will decrease. If the cell size is fixed, the average signal-to-cochannel interference ratio will be independent of the transmitted power of each cell (discussed in Section 5.4). The distance between any two cochannel cells can be examined by making use of the geometry of hexagonal cells.

Example 5.2 Number of Frequency Channels

Consider a cellular system with a total bandwidth of 30 MHz which uses two 25 kHz simplex channels to provide full duplex voice and control channels. Assuming that the system uses a nine-cell reuse pattern and 1 MHz of the total bandwidth is allocated for control channels,

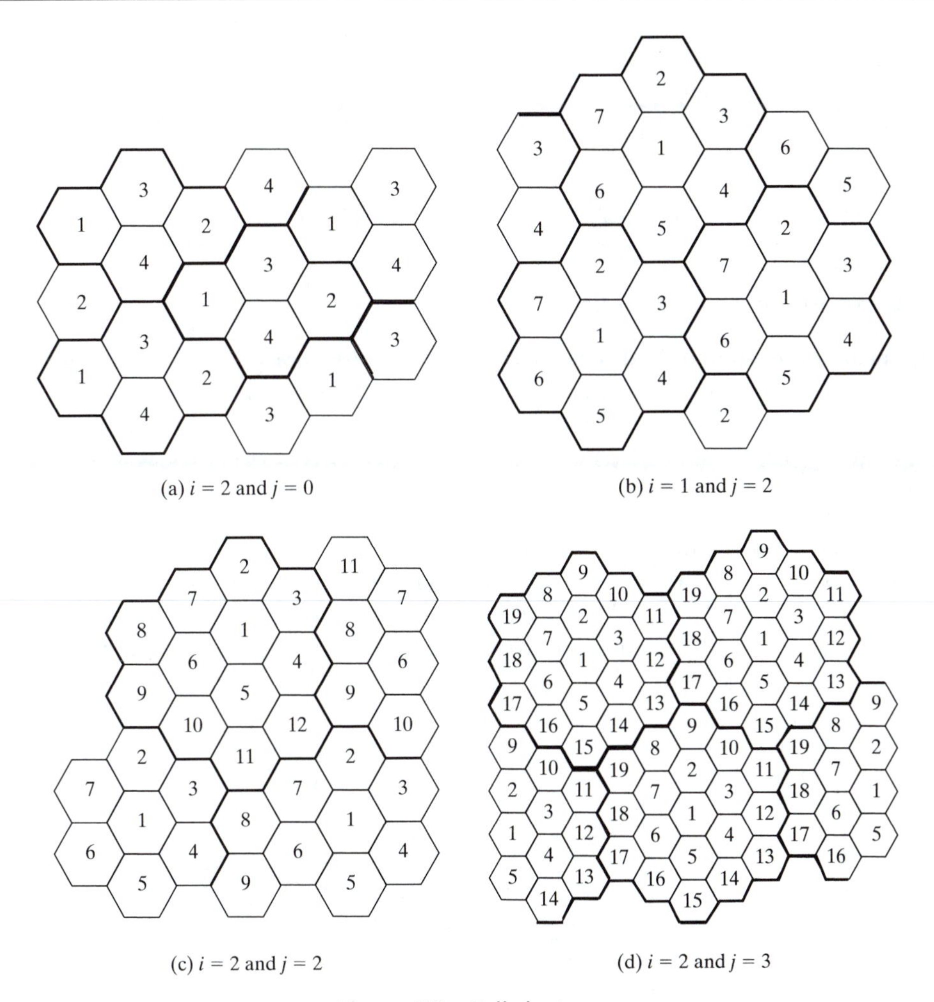

(a) $i = 2$ and $j = 0$ (b) $i = 1$ and $j = 2$

(c) $i = 2$ and $j = 2$ (d) $i = 2$ and $j = 3$

Figure 5.2 Cell clusters.

a. calculate the total available channels,
b. determine the number of control channels,
c. determine the number of voice channels per cell, and
d. determine an equitable distribution of control channels and voice channels in each cell.

Solution Given:

Total bandwidth = 30 MHz
Channel bandwidth = 25 kHz × 2 = 50 kHz/duplex channel

a. The total number of available channels = $\frac{30000}{50} = 600$.
b. The number of control channels = $\frac{1000}{50} = 20$.

c. The number of voice channels per cell $= \frac{600-20}{9} \simeq 64$.
d. Since only a maximum of 20 channels can be used as control channels, for $N = 9$, one way to allocate is 7 cells with 2 control channels and 64 voice channels each, and 2 cells with 3 control channels and 66 voice channels each.

Note that the channel allocation performed in part (*d*) is not unique.

5.3.3 Geometry of Hexagonal Cells

The geometry of an array of hexagonal cells is depicted in Figure 5.3, where R is the radius of the hexagonal cell (from the center to a vertex). A hexagon has exactly six equidistant neighbors. As can be seen from Figure 5.3, in a cellular array the lines joining the centers of any cell and each of its neighbors are separated by multiples of 60 degrees. Note that in Figure 5.3 the 60° angle is bounded by the vertical line and the 30° line, both of which join centers of hexagonal cells.

The distance between the nearest cochannel cells in a hexagonal area can be calculated from the geometry shown in Figure 5.3. For notational convenience, the cell under consideration will be referred to as the candidate cell. The distance between the centers of two adjacent hexagonal cells is $\sqrt{3}R$. Let D_{norm} be the distance from the center of the candidate cell to the center of a

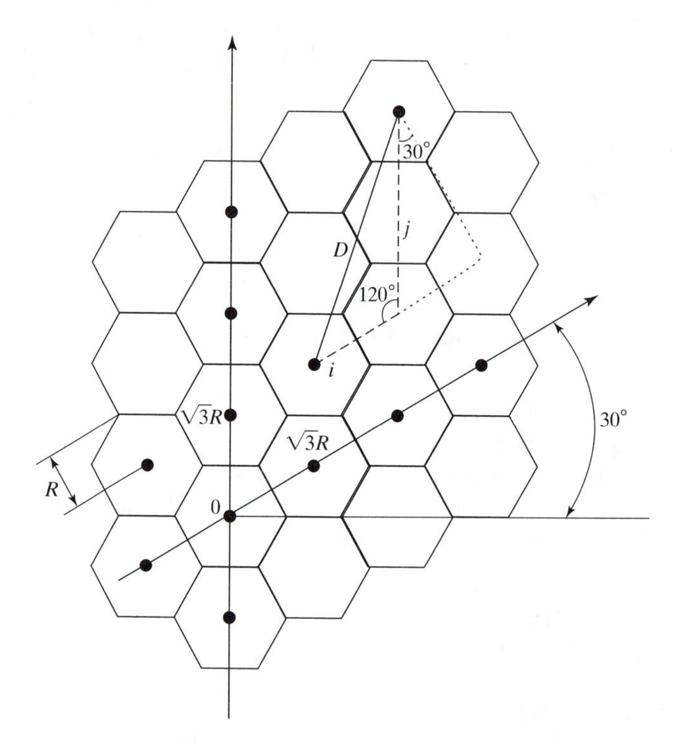

Figure 5.3 Distance between nearest cochannel cells.

nearest cochannel cell, normalized with respect to the distance between the centers of two adjacent cells, $\sqrt{3}R$. Note that the normalized distance between two adjacent cells (either with $i = 1$ and $j = 0$ or with $i = 0$ and $j = 1$) is unity. Let D be the actual distance between the centers of two adjacent cochannel cells. D is a function of D_{norm} and R.

From the geometry shown in Figure 5.3 we readily have

$$\begin{aligned} D_{norm}^2 &= j^2 \cos^2(30°) + (i + j \sin(30°))^2 \\ &= i^2 + j^2 + ij. \end{aligned} \tag{5.3.4}$$

From Eqs. (5.3.4) and (5.3.3),

$$D_{norm} = \sqrt{N}.$$

With the actual distance $\sqrt{3}R$ between the centers of two adjacent hexagonal cells, the actual distance between the center of the candidate cell and the center of a nearest cochannel cell is

$$D = D_{norm} \times \sqrt{3}R = \sqrt{3N}R. \tag{5.3.5}$$

For hexagonal cells, there are six nearest cochannel neighbors to each cell. Cochannel cells are located in tiers. In general, a candidate cell is surrounded by $6k$ cells in tier k. For cells with the same size, the cochannel cells in each tier lie on the boundary of the hexagon that chains all the cochannel cells in that tier. As D is the radius between two nearest cochannel cells, the radius of the hexagon chaining the cochannel cells in the kth tier is given by kD. For the frequency reuse pattern with $i = 2$ and $j = 1$ so that $N = 7$, the first two tiers of cochannel cells are illustrated in Figure 5.4. It can be readily observed from Figure 5.4 that the radius of the first tier is D and the radius of the second tier is $2D$.

Example 5.3 Number of Cells in a Cluster

Verify that the cell cluster size is $N = i^2 + ij + j^2$, where i and j are the integer parameters determining the cochannel cells as illustrated in Figure 5.1.

Solution A candidate cell has 6 nearest cochannel cells. By joining the centers of the 6 nearest neighboring cochannel cells, we form a large hexagon, as shown in Figure 5.5. This large hexagon has radius equal to D, which is also the cochannel cell separation. With the cell radius R, from Eq. (5.3.4), we have

$$D = \sqrt{3}RD_{norm} = \sqrt{3(i^2 + ij + j^2)}R.$$

In general, the area of a hexagon is proportional to the square of its radius. Let $\beta(\approx 2.598)$ be the proportional constant. Then the area of the large hexagon with radius D is

$$A_{\text{large}} = \beta D^2 = \beta[3(i^2 + ij + j^2)R^2]$$

and the area of a cell (the small hexagon) with radius R is

$$A_{\text{small}} = \beta R^2.$$

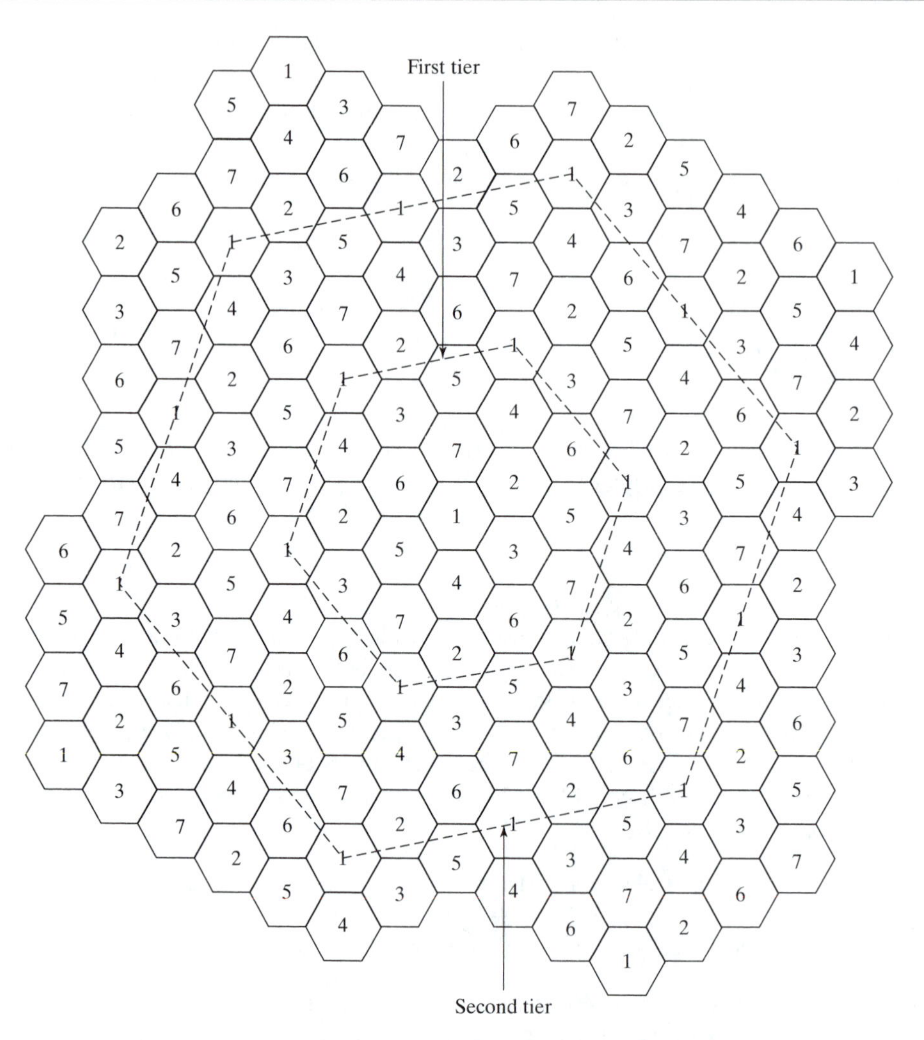

Figure 5.4 Two tiered interfering cells with $N = 7$.

The number of cells in the large hexagon is then

$$\frac{A_{\text{large}}}{A_{\text{small}}} = 3(i^2 + ij + j^2). \tag{5.3.6}$$

On the other hand, from the geometry (e.g., Figure 5.4) it can be seen that, in general, the large hexagon encloses the center cluster of N cells plus 1/3 the number of the cells associated with six other peripheral large hexagons. Hence, the total number of cells enclosed by the large hexagon is

$$N + 6\left(\frac{1}{3}N\right) = 3N. \tag{5.3.7}$$

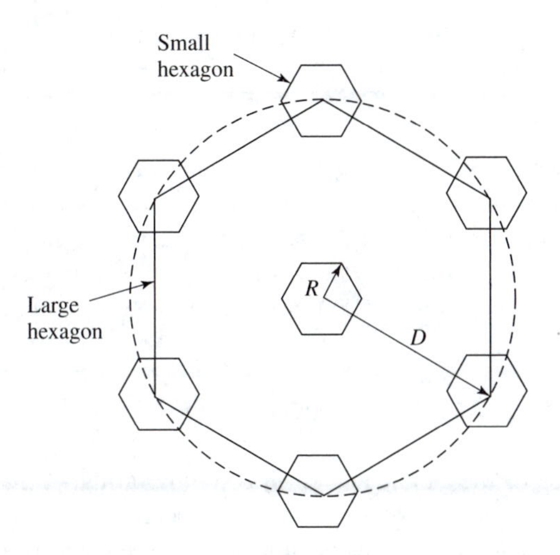

Figure 5.5 First tier cochannel interfering cells.

From Eqs. (5.3.6) and (5.3.7), we have

$$N = i^2 + ij + j^2.$$

5.3.4 Frequency Reuse Ratio

The frequency reuse ratio, q, is defined as

$$q \triangleq \frac{D}{R}. \tag{5.3.8}$$

Because frequency reuse leads to cochannel cell operation, q is also referred to as the cochannel reuse ratio.

Table 5.1 Frequency Reuse Ratio and Cluster Size

Frequency Reuse Pattern (i, j)	Cluster Size N	Frequency Reuse Ratio q
(1, 1)	3	3.00
(2, 0)	4	3.46
(2, 1)	7	4.58
(3, 0)	9	5.20
(2, 2)	12	6.00
(3, 1)	13	6.24
(3, 2)	19	7.55
(4, 1)	21	7.94
(3, 3)	27	9.00
(4, 2)	28	9.17
(4, 3)	37	10.54

Substituting Eq. (5.3.5) into Eq. (5.3.8), the frequency reuse ratio q is related to the cluster size (or frequency reuse factor) N by

$$q = \sqrt{3N}. \tag{5.3.9}$$

Since q increases with N and since a smaller value of N has the effect of increasing the capacity of the cellular system (see Subsection 5.3.1) and, at the same time, increasing cochannel interference, the choice of q or N has to be made such that the signal-to-cochannel interference ratio is at an acceptable level. Several frequency reuse patterns, together with the corresponding cluster sizes and frequency reuse ratios, are tabulated in Table 5.1 for easy reference.

5.4 COCHANNEL AND ADJACENT CHANNEL INTERFERENCE

Consider the performance of the candidate cell in a cellular array. A given base station provides the capacity to handle the services of many mobile users. The signal received from a target user at the cell-site (base station) receiver is subject to interference from transmissions of other mobiles in the same cell, background noise, and interference from transmissions by mobiles in neighboring cells. Under the assumption of a proper separation between uplink transmissions and downlink transmissions, either in the time domain (i.e., time division duplexing) or in the frequency domain (i.e., frequency division duplexing), interference from transmissions in the other link can be neglected. Interference from other mobiles at the cell-site receiver in the same cell is *intracell interference*. Interference from other cells is *intercell interference*. Intercell interference in the downlink that affects the reception at the individual mobile hosts may be more of a problem than uplink interference at the cell-site receiver. The reason for this can be attributed to the fact that the cell-site receiver may be more sophisticated than the receivers of the individual mobile users.

If the different cells in the entire cellular system were to use different sets of frequencies, intercell interference would be kept at a minimum. However, the system capacity would be limited; deployment of frequency reuse (see Subsection 5.3.1) is necessary to enlarge the system capacity. On the other hand, frequency reuse will introduce cochannel interference from cells using the same set of frequencies. Therefore, frequency reuse needs to be carefully planned so that cochannel interference is kept at an acceptable level.

5.4.1 Cochannel Interference

As mentioned in Chapter 1, wireless channels are interference limited. Except for the cochannel cells, other neighboring cells will operate at frequencies different from those of the candidate cell so that interference from noncochannel cells will be minimal. Intercell interference is thus dominated by cochannel interference. It is thus of interest to assess system performance, taking into consideration interference from cochannel cells. For simplicity in the following analysis, we will consider only the average channel quality as a function of the distance dependent path loss, without going into details of channel statistics due to propagation shadowing and multipath fading.

We will use the symbols S and I to denote respectively the power of the desired signal and the power of the cochannel interference at the output of the receiver demodulator. Let N_I be the number of cochannel interfering cells and I_i be the interference power caused by transmissions from the ith interfering cochannel cell base station. The signal-to-cochannel interference ratio (S/I) at the desired mobile receiver is given by

$$\frac{S}{I} = \frac{S}{\sum_{i=1}^{N_I} I_i}.$$

The average received signal strength at any point decays as a power law of the distance between the transmitter and the receiver, as discussed in Section 2.4.

Let D_i be the distance between the ith interferer and the mobile. The received interference, I_i, at a given mobile due to the ith interfering cell is proportional to $(D_i)^{-\kappa}$, where κ is the path loss exponent. The path loss exponent κ is normally determined by measurement. In many cases it is in the range $2 \leq \kappa \leq 5$.

In addition to cochannel interference, there is always the inherent background noise. However, in an interference dominated environment, we may neglect the background noise. It is noted that the desired received signal power S is proportional to $r^{-\kappa}$, where r is the distance between the mobile and the serving base station. When the transmit powers from all base stations are equal, and the path loss exponent is the same throughout the geographical coverage area, the cochannel interference from the ith cochannel cell, I_i, for all i, depends on D_i and κ only. The S/I at a typical mobile receiver can be approximated by

$$\frac{S}{I} = \frac{r^{-\kappa}}{\sum_{i=1}^{N_I} D_i^{-\kappa}}. \tag{5.4.1}$$

The degree of cochannel interference is a function of the location of the mobile within the cell of the serving base station. When the mobile is located at the cell boundary (i.e., $r = R$), the worst case cochannel interference occurs as the power of the desired signal is minimum. With hexagon shaped cellular systems, there are always six cochannel interfering cells in the first tier. If we neglect cochannel interference from the second and other higher tiers, this means that $N_I = 6$. In the case that $r = R$ and using $D_i \approx D$ for $i = 1, 2, \ldots, N_I$,

$$\frac{S}{I} = \frac{(D/R)^{\kappa}}{N_I} = \frac{q^{\kappa}}{N_I} = \frac{(\sqrt{3N})^{\kappa}}{N_I}. \tag{5.4.2}$$

Thus, the frequency reuse ratio can be expressed as

$$q = \left(N_I \times \frac{S}{I}\right)^{1/\kappa} = \left(6 \times \frac{S}{I}\right)^{1/\kappa}. \tag{5.4.3}$$

For the U.S. AMPS analog FM system, a value of $S/I = 18$ dB or greater is acceptable. With a path loss exponent of $\kappa = 4$, the frequency reuse ratio q is determined as

$$q = (6 \times 10^{1.8})^{1/4} = (6 \times 63.1)^{0.25} \simeq 4.41.$$

Therefore, the cluster size N should be

$$N = q^2/3 = 6.49 \simeq 7.$$

Example 5.4 S/I Ratio versus Cluster Size

Suppose the acceptable signal-to-cochannel interference ratio in a certain cellular communications situation is $S/I = 20$ dB or 100. Also, from measurements, it is determined that $\kappa = 4$. What is the minimum cluster size?

Solution The frequency reuse ratio can be calculated, using Eq. (5.4.3), as

$$q = (6 \times 100)^{1/4} = 4.9492.$$

Then, from Eq. (5.3.9), the cluster size is given by

$$N = q^2/3 = 8.165 \simeq 9.$$

In this case, a 9-reuse pattern is needed for an S/I ratio of at least 20 dB. Since

$$q = D/R \quad \text{or} \quad D = qR,$$

D can be determined, given the cell radius R, and vice versa. Note that if N is less than 9, the S/I value would be below the acceptable level of 20 dB.

Consider that the mobile is at the cell boundary, where it experiences worst case cochannel interference on the forward channel. If we use a better approximation of the distance between the mobile and the first tier interfering base stations as illustrated in Figure 5.6, then from Eq. (5.4.1) the S/I ratio can be expressed as (see Problem 5-3)

$$\frac{S}{I} = \frac{R^{-\kappa}}{2(D-R)^{-\kappa} + 2D^{-\kappa} + 2(D+R)^{-\kappa}}. \tag{5.4.4}$$

Recall that $D/R = q$. With a path loss exponent of $\kappa = 4$, Eq. (5.4.4) can be written as

$$\frac{S}{I} = \frac{1}{2(q-1)^{-4} + 2q^{-4} + 2(q+1)^{-4}}. \tag{5.4.5}$$

Example 5.5 Worst Case Cochannel Interference

Consider a cellular system that requires an S/I ratio of 18 dB.

a. For a frequency reuse factor of 7, calculate the worst-case signal-to-cochannel interference ratio.
b. Is a frequency reuse factor of 7 acceptable in terms of cochannel interference? If not, what would be a better choice of frequency reuse factor?

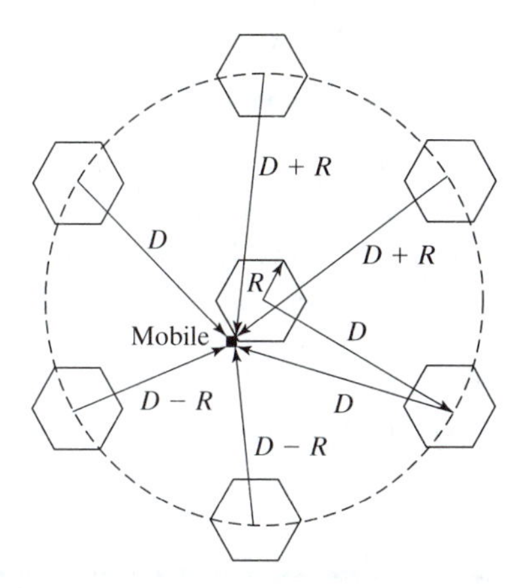

Figure 5.6 Worst case cochannel interference scenario for $N = 7$.

Solution

a. For $N = 7$, the frequency reuse ratio is $q = \sqrt{3N} = 4.6$. Assuming a path loss exponent of $\kappa = 4$, then, from Eq. (5.4.5), the worst-case signal-to-cochannel interference ratio is

$$\frac{S}{I} = 54.3 \quad \text{or} \quad 17.3 \text{ dB}.$$

b. The value of S/I for a 7 frequency reuse factor is below the acceptable level of 18 dB. To increase S/I, we need to decrease I. This can be achieved by increasing the frequency reuse factor, N. For $N = 9$, we have $q = \sqrt{3N} = 5.20$, so that $S/I = 95.66$ or 19.8 dB. This value of S/I is above the acceptable level of 18 dB. Therefore, a frequency reuse factor of 9 is a suitable choice.

Although an increase of frequency reuse factor from 7 to 9 in Example 5.5 yields an acceptable S/I level, this increase in N is accompanied by a decrease in system capacity, since a 9-cell reuse offers a spectrum utilization of 1/9 within each cell, whereas a 7-cell reuse offers a spectrum utilization of 1/7. The capacity reduction may not be tolerable. From an operational point of view, catering to the worst-case scenario, which rarely occurs, may not be desirable. The designer may want to seek the best tradeoff by accepting the fact that, with a small but nonzero probability, the worst-case scenario may occur and, hence, degrade the performance to a level below specification during some interval of the holding time of a call.

From the preceding analysis, it is clear that cochannel interference determines link performance which, in turn, dictates the frequency reuse plan and the overall capacity of cellular systems.

5.4.2 Adjacent Channel Interference

Adjacent channel interference (ACI) results from signals which are adjacent in frequency to the desired signal. ACI is mainly due to imperfect receiver filters which allow nearby frequencies to leak into the passband. Consider the uplink transmissions from two mobile users using adjacent channels, one very close to the base station and the other very close to the cell boundary. Without proper transmission power control, the received power from the mobile close to the base station is much larger than that from the other mobile far away. This near–far effect can significantly increase the ACI from the strong received signal to the weak received signal. To reduce ACI, we should: (*a*) use modulation schemes which have low out-of-band radiation (e.g., MSK is better than QPSK and GMSK is better than MSK); (*b*) carefully design the bandpass filter at the receiver front end; (*c*) use proper channel interleaving by assigning adjacent channels to different cells; (*d*) avoid using adjacent channels in adjacent cells to further reduce ACI if the cell cluster size is large enough; and (*e*) separate the uplink and downlink properly by TDD or FDD.

5.5 CALL BLOCKING AND DELAY AT THE CELL-SITE

Signal-to-interference ratio, which determines the transmission bit error rate, is a QoS factor at the output of the cell-site receiver. From the user's perspective, quality of service is more than an acceptable transmission accuracy. In fact, there are two crucial questions: (*a*) How successfully can a new user get a connection established? (*b*) After connection establishment, how successfully will the connection be maintained as the user moves from one cell to another? The first question refers to the admission of new calls, while the second question refers to the admission of handoff calls. The performance measure is the probability that a call (new or handoff) is blocked.

To formulate the probability of call blocking, consider a radio cell which has been allocated J channels. Assume a large population size of mobile users in the cell. Suppose during the connection time of a call, each user occupies one channel. If the number of active users during any epoch equals J, all available channels will be occupied. Then, with probability 1, a call request will be denied (i.e., blocked). If the number of ongoing calls is fewer than J, a call will be blocked with probability smaller than 1. This is equivalent to the condition that the trunk traffic load in Erlangs is less than J, the number of available channels. One Erlang represents the amount of traffic load carried by a channel that is completely occupied, such as one call-hour per hour. If a channel is busy for 30 minutes during a one hour period, then the channel is said to carry 0.5 Erlangs of traffic. Offered traffic refers to the amount of traffic sent by the users, while carried traffic refers to the amount of traffic served.

To obtain an expression for call blocking probability, we model the cell-site as a bufferless system. With no buffer, blocked calls are lost. Assume that (*a*) there are $L (\gg 1)$ users in the system; (*b*) the aggregate arrival traffic is Poisson distributed with rate λ; (*c*) the duration of a call is exponentially distributed with parameter μ_1; and (*d*) the residence time of each user in a cell is exponentially distributed with parameter μ_2. As an exponential random variable is memoryless, the channel holding time is the minimum of the call duration and the cell residence time, which is also exponentially distributed with parameter $\mu = \mu_1 + \mu_2$. That is, the mean channel holding time of the call is μ^{-1}, corresponding to a mean service rate of μ for the call. Since the channel holding time is exponentially distributed, the service time of each of the servers is also exponentially distributed. With Poisson arrivals and exponential service times, the underlying queueing process is Markovian.

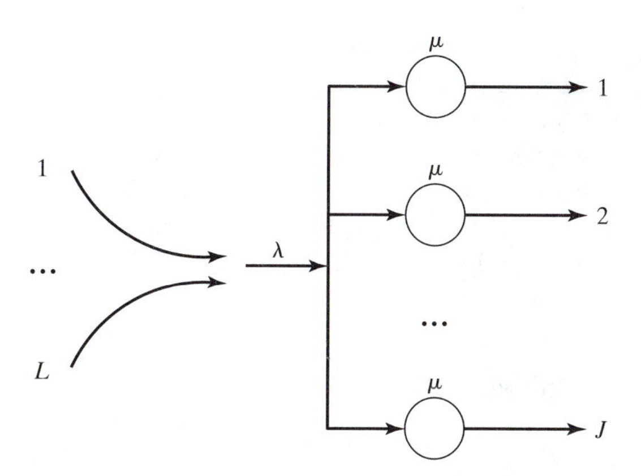

Figure 5.7 *J*-server bufferless model.

The cell-site receiver can be modeled as a J-server system, in which each server serves traffic at a mean rate μ. The J-server system for a population of size L and aggregate arrival rate λ is illustrated in Figure 5.7. The state transition rate diagram of the multiserver system with Poisson arrival and exponential service time is shown in Figure 5.8. As shown in the diagram, when the system is in state j, for $j = 0, 1, \cdots, J$, there are j ongoing calls, and j servers, each with mean service rate μ, are being engaged. When the system is in state $j = J$, all J servers are engaged and new requests will be blocked. Note that, if $L \leq J$, the number of users is fewer than the number of servers, there will be no blocking. But this would not be the best way to utilize the available resources.

The traffic intensity is defined as $\rho \triangleq \lambda/\mu$. In terms of the traffic intensity ρ, the probability of blocking is given by the *Erlang loss formula*

$$P(\text{blocking}) = \frac{\rho^J/J!}{\sum_{j=0}^{J} \rho^j/j!}. \tag{5.5.1}$$

Equation (5.5.1) is also known as the Erlang-B loss formula. The call blocking probability is considered to be the grade of service (GoS) parameter for the Erlang-B system. The traffic intensity in Erlangs for a prescribed GoS is tabulated in Appendix F.

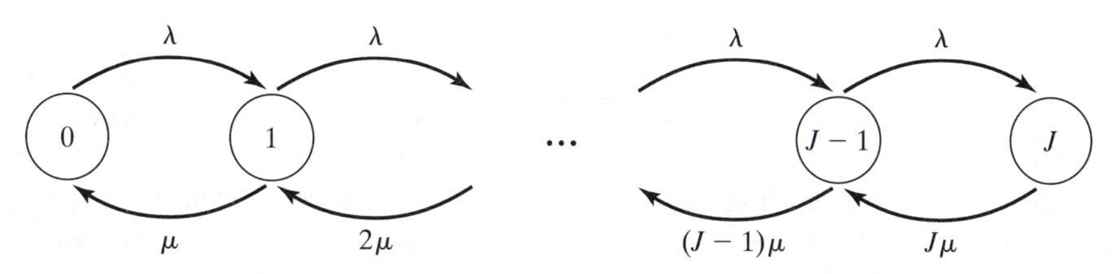

Figure 5.8 State transition rate diagram of a *J*-server system.

In the bufferless J-server system, the delay for any request is zero. The request is either granted or blocked. If some queueing delay can be tolerated, the J-server system in Figure 5.7 can also have a common buffer for queueing requests when all J servers are fully engaged. Upon arrival, a request (new or handoff), finding all J servers are busy, can join the queue. When a server becomes available, the head of the queue will then receive service. In this case, the queueing delay would be nonzero. The reason for queueing requests is to decrease the call blocking probability. The probability of queueing is given by [31, 158]

$$P(\text{queueing}) = \frac{\dfrac{J\rho^J}{J!(J-\rho)}}{\left[\dfrac{J\rho^J}{J!(J-\rho)}\right] + \displaystyle\sum_{j=0}^{J-1}\left(\frac{\rho^j}{j!}\right)}. \tag{5.5.2}$$

Queueing gives rise to delay. The probability of nonzero delay is given by

$$P(\text{delay} > 0) = \frac{\rho^J}{\rho^J + J!\left(1 - \dfrac{\rho}{J}\right)\displaystyle\sum_{j=0}^{J-1}\frac{\rho^j}{j!}}. \tag{5.5.3}$$

Equations (5.5.2) and (5.5.3) are basically the same. Both equations are known as the Erlang-C formula. The probability of nonzero delay that can be tolerated is also considered to be the GoS parameter for the Erlang-C system. The offered loads in Erlangs corresponding to the number of channels available for different GoS, given by the Erlang-C formula, are tabulated in Appendix F.

A given request for network access can only tolerate a finite amount of delay. It is of interest to know the probability that the delay exceeds a given delay threshold. Let t be the delay threshold. Given that the mean channel holding time of a call is μ^{-1}, the probability that the delay exceeds t is given by

$$\begin{aligned} P(\text{delay} > t) &= P(\text{delay} > 0) \times P(\text{delay} > t|\text{delay} > 0) \\ &= P(\text{delay} > 0)\exp[-(J-\rho)\mu t]. \end{aligned} \tag{5.5.4}$$

The average delay, $\bar{D}$, for all calls in the queueing system is given by

$$\bar{D} = P(\text{delay} > 0) \times \frac{1}{\mu(J-\rho)}. \tag{5.5.5}$$

In the Erlang-B or Erlang-C system, a channel is allocated to a user on a call by call basis, thus allowing the system to serve a number of subscribers much greater than the number of available channels. The system performance can be described by *trunking efficiency*. Trunking efficiency is defined as the carried traffic intensity in Erlangs per channel, which is a value between zero and one, and is a function of the number of channels per cell and the specified GoS parameters, such as call blocking rate in the Erlang-B system and average queueing delay in the Erlang-C system.

For a comparison of the Erlang-B and Erlang-C systems, given (*a*) the same number of channels per cell and (*b*) the call blocking rate in the Erlang-B system being the same as the probability of nonzero queueing delay in the Erlang-C system, the offered traffic intensity of the Erlang-B system is larger than that of the Erlang-C system. For example, with $J = 20$, $P(\text{blocking}) = P(\text{queueing}) = 0.1$, from Appendix F, the offered traffic intensity is 17.613 Erlangs in the Erlang-B system and is 14.116 Erlangs in the Erlang-C system. This is because, in the Erlang-B system, the blocked calls are cleared and are not served by the system; while in the Erlang-C system, the new calls are buffered (not dropped) when all the channels are busy and are then served when a channel becomes available. As a result, some offered traffic (the blocked calls) in the Erlang-B system is not carried by the system, but all the offered traffic in the Erlang-C system is carried by the system, under the assumption that the incoming calls can tolerate any queueing delay.

Example 5.6 Traffic Load and Call Blocking Probability

Consider a cellular system with 416 radio channels available for handling traffic. Suppose 21 of these channels are designated as control channels. Let the average channel holding time of a call be 3 minutes, the blocking probability during busy hours be 2%, and the frequency reuse factor be 9.

a. Determine the number of calls per cell per hour.
b. Determine the signal-to-cochannel interference ratio, S/I, in dB.

Solution

a. In a cellular system with frequency reuse, all the available radio channels are allocated to handle services in a single cluster of cells. Therefore, the number of voice channels in a cluster is $416 - 21 = 395$. With a frequency reuse factor of 9, the effective number of voice channels per cell $= \frac{395}{9} \simeq 44$. With 44 available channels and a 2% blocking probability, the traffic load is (from the Erlang-B table in Appendix F) 34.683 Erlangs.

$$\text{The number of calls per cell per unit time} = \frac{\text{number of calls per cell}}{\text{average holding time of a call}}.$$

Therefore,

$$\begin{aligned}\text{the number of calls per cell per unit time} &= \frac{34.683}{3} \text{ calls per cell per minute}\\ &= \frac{34.683}{3} \times 60 \approx 693 \text{ calls per cell per hour.}\end{aligned}$$

b. The frequency reuse ratio $q = \sqrt{3N} = \sqrt{3 \times 9} = 5.1962$. For $\kappa = 4$, the signal-to-cochannel interference ratio is

$$S/I = \frac{1}{6} \times q^{\kappa} = \frac{1}{6} \times (5.1962)^4 = 121.5 \quad \text{or} \quad 20.845 \text{ dB}.$$

5.6 OTHER MECHANISMS FOR CAPACITY INCREASE

As discussed in Subsection 5.3.1, the capacity of a cellular system can be enlarged through frequency reuse. The capacity can also be improved based on cellular layout and antenna design using

a. cell splitting, and
b. antenna sectoring.

5.6.1 Cell Splitting

One way to perform cell splitting, as illustrated in Figure 5.9, is to subdivide a congested cell into smaller cells, each with its own base station and a corresponding reduction in antenna height and transmit power. With more cells, there will be more clusters in the same coverage area. This is equivalent to replicating a cell cluster more times. In the context of Subsection 5.3.1, the replication factor M is increased. Hence, cell splitting increases the capacity of a cellular system since it increases the number of times that channels are reused. In Figure 5.9, the central area is assumed to be saturated with traffic (i.e., the call blocking probability in the area exceeds an acceptance level). The original large cell with radius R in the center is split into the medium cells with radius $R/2$ and the medium cell in the center is further split into the small cells with radius $R/4$. The cell splitting reduces the call blocking probability in the area, and increases the frequency with which mobiles hand off from cell to cell.

Let d be the distance between the transmitter and the receiver, and d_0 be the distance from the transmitter to a close-in reference point. Let P_0 be the power received at the close-in reference point. From Section 2.4, the average received power, P_r, is proportional to P_0, and is given by

$$P_r = P_0 \left(\frac{d}{d_0}\right)^{-\kappa} \tag{5.6.1}$$

where $d \geq d_0$ and κ, as defined earlier, is the path loss exponent. Taking the logarithm, Eq. (5.6.1) can be expressed as

$$P_{r(\text{dBW})} = P_{0(\text{dBW})} - 10\kappa \log_{10}\left(\frac{d}{d_0}\right), \quad d \geq d_0. \tag{5.6.2}$$

Let P_{t1} and P_{t2} be the transmit power of the large cell base station and the medium cell base station, respectively. The received power, P_r at the large (old) cell boundary is proportional to $P_{t1}R^{-\kappa}$, and P_r at the medium (new) cell boundary is proportional to $P_{t2}(R/2)^{-\kappa}$. On the basis of equal received power, we have

$$P_{t1}R^{-\kappa} = P_{t2}(R/2)^{-\kappa} \quad \text{or} \quad P_{t1}/P_{t2} = 2^{\kappa}.$$

Taking the logarithm, the above can be expressed as

$$10 \log_{10}\left(\frac{P_{t1}}{P_{t2}}\right) = 10\kappa \log_{10} 2 \simeq 3\kappa \text{ dB}.$$

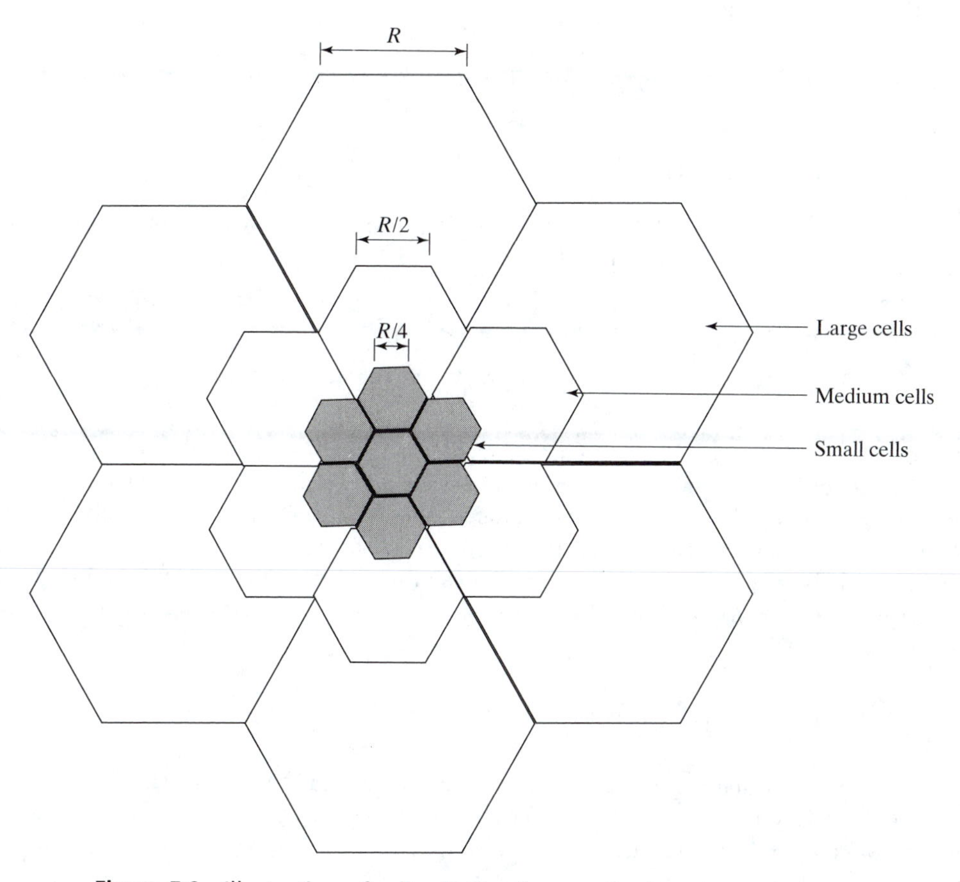

Figure 5.9 Illustration of cell splitting from radius R to $R/2$ and to $R/4$.

For $\kappa = 4$, $P_{t1}/P_{t2} = 12$ dB. Thus, with cell splitting, where the radius of the new cell is one-half of that of the old cell, we can achieve a 12 dB reduction in the transmit power.

How cell splitting increases system capacity is best illustrated by an example.

Example 5.7 Cell Splitting and Capacity Increase

Consider the cellular system shown in Figure 5.10, where the original cells have radius R. These cells are split into smaller cells, each with radius $R/2$. Suppose each base station is allocated 60 channels regardless of the cell size. There are obviously more small cells than original cells in the same coverage area. Since the number of channels allocated in a small cell is the same as that in a large cell, it is obvious that cell splitting increases the number of channels within the same coverage area. Find the number of channels contained in a 3×3 km^2 area centered around (small) cell "A" for the following cases:

a. without cell splitting (i.e., just the original large cells), and
b. with cell splitting (i.e., using the small cells (microcells)).

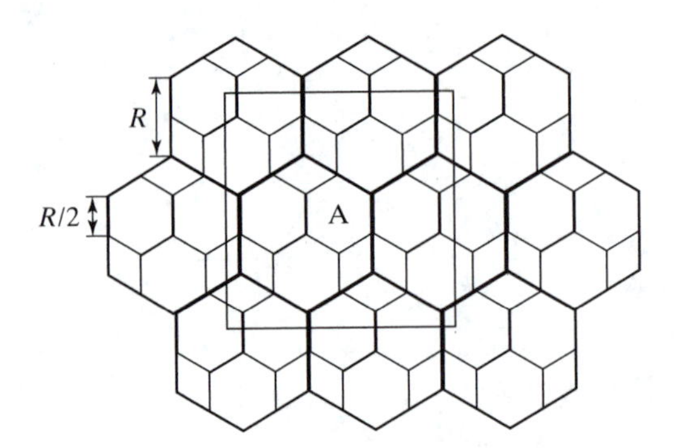

Figure 5.10 A cell splitting example with $R = 1$ km.

Solution To cover an area of 3×3 km^2 centered around cell A, we need to cover 1.5 km to the right, left, top and bottom of base station A, as shown in Figure 5.10. From Figure 5.10, it is observed that the 3×3 km^2 square centered around cell "A" contains more small cells than large cells. However, because of edge effect, the number of either type of cells contained within the square can only be an estimate. A reasonable estimate is that there are approximately 4 large cells (from visual observation). With a 1/2 radius split, the number of small cells within the square would be at most

$$\left(\frac{R}{R/2}\right)^2 \times \text{ the number of large cells } = 4 \times 4 = 16 \text{ small cells.}$$

The preceding calculation would be correct if the enclosed area is infinitely large. With a finite area, it is necessary to take edge effect into consideration, so that the number of small cells contained within the 3×3 km^2 area would be less than 16. A reasonable estimate would be 15 small cells.

a. With an estimate of 4 base stations contained within the square, the number of channels equals $4 \times 60 = 240$.
b. With an estimate of 15 small cells, the number of channels contained in the square, with cell splitting, is $15 \times 60 = 900$ channels, which is 3.75 times the channel number in the nonsplitting case.

Note that the upper bound is a 4-fold increase.

5.6.2 Directional Antennas (Sectoring)

In the basic form, antennas are omnidirectional. Directional antennas can increase the system capacity relative to that of omnidirectional antennas. From Eq. (5.4.1) the worst-case S/I is given by

$$\frac{S}{I} = \frac{R^{-\kappa}}{\sum_{i=1}^{N_I} (D_i)^{-\kappa}},$$

where the value of N_I depends on the form of antenna used. In the omnidirectional case, $N_I = 6$ for the first tier of cochannel cells. Assuming $D_i \approx D$, $i = 1, 2, \ldots, N_I$,

$$\left(\frac{S}{I}\right)_{omni} = \frac{1}{6} \times q^{\kappa},$$

where $q = D/R$. In terms of capacity increase through sectorization, the omnidirectional case can be used as the benchmark.

With hexagonal cells, as illustrated in Figure 5.11, sectorization can be done in multiples of 60°. Assuming a 7-cell reuse, for the 3-sector case (with 120° in each sector) the number of interferers in the first tier is reduced from 6 to 2.

With $D_i \approx D$,

$$\left(\frac{S}{I}\right)_{omni} = \frac{1}{6} \times q^{\kappa} \quad \text{and} \quad \left(\frac{S}{I}\right)_{120^\circ} = \frac{1}{2} \times q^{\kappa}.$$

The increase in the signal-to-interference ratio is then

$$\frac{(S/I)_{120^\circ}}{(S/I)_{omni}} = 3.$$

That is, theoretically, the capacity increase from the omnidirectional antenna case is given by the number of sectors in each cell resulting from the use of directional antennas. Note that, within each cell, mobiles may need to hand off from sector to sector. However, the handoff process can be easily managed by the base station. If the total number of channels available to each cell needs to be partitioned for the sectors, the trunking efficiency for each cell is reduced from that without sectoring.

A worst-case scenario in a 120° sectorization is shown in Figure 5.12, where the mobile is located at the corner of the cell, R is the cell radius, and D is the distance between the adjacent cochannel cells. In the 3-sector case, the mobile experiences interference from one sector of each of the two interfering cells. With the distance approximation shown and a path loss exponent of $\kappa = 4$, we have

$$\left(\frac{S}{I}\right)_{120^\circ} = \frac{R^{-4}}{D^{-4} + (D + 0.7R)^{-4}} = \frac{1}{q^{-4} + (q + 0.7)^{-4}}. \tag{5.6.3}$$

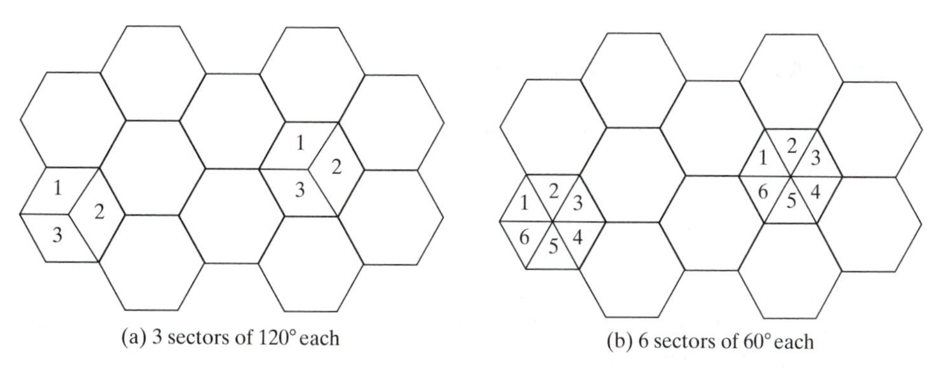

Figure 5.11 Antenna sectorization.

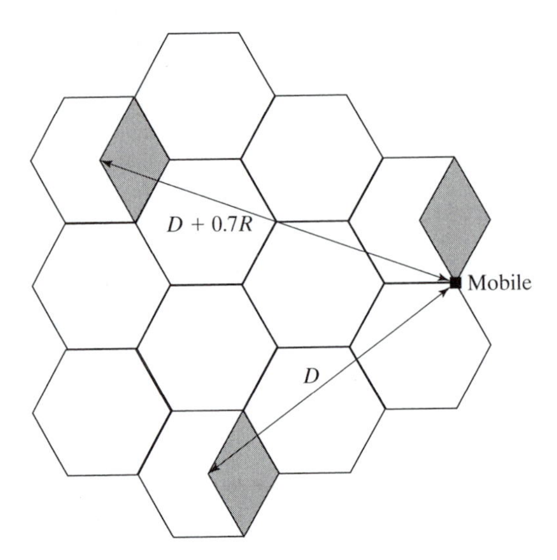

Figure 5.12 Worst-case scenario in 120° sectoring.

Example 5.8 Cochannel Interference with Sectoring

In Example 5.6, it is shown that, with a frequency reuse factor of 7, base stations using omnidirectional antennas cannot satisfy the 18 dB signal-to-cochannel interference ratio requirement. Determine whether the use of 120° sectoring and 7-cell frequency reuse would satisfy the 18 dB requirement.

Solution For a 7-cell reuse, we have $q = \sqrt{3 \times 7} = 4.6$. Substituting for q in Eq. (5.6.3), we get

$$\left(\frac{S}{I}\right)_{120^\circ} = 285 \quad \text{or} \quad 24.5 \text{ dB}.$$

Since this is greater than 18 dB, the 3-sector worst case for a 7-cell reuse is acceptable.

5.7 CHANNEL ASSIGNMENT STRATEGIES

There are essentially two channel assignment approaches: (*a*) fixed channel assignment, and (*b*) dynamic channel assignment.

Fixed Channel Assignment (FCA). In FCA, each cell is allocated a predetermined set of voice channels. Any call attempt within the cell can only be served by the unused channels in that particular cell. To improve utilization, a borrowing option may be considered. With the borrowing option, a cell is allowed to borrow channels from a neighboring cell if all of its own channels are already occupied and the neighboring cell has spare channels. Borrowing is normally supervised by the MSC.

As mentioned in Subsection 5.2.2, since handoff is to be performed by the MSC, the MSC has full knowledge of the capacity usage of the cluster of cells within its jurisdiction. Therefore, the MSC is the natural subsystem to oversee functions such as channel borrowing.

Dynamic Channel Assignment (DCA). In DCA, voice channels are not allocated to different cells on a permanent basis [33]. Each time a call request is made, the serving base station requests a channel from the MSC. The MSC determines (dynamically) the availability of a channel and executes its allocation procedure accordingly. The MSC only allocates a given frequency (radio channel) if that frequency (radio channel) is not presently in use in the cell, or any other cell which falls within the minimum restricted distance of frequency reuse to avoid cochannel interference.

Dynamic channel assignment reduces the likelihood of call blocking, which increases the trunking capacity of the system, since all available channels under the control of the MSC are accessible to all of the cells. Dynamic channel assignment strategies require the MSC to collect real-time data on channel occupancy, traffic distribution, and radio signal quality of all channels on a continuous basis. In any case, the MSC needs to do this data collection in order to manage handoff.

SUMMARY

Cellular communication is designed to enhance system capacity while allowing for the use of low power transmitters and frequency reuse. The cellular concept is to divide a large geographical area into many small geographical coverage areas, and to reuse the frequencies used in a given small area elsewhere. Since the geographical coverage areas are now smaller, a mobile will move out of one base station coverage area and into a neighboring area more frequently. Therefore, the tradeoff in attaining capacity expansion is the need to handle handoff when the mobile roams from cell to cell, in order to maintain continuous service of a connection. This chapter has examined the basic characteristics of cellular systems, including system capacity and frequency reuse properties. Handoff management issues will be considered in Chapters 7 and 8.

ENDNOTES

1. For the cellular concept, see the special issue of the *Bell System Technical Journal* [19] which includes papers by MacDonald [92] and Young [163]. For system aspects of cellular radio, see the paper by Steele, Whitehead and Wong [148].
2. The analysis on the cochannel interference (CCI) given in Section 5.4 considers the average performance versus the propagation distance. In practice, the interference changes randomly, depending on channel fading (due to multipath propagation) and shadowing (depending on the propagation terrain), power control mechanism, receiver structure, traffic load and distribution, etc., in addition to the propagation distance. As a result, the interference should be modeled as a random process as the mobile user moves. The main probabilistic performance measure is the probability of CCI (also called the outage probability), defined as the probability that the instantaneous signal-to-cochannel interference ratio is below a given threshold. The threshold

value depends on the link quality requirement. The book by Stüber [147] provides an in-depth analysis on CCI modeling, the probability of CCI, and related references such as [63, 50, 88, 120, 146, 161].

3. For the fundamentals of queueing theory and detailed analysis of the Erlang systems, see the books by Cooper [31] and by Wolff [158].
4. For cellular system layout, performance measures, and traffic load analysis, see the paper by Everitt [48], Chapter 7 of the book edited by Jakes [42], and Chapter 2 of the book by Rappaport [128].
5. For effects of cell sectorization on the spectrum efficiency of cellular radio systems, see the paper by Chan [26].
6. For dynamic channel assignment, see the papers by Cox and Reudnik [33, 34, 35].

PROBLEMS

P5-1 Consider a cellular system with hexagonal cells. The cell radius is R. The service area is partitioned into cell clusters. Frequency is reused from cell cluster to cell cluster. The geometric relation between adjacent cochannel cells can be described by the two nonnegative integers i and j.

a. With pictorial illustration, explain what i and j represent.
b. Determine the distance D between the centers of two adjacent cochannel cells.
c. Derive an expression for the number of cells, N, in each cell cluster.
d. What should be considered in choosing a value for N?

P5-2 Show that the frequency reuse factor for a cellular system is given by K/J, where J is the average number of channels per cell and K is the total number of channels available to the cellular network provider without frequency reuse. The cell cluster can be replicated M times to provide a total capacity of C channels. Discuss the changes in the value of C when you increase or decrease the frequency reuse factor while keeping K constant.

P5-3 Verify the distance approximations shown in Figure 5.6 and Eq. (5.4.4).

P5-4 In the radio cell layout, in addition to the hexagonal topology, a square or an equilateral triangle topology can also be used.

a. Given the same distance between the cell center and its farthest perimeter points, compare the cell coverage areas among the three regular polygons (hexagon, square, and triangle);
b. Discuss the advantages of using the hexagonal cell shape over the square and triangle cell shapes.

P5-5 A cellular system has a total of 500 duplex voice channels (without frequency reuse). The service area is divided into 150 cells. The required signal-to-cochannel interference ratio is 18 dB. Consider the path loss exponent κ equal to 3, 4, and 5, respectively. Based on Eqs. (5.3.9) and (5.4.3), determine

a. the cell cluster size;
b. the number of cell clusters in the service area; and
c. the maximum number of users in service at any instant.

Discuss effects of the path loss exponent on the frequency reuse and on the transmit power (when the cell size is fixed).

P5-6 Consider a cellular system with a total bandwidth of 30 MHz. Each full duplex voice or control channel uses two 25 kHz simplex channels. It is assumed that (1) the system uses a 9-cell reuse pattern and 0.75 MHz of the total bandwidth is allocated for control channels; (2) the system service area consists of 50 cells; (3) the call blocking probability bound is 2%, as given by the Erlang-B formula. If the offered traffic per user is 0.025 Erlangs, calculate

a. the traffic load of each cell and the trunking efficiency,
b. the total number of users in each cell and in the system, respectively,
c. the number of mobile users per channel in each cell and in the system, respectively, and
d. the maximum number of users in service at any instant in the system.

P5-7 Consider the cellular system in Example 5.1 with the cell cluster of size 7. Given that the traffic load per user is 0.03 Erlangs and the average number of calls per hour per user is 1.5, for an Erlang-C system with a probability of delaying a call being 5%, determine

a. the traffic load per cell,
b. the number of users per km^2 that can be supported by this system,
c. the mean duration of a call,
d. the probability that a delayed call will have to wait for more than 10 s, and
e. the probability that a call will be delayed for more than 10 s.

P5-8 Repeat the calculations in Problem 5-7 for the case where the cell cluster size is increased to nine. Discuss effects of the cluster size on the service quality.

P5-9 Consider a cellular system with a total of 395 traffic channels and a 7-cell frequency reuse. Suppose the probability of call blocking is to be no more than 1%. Assume that every subscriber makes 1 call per hour and each call lasts 3 minutes, on average.

a. For omnidirectional antennas, determine the traffic load in Erlangs per cell and the number of calls per cell per hour.
b. Repeat part (a) for a 120° sectoring.
c. Repeat part (a) for a 60° sectoring.
d. Discuss the effect of sectoring, using the information from parts (*b*) and (*c*), on the S/I ratio and on the trunk efficiency.

P5-10 Consider the system in Problem 5-9.

a. Determine the minimum frequency reuse factors for no sectoring, 120° sectoring and 60° sectoring, respectively, taking into consideration that a S/I value of 18 dB or better is satisfactory.
b. Determine the traffic loads per cell and trunk efficiencies with no sectoring, 120° sectoring, and 60° sectoring.

P5-11 A cellular network provider uses a digital TDMA scheme that can tolerate a S/I ratio of 15 dB in the worst case, in a propagation environment with path loss exponent $\kappa = 4$.

a. Find the best value of N for (1) omnidirectional antennas, (2) 120° sectoring, and (3) 60° sectoring.

b. Should sectoring be used?
c. If sectoring is used, should you use 120° or 60° sectoring? Explain.

P5-12 Consider a cellular system that employs omnidirectional antennas.

a. If the path loss exponent is $\kappa = 4$, show that a cell can be split into 4 smaller cells, each with a radius that is one-half of the radius of the original cell, and 1/16 of the transmitter power of the original base station.
b. If it is determined, through careful experimentation, that $\kappa = 3$ is the correct value, how should the transmitter power be changed in order to split a cell into 4 smaller cells?
c. Provide drawings that show how the new cells would fit within the original macrocells, so that all the base stations before splitting are also used after splitting.

P5-13 One method to increase the capacity of a cellular system is cell splitting, in which a large cell is divided into several smaller cells. This normally takes place in regions where there are heavy concentrations of users. Cell splitting needs to preserve the frequency reuse plan of the original large cells.

a. While cell splitting increases system capacity, this gain comes with the introduction of other networking problems. Describe two main problems arising from cell splitting and explain how you propose to address these problems.
b. The base station can be placed anywhere within the cell. Assuming that base stations are located at the centers of the large cells and using appropriate labeling, construct a cell splitting scenario starting from the center of a large cell such that the original $N = 7$-cell frequency reuse pattern is preserved.

P5-14 Here we want to investigate the cochannel interference in a more practical environment based on computer simulation. Consider a cellular system with hexagonal cells and with a frequency reuse factor of 7. Consider the forward link transmission with the cochannel interference resulting only from the 6 cochannel cells (base stations) in the first tier. The propagation environment is characterized by the lognormal shadowing as described by Eqs. (2.4.15)–(2.4.16), with path loss exponent κ and the standard deviation σ_ϵ. All the base stations have the same transmit signal power and the same values for d_0 and $\bar{L}_p(d_0)$, respectively. The required instantaneous signal-to-cochannel interference ratio (S/I) is 18 dB.

a. Assume that the mobile user location is uniformly distributed in the cell with $d > d_0$, find the probability that the instantaneous (S/I) value is below 18 dB, (1) for $\sigma_\epsilon = 8$ dB and $\kappa = 2, 2.5, 3, 3.5$, and 4, respectively, and (2) for $\kappa = 4$ and $\sigma_\epsilon = 7, 8$, and 9 dB, respectively.
b. Consider the worst-case scenario where the mobile user is at the cell boundary. Find the probability that the instantaneous (S/I) value is below 18 dB, (1) for $\sigma_\epsilon = 8$ dB and $\kappa = 2, 2.5, 3, 3.5$, and 4, respectively, and (2) for $\kappa = 4$ and $\sigma_\epsilon = 7, 8$, and 9 dB, respectively.
c. From the simulation results, comment on the effects of the user location and the channel parameters κ and σ_ϵ on the (S/I) ratio.

6

Multiple Access Techniques

Multiple access is a signal transmission situation in which two or more users wish to simultaneously communicate with each other using the same propagation channel. This is precisely the uplink transmission situation in a wireless communications system. In the uplink or reverse channel, multiple users will want to transmit information simultaneously. Without proper coordination among the transmitting users, collisions will occur when two or more users transmit simultaneously. Access methods that incur collision are referred to as random access and variants of random access. This chapter discusses the throughput characteristics of two popular random access methods: Aloha and carrier-sense multiple access (CSMA). Multiple access strategies based on orthogonality among the competing transmissions are collision-free. Orthogonality can be in the form of frequency division, time division or code division. Techniques with built-in conflict resolution capability presented in this chapter are frequency-division multiple access (FDMA), time-division multiple access (TDMA) and code-division multiple access (CDMA). Performance analysis and evaluation of these conflict-free multiple access methods in terms of spectral efficiency and system capacity are described and discussed.

6.1 MULTIPLE ACCESS IN A RADIO CELL

In each radio cell, the transmission from the base station in the downlink can be heard by each and every mobile user in the cell. For this reason, this mode of transmission is referred to as *broadcasting*. On the other hand, transmissions from the mobile users in the uplink to the base station is many-to-one, and is referred to as *multiple access*. Figure 6.1 illustrates the uplink/downlink transmission scenarios.

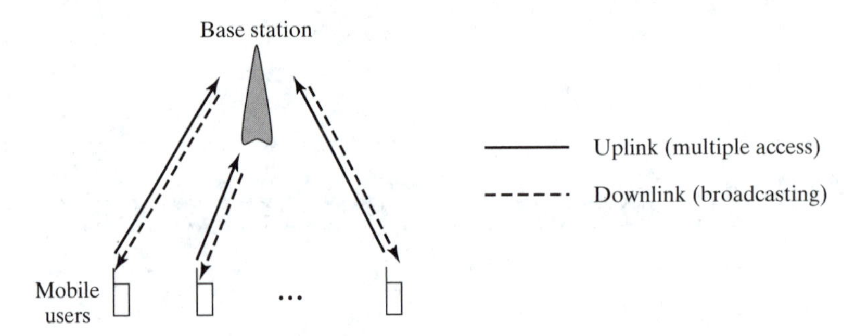

Figure 6.1 The uplink and downlink transmissions.

Transmissions in the uplink have the following attributes:

a. Multiple mobile users want to access the common resource (base station) simultaneously;
b. If the transmissions from two or more users arrive at the base station receiver at the same time, there will be destructive interference, unless the multiple arriving signals are mutually orthogonal;
c. Orthogonality between two signals $x_i(t)$ and $x_j(t)$, $t \in [0, T]$, means that their inner product over the signaling interval vanishes. That is

$$\int_0^T x_i(t)x_j(t)dt = 0, \quad \text{for } i \neq j. \tag{6.1.1}$$

The key element in multiple access is to make the transmitted signals from the different users orthogonal to each other. This raises the fundamental question of how this orthogonality condition should be mechanized.

Conflict-Free Multiple Access. Orthogonality can be mechanized using

a. space division multiple access (SDMA),
b. frequency division multiple access (FDMA),
c. time division multiple access (TDMA), or
d. code division multiple access (CDMA).

In theory, SDMA, FDMA, TDMA and CDMA are conflict-free multiple access techniques. The conflict-free property is achieved through coordination among all the participating users. In the case of SDMA, FDMA and TDMA, the coordination among all participating users is performed through fixed assignment. For example, in FDMA the system bandwidth, B_s, is partitioned into frequency bands and each user is assigned a unique frequency band for information transmission for the entire duration of the connection. CDMA is a spread spectrum technique. Each user is assigned a unique spreading function from a set of wideband orthogonal functions. Based on the orthogonality property given in Eq. (6.1.1), an individual user can transmit using the entire system bandwidth, B_s, during one use of the channel. Thus, FDMA is a narrowband multiple access plan while CDMA is wideband.

Random Access and Variants. In certain situations, depending on the traffic load and mixture, it may be advantageous to employ a non-conflict-free multiple access scheme. In non-conflict-free multiple access, transmissions by the different users are either uncoordinated or are only partially coordinated. A completely uncoordinated scheme is referred to as *random access*. In random access, a user contends for usage of the same resource, independent of any other users. For this reason, random access is also referred to as *contention access*. In a random access scheme, a user transmits whenever it has information to be transmitted, independent of the status of any other users.

In conflict-free multiple access systems, random access is often used for users to gain the initial access to the systems. For example, in GSM systems, there is a random access channel (RACH) among the control channels which provide the necessary control functions. The RACH is used by a mobile user to originate a call or to respond to a paging signal in the reverse link. The RACH uses a slotted Aloha access scheme. In responding to a call request from a mobile user via the RACH, the base station allocates a conflict-free channel to the user during the call connection.

6.2 RANDOM ACCESS

While we will not be dwelling much on non-conflict-free multiple access techniques in this chapter, it seems appropriate to briefly study some of the popular methods for non-conflict-free multiple access.

6.2.1 Aloha Systems

Random access was used by a research group from the University of Hawaii in the late 1960s and early 1970s for its satellite communications with the U.S. mainland at a transmission speed of 50 kbps [1, 2]. This system was called Aloha and the term Aloha has been used as a general name for random access. Aloha is a packet-switching system. The time interval required to transmit one packet is called a *slot*. When transmissions from two or more users overlap, they destroy each other, whether it is complete overlap or partial overlap. The maximum interval over which two packets can overlap and destroy each other is called the *vulnerable period*. The mode of random access in which users can transmit at anytime is called *pure Aloha*. In a pure Aloha system, where the packet length is a fixed constant, the vulnerable period is two slot times. A version in which users are restricted to transmit only from the instant corresponding to the slot boundary is referred to as *slotted Aloha*. The alignment of transmissions to coincide with the slot boundary means that packets can only experience complete overlap, so that the vulnerable period in slotted Aloha is one slot time. This means that the maximum throughput rate of slotted Aloha doubles that of pure Aloha.

Throughput of Aloha Systems. In the Aloha systems, a user can hear its own transmission or the transmissions by other users within the footprint of the serving satellite. A transmitting user, upon hearing a collision, backs off for a random delay interval and transmits again, until success is achieved. The transmission is successful when there are no other packet transmissions during the vulnerable period. Thus, the probability of successful transmission is defined as

$$\begin{aligned} \text{P[success]} &= \text{P[no other packet transmission occurs} \\ &\quad \text{within a vulnerable period].} \end{aligned} \tag{6.2.1}$$

Let S be the throughput, defined as the successfully transmitted traffic load, and G be the total offered channel traffic load. Assuming that traffic generated for transmission obeys a Poisson distribution, then

$$\text{P[no other packet transmission occurs]} = e^{-\tau G}, \tag{6.2.2}$$

where τ is the vulnerable period. Using Eq. (6.2.2) in Eq. (6.2.1), we have

$$\text{P[success]} = e^{-\tau G}. \tag{6.2.3}$$

By definition, we also have

$$\text{P[success]} = \frac{S}{G}. \tag{6.2.4}$$

Combining Eqs. (6.2.3) and (6.2.4), we have the throughput equation of the Aloha systems given by

$$S = Ge^{-\tau G}. \tag{6.2.5}$$

From Eq. (6.2.5), we note that $S \longrightarrow 0$ in the limit as $G \longrightarrow \infty$. That is, the negative exponential decays faster than G increases.

The maximum value of S occurs when the slope of the throughput curve is 0 (i.e., $\frac{dS}{dG} = 0$). Taking the derivative of the right-hand side of Eq. (6.2.5) with respect to G and setting the result equal to zero, we get $G = 1/\tau$. For pure Aloha, the vulnerable period is $\tau = 2$ slots, while for slotted Aloha, it is $\tau = 1$ slot. The maximum throughput, for pure Aloha, is therefore

$$S_{\text{max}} = \frac{1}{2e} \approx 0.184, \tag{6.2.6}$$

and that for slotted Aloha is

$$S_{\text{max}} = \frac{1}{e} \approx 0.368. \tag{6.2.7}$$

Delay Throughput Characteristics of Aloha. Intuitively, the more traffic one tries to push through a system, the longer it will take to get through the system. That is, delay and throughput have paradoxical requirements. The delay experienced by a packet in the system is measured from the instant of the packet's arrival until the instant the sender receives confirmation. The packet delay is thus a function of the number of transmissions, the retransmission delay, and the time required for the sender to receive confirmation of successful transmission. Let

R be the number of slots to receive an acknowledgment,
$\bar{D}_{ret}$ be the mean retransmission delay, and
E be the mean number of transmissions until success.

The mean number of transmissions until success, E, is then

$$E = \frac{G}{S} = e^{\tau G}.$$

Hence, the average packet delay, $\bar{D}$, is given by

$$\bar{D} = R + (E - 1)(R + \bar{D}_{ret}) \tag{6.2.8}$$

for pure Aloha, and

$$\bar{D} = R + 0.5 + (E - 1)(\lceil R \rceil + \bar{D}_{ret}) \tag{6.2.9}$$

for slotted Aloha. In both Eqs. (6.2.8) and (6.2.9), R represents the time taken for the initial transmission and confirmation, and $(E - 1)$ represents the mean number of retransmissions. In Eq. (6.2.9), the term 0.5 represents the fact that, on average, an arrival is 1/2 slot time to the slot boundary, and the symbol $\lceil x \rceil$ denotes the smallest integer equal to or greater than x. Substituting $\tau = 2$ for pure Aloha and $\tau = 1$ for slotted Aloha, we have

$$\bar{D}_{pure} = R + (e^{2G} - 1)(R + \bar{D}_{ret}) \tag{6.2.10}$$

and

$$\bar{D}_{slotted} = R + 0.5 + (e^{G} - 1)(\lceil R \rceil + \bar{D}_{ret}). \tag{6.2.11}$$

In a random access environment, when two or more users transmit packets simultaneously, collisions will take place. An efficient scheme to resolve collisions will help to improve system throughput. One approach to constructing a collision resolution algorithm is by means of a *tree protocol* to resolve collisions using a divide-and-conquer approach. We discuss the tree collision resolution algorithm using the following example.

Example 6.1 Tree Protocol for Collision Resolution

Consider the situation in which there are 8 users in the radio cell. Suppose we number the users from 0 to 7. At a given epoch, users 0, 1, 2, 5 and 7 have packets ready for transmission, while users 3, 4 and 6 are idle. When the ready users transmit simultaneously, collisions will occur. The interval of time within which the collisions are resolved is referred to as the collision resolution interval (CRI). In the divide-and-conquer approach, the binary tree is divided into two halves; one half is searched to completion and then the other half is searched to resolve collisions. Collision resolution can be performed on a per collision resolution interval basis.

The channel states are described by the 3-tuple {idle, collision, success}. Assume that the users can detect the channel states.

a. Draw a binary tree showing that the users are located at the leaves of the tree, and label the intermediate nodes using the letters A through G, with A representing the root node.
b. Using the tree structure, describe, with diagrammatic illustration, a static approach to resolve collisions. What is the length of the collision resolution interval?

Solution

a. With 8 users, the binary tree has 4 levels, with the root at level 0 and the leaves at level 3, as shown in Figure 6.2, where ready users are indicated by the symbol ⨀.
b. Consider that, at the end of the $(i-1)$th collision resolution interval, and hence the start of the ith collision resolution interval, users 0, 1, 2, 5 and 7 are ready for packet transmission. At the root node A (level 0), collisions occur. Split the tree into two halves and resolve the left half first. Then at intermediate node B (level 1), users 0, 1, and 2 transmit, resulting in collision. Divide the subtree into two halves and search the left half. Users 0 and 1 transmit and collision occurs. Further divide the subtree, with node D as the root, into two halves. Since each of the halves only has the leaf node, transmissions by user 0, and then by user 1, will both be successful. This completes the search of the left half of the subtree with node D as the root. The algorithm next searches the right half of the subtree with node B as the root. The procedure is repeated until the entire tree has been searched and all ready users have successfully transmitted.
The ith collision resolution interval (CRI_i) is illustrated in Figure 6.3. The length of CRI_i is 9.

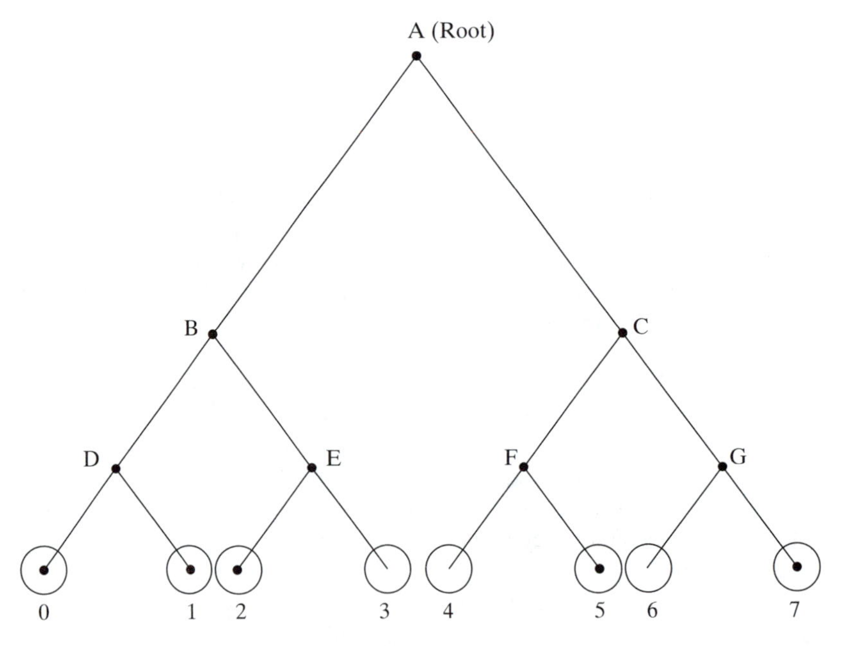

Figure 6.2 Labeled binary tree for 8 users.

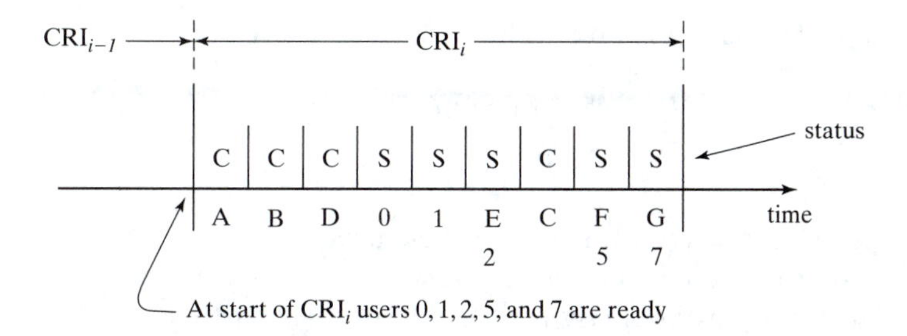

Figure 6.3 Collision resolution interval.

6.2.2 Carrier Sense Multiple Access (CSMA)

In CSMA, users listen before transmission. This listening is referred to as sensing the channel. If a user senses the channel idle, it transmits. Otherwise, the ready user takes one of the following actions [73]:

a. Defers transmission and tries again after a random delay. This mode of retry is called non-persistent CSMA;
b. Waits until the channel becomes idle and then transmits again. This mode of retry is called 1-persistent CSMA.

Throughput of CSMA. Here we derive the throughput for unslotted CSMA. To facilitate the derivation, we quantify the parameters as follows. For fixed packet lengths, the packet transmission time is one slot. The propagation delay between two users is τ s, and the normalized propagation delay, a, is given by

$$a = \frac{\tau}{l/C} = \frac{\tau C}{l},$$

where C is the channel capacity in bits/s and l is the packet length in bits. The status of the channel alternates with busy and idle periods. Let

B_n be the busy period in the nth cycle,
I_n be the idle period in the nth cycle, and
U be the time that the channel is used without collision.

Let $\bar{X}$ denote the mean value of X. The success rate S is then given by

$$S = \frac{\bar{U}}{\bar{B} + \bar{I}}. \tag{6.2.12}$$

The parameters $\bar{U}$, $\bar{I}$ and $\bar{B}$ are determined as follows. Let

$$\begin{aligned} P_s &= P[\text{a busy period has a single packet transmission}] \\ &= P[\text{no transmission in normalized interval } a] \\ &= e^{-aG}. \end{aligned} \tag{6.2.13}$$

The mean time that the channel is used without collision is given by

$$\begin{aligned}\bar{U} &= \text{packet transmission time} \times P_s \\ &= 1 \times P_s.\end{aligned}$$

The mean idle period is $\bar{I} = 1/G$. It remains to determine $\bar{B}$.

A busy period can have multiple packet transmissions. Let Y be the time between the first and last packet transmissions in a busy period. $Y = 0$ means that there is only a single packet transmission in the busy period so that the transmission is successful. If Y is equal to or longer than the normalized propagation delay a, there is no collision. Collisions take place when $0 < Y < a$. For this case, the cumulative distribution function of Y is

$$\begin{aligned}F_Y(y) &= P[Y \leq y] \\ &= P[\text{no packet arrives in } (y, a)] \\ &= e^{-(a-y)G} \qquad 0 \leq y \leq a.\end{aligned} \tag{6.2.14}$$

The probability density function of Y, $f_Y(y)$, is obtained by differentiating Eq. (6.2.14)

$$f_Y(y) = e^{-aG}\delta(y) + Ge^{-(a-y)G}, \quad 0 \leq y \leq a. \tag{6.2.15}$$

The mean value of Y is thus given by

$$\begin{aligned}\bar{Y} &= \int_0^{\infty} y f_Y(y) dy \\ &= a - \frac{1 - e^{-aG}}{G}.\end{aligned} \tag{6.2.16}$$

The mean busy period is given by the sum of the packet transmission time, the normalized propagation delay, and the mean interval between the first and last packet transmissions instants. That is

$$\bar{B} = 1 + a + \bar{Y}. \tag{6.2.17}$$

The throughput for unslotted non-persistent CSMA is obtained by substitution of $\bar{U}$, $\bar{I}$ and $\bar{B}$ in Eq. (6.2.12), yielding

$$S_{np-CSMA} = \frac{Ge^{-aG}}{G(1 + 2a) + e^{-aG}}. \tag{6.2.18}$$

In the limit when $a \longrightarrow 0$, $S \longrightarrow G/(1 + G)$. The throughput for unslotted 1-persistent CSMA can be derived in a similar manner.

$$S_{1p-CSMA} = \frac{Ge^{-G(1+2a)}[1 + G + aG(1 + G + aG/2)]}{G(1 + 2a) - (1 - e^{-aG}) + (1 + aG)e^{-G(1+a)}}. \tag{6.2.19}$$

In the limit as $a \longrightarrow 0$, $S \longrightarrow \frac{Ge^{-aG}(1+G)}{G+e^{-aG}}$. The mean packet delay of CSMA is approximately given by

$$\bar{D}_{CSMA} = R + \left(\frac{G}{S} - 1\right)(R + \bar{D}_r), \qquad (6.2.20)$$

where the parameters R and $\bar{D}_r$ are as defined earlier.

6.3 CONFLICT-FREE MULTIPLE ACCESS TECHNOLOGIES

The methods commonly used in mobile wireless cellular systems are FDMA, TDMA and CDMA. As discussed in Chapter 1, the first generation wireless systems use FDMA while those in the second generation use TDMA and CDMA. CDMA is the targeted multiple access technology for the third generation (3G) wireless communications systems. FDMA and TDMA are fixed capacity allocation schemes in that an individual user is assigned a frequency band (in FDMA) or a time slot (in TDMA) for the duration of the connection. With sufficiently well designed filters (in the FDMA case) and slot synchronizers (in the TDMA case), there should be no, or a minimal amount of, spectral overlap (in FDMA) or timing jitter (in TDMA). In this way, FDMA and TDMA would be conflict-free multiple access schemes.

CDMA is a spread spectrum technique. Orthogonality between any pair of transmitted signals in CDMA is based on algebraic properties. However, practically generated wideband spreading functions are not truly orthogonal. The cross correlation between any pair of transmitted signals represents interference. Hence, CDMA is an interference limited multiple access strategy.

6.3.1 FDMA

In FDMA, the total bandwidth is divided into non-overlapping frequency subbands. Each user is allocated a unique frequency subband for the duration of the connection, whether the connection is in an active or idle state. Orthogonality among transmitted signals from different mobile users is achieved by bandpass filtering in the frequency domain. This type of multiple access support is narrowband, and is not suitable for multimedia communications with various transmission rates. In addition, it incurs a waste of bandwidth when the user is in a dormant state.

FDMA is relatively simple to implement. However, the power amplifiers and the power combiners used are nonlinear, and tend to generate intermodulation frequencies, resulting in intermodulation distortion. To minimize the effects of intermodulation distortion, stringent RF filters are required to reject intermodulation distortion. RF filters are heavy, cumbersome, and costly.

To provide interference-free transmissions between the uplink and the downlink channels, the frequency allocations have to be separated by a sufficient amount. The frequency separation can be achieved using two antennas operating at different frequencies, or one antenna with frequency division duplexing. That is, the uplink and downlink channels of FDMA operate at distinctly different frequency bands. Therefore, the channel impairments seen at the cell-site receiver are different from those seen at the receiver of each of the mobile users.

6.3.2 TDMA

In a TDMA system, the channel time is partitioned into frames. The length of a frame is long enough so that every user in service has an opportunity to transmit once per frame. To achieve this, a TDMA frame is further partitioned into time slots. Users have to transmit in their assigned slots from frame to frame. The slot assignment can be fixed or dynamic. If the assigned slot is fixed from frame to frame for the duration of the connection, the users have to synchronize to their respective assigned slots. This mode of TDMA is referred to as synchronous TDMA (STDMA). With packet-switched transmission, it is more efficient to allow a user to transmit only when it has a packet to send. In this case, a user is not assigned a fixed time slot for the duration of its connection. Transmission slots are dynamically assigned from frame to frame. This mode of TDMA is referred to as asynchronous TDMA (ATDMA). In STDMA, the frame length is fixed by the number of users, whether or not they are active. In ATDMA, the frame length varies from frame to frame, depending on the number of active users in the frame. In ATDMA, dynamic assignment of slots is performed through a reservation access procedure. This subsection will discuss the ramifications of both STDMA and ATDMA methods.

STDMA. Depending on the manner in which frequency is allocated, STDMA can be wideband or narrowband. It is called wideband TDMA if the channel time is divided into slots, and an individual user is allowed to use the entire available channel bandwidth to transmit its information in the assigned slot, as shown in Figure 6.4(a) where B_s is the total frequency band allocated to the uplink transmission. It is called narrowband TDMA if the overall bandwidth is first divided into frequency bands, and the channel time corresponding to each frequency band is divided into time slots for packet transmission, as shown in Figure 6.4(b). In this way an individual user can only transmit at a rate governed by the allocated frequency subband.

In STDMA, the channel time is divided into contiguous slots, each of which is long enough to transmit or receive one information unit. STDMA operates on a frame-by-frame basis. A group of N_{slot} slots plus a header and a trailer form a TDMA frame, as illustrated in Figure 6.5. In addition to carrying the information data, a slot also includes other fields, such as trailer bits, synchronization (sync) bits, guard bits, etc. A slot can only be used by one user to transmit or receive during one use of the transmission channel.

With TDMA, the receiver must be able to synchronize to the received signal within a slot time. This means that timing information has to be extracted from the observed signal. A conventional approach to extract timing information is to use matched filtering or correlation detection to achieve synchronization within a time slot.

To provide the required separation between the transmissions in the uplink and downlink channels, TDMA can use TDD or FDD. FDD provides two simplex channels at the same time, while TDD provides two simplex time slots on the same frequency band. The manner in which FDD and TDD provide duplex operation is shown in Figure 6.6. As mentioned previously, in FDD systems, the channel disturbances in the uplink and downlink channels are different. On the other hand, in a TDD system, the uplink and downlink channels operate at the same frequency band. In this case, the cell-site receiver and the user's receiver see approximately the same propagation channel if the channel coherence time is much larger than the frame duration. From Figure 6.6, it

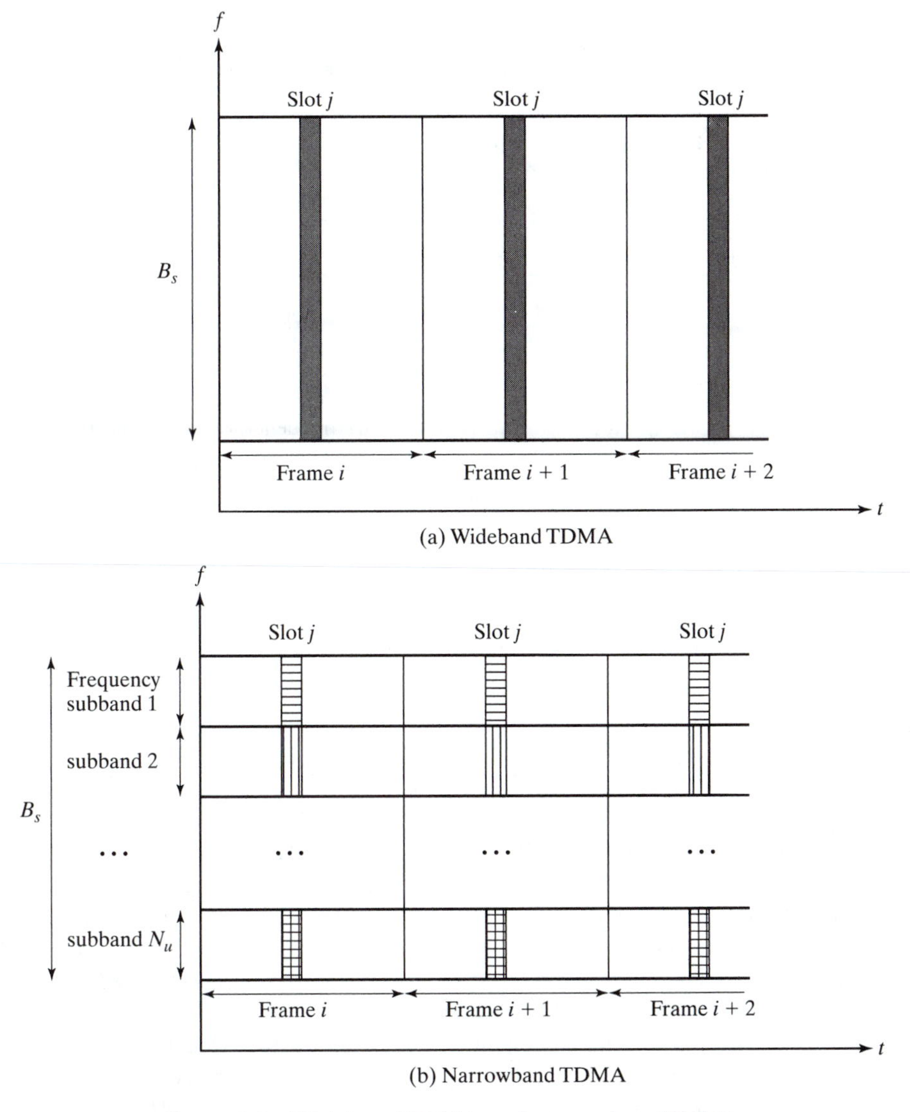

Figure 6.4 Wideband TDMA and narrowband TDMA.

is observed that uplink and downlink propagations are separated by an *off* interval (either in time or in frequency).

In summary, STDMA has a wideband version and a narrowband version. In *wideband TDMA*, transmission in each slot uses the entire frequency band, while in *narrowband TDMA*, since the whole frequency band is divided into subbands, transmission in each slot only uses the frequency width of one subband. The number of contiguous slots during one use of the channel (whether it is the entire band or a subband) constitute a frame.

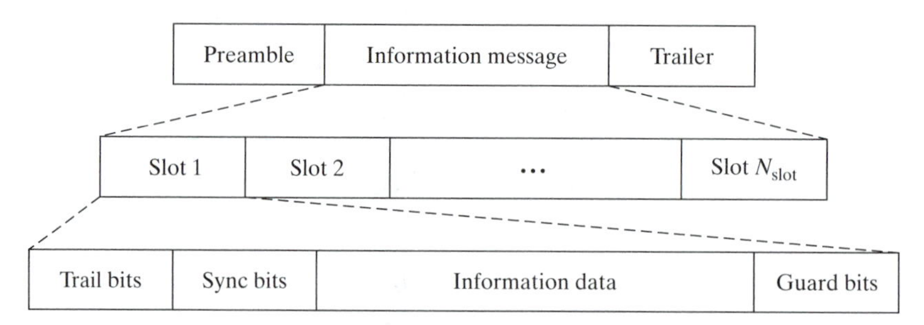

Figure 6.5 TDMA frame structure.

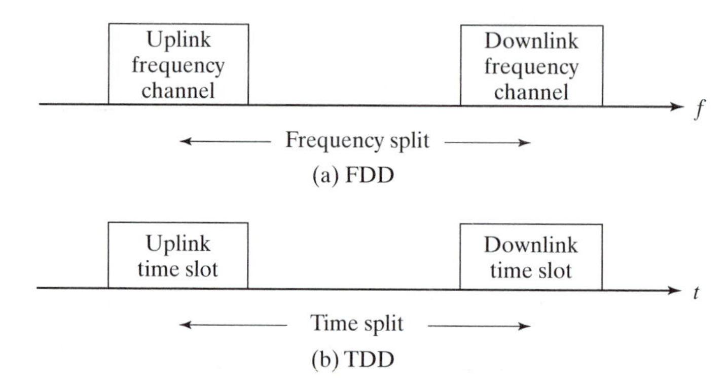

Figure 6.6 FDD and TDD methods for duplex operation.

ATDMA. Asynchronous TDMA is implemented using a reservation access mechanism. An ATDMA frame contains two segments: a leading segment and a trailing segment. The leading segment contains mini request slots for active users to submit requests for slot allocation. The trailing segment contains information slots for packet transmission. Since the leading segment is used for reservation, it represents an overhead on the system. Requests submitted in the request slots in the current uplink frame will be received by the base station. Based on the information contained in the request vector, the base station assigns information slots to requesting users in the next downlink frame. Thus, in an ATDMA system, requests submitted in the current frame will be accommodated in the next frame.

The simplest way to structure an ATDMA frame is to allocate a request slot to each user on a permanent basis. If there are N users, the request segment has to contain N request slots. In this way, each user owns a request slot and requests can be submitted on a conflict-free basis. However, if the user population is overly large, the overhead can be quite large. If the traffic load is very heavy such that every user is active, this way of structuring the request slot segment would be fine. But, if the system were not that heavily loaded, it would be more efficient to use STDMA.

Under normal operating conditions, the average traffic load would never be 100%. Consider voice transmission for example. On average, a voice process is in a talk state only 40% of the time; the other 60% of the time, it is in a silent state. So, the average load would be 0.4. In this case, the number of request slots in the ATDMA frame can be fewer than the number of users. However, if the number of request slots is fewer than the number of users, then how should request submissions be performed? One approach is to let active users choose a request slot randomly to submit requests. This can lead to collision in transmission of requests. The uncollided requests will be successful and those users will receive slot allocation in the next frame. A collided request will have to be retransmitted, again by randomly choosing a request slot in the next uplink frame.

In TDMA, each user collects the low-rate source information data in the previous frame and transmits the data at a much higher rate in the allocated time slot of the current time frame. As the frame duration is usually very small (e.g., 5 ms or 10 ms), the small delay in the transmission may not be noticed by the end user.

6.3.3 CDMA

CDMA is a spread spectrum multiple access method [155, 80]. The principle of spread spectrum communications is that the bandwidth of the baseband information-carrying signals from the different users is spread by different signals with a bandwidth much larger than that of the baseband signals. Ideally, the spreading signals used for the different users are orthogonal to each other. Thus, at the receiver, the same spreading signal is used as the despreading signal to coherently extract the baseband signal from the target user, while suppressing the transmissions from any other users. In spread spectrum communications, the spreading signals have to be derived from a set of orthogonal functions. Orthogonal functions with an infinitely large bandwidth will look like white noise to each other. However, white noise cannot be practically generated. Thus, truly orthogonal functions with an infinitely large bandwidth are difficult, if not impossible, to generate in practice.

Sequences, or functions, that can be generated have a deterministic feature. These sequences are referred to as pseudorandom noise (PN) sequences. The generation of PN sequences using linear shift registers is shown in Appendix E. CDMA in which the spread spectrum is achieved by directly multiplying the user's baseband signal with a high rate PN sequence is referred to as *direct sequence CDMA* (DS-CDMA). Different users in a DS-CDMA system use different spreading signals, therefore, they may use the same carrier frequency, f_c, and transmit the spread signals simultaneously. In a DS-CDMA system, each user is assigned its own PN sequence, which is approximately orthogonal to all other sequences assigned to other users. The users may be randomly located within the footprint of the base station. Even if the different users transmit at the same power level, because the distances from the various users to the cell-site differ, the power levels received from the different users at the cell-site receiver will differ. This phenomenon is called the near–far problem. The near–far problem can be avoided through power control. In a CDMA-based system, power control is implemented to provide the same received signal power level (or desired power levels) at the base station receiver from the different users, independent of the location of each mobile user within the cell area [8, 134, 94].

Since the spreading PN sequences are not truly orthogonal, as a multiple access technology, CDMA is interference limited. This means that the capacity of CDMA has a soft limit, and is a function of the service quality required.

Direct Sequence Spread Spectrum. Consider a radio cell with a population of K users. Each of the mobile users is assigned a unique spreading sequence. Each symbol in the PN sequence is called a chip. Figure 6.7 shows the functional block diagram of the transmitter and receiver for the kth user, $k = 1, 2, \ldots, K$. For simplicity, consider real-valued binary spreading waveforms. The information-carrying baseband signal $d_k(t)$ is

$$d_k(t) = \sum_i s_{k,i} \Pi\left(\frac{t - iT_b}{T_b}\right),$$

where $s_{k,i} \in \{-1, +1\}$ is the ith binary information bit, T_b is the information bit interval, and $\Pi(t/T_b)$ is the rectangular pulse

$$\Pi\left(\frac{t}{T_b}\right) = \begin{cases} 1, & 0 \le t \le T_b \\ 0, & \text{otherwise} \end{cases} .$$

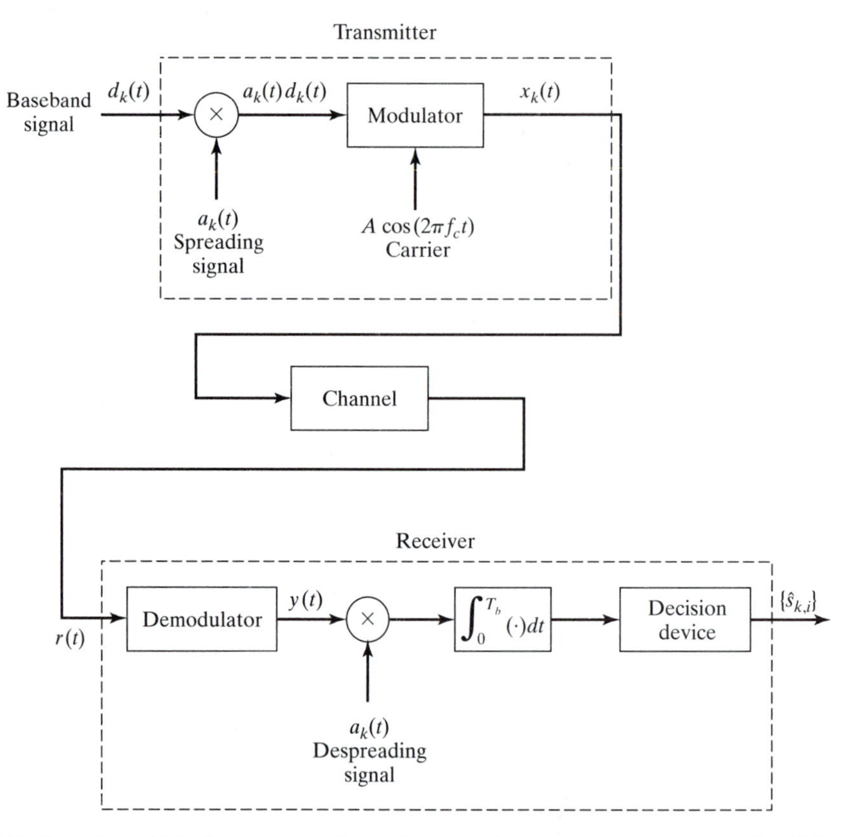

Figure 6.7 Functional block diagram of user k transmitter and receiver in a DS-CDMA system.

The spreading signal of the kth user is $a_k(t)$ and can be represented as

$$a_k(t) = \sum_l a_{k,l} P_{T_c}(t - lT_c),$$

where $a_{k,l} \in \{-1, +1\}$ is the lth chip of the binary PN sequence assigned to user k, $P_{T_c}(t)$ is the chip pulse waveform depending on baseband pulse shaping, and T_c is the chip interval, corresponding to a chip rate of $1/T_c$. The spreading process is to modulate $a_k(t)$ onto $d_k(t)$, which gives the spread signal $a_k(t)d_k(t)$. The spreading function $a_k(t)$ is often phase shift keyed onto the baseband signal $d_k(t)$ (i.e., the product operation, $a_k(t)d_k(t)$, is performed by the phase shift keying process, as discussed in Chapter 3). The spread signal $a_k(t)d_k(t)$ is then modulated with carrier frequency f_c ($\gg 1/T_c$), resulting in the bandpass signal $x_k(t)$ with amplitude A_c and bit interval T_b, for transmission. Normally, $T_b = LT_c$, where L is an integer. The principle of signal spreading is that $L \gg 1$ (i.e., $T_b \gg T_c$). Since the spreading signal has a chip rate much larger than the transmitted information symbol rate[1], the bandwidth of the spread signal is much larger than the bandwidth of the baseband information signal, hence, the name spread spectrum modulation. In the following, for simplicity, we consider $P_{T_c}(t) = \Pi(t/T_c)$ and BPSK for the passband modulation with coherent demodulation. If the PN sequences are periodic, with period L, then the transmitted signal is

$$x_k(t) = A_c \left[\sum_i s_{k,i} \sum_{l=1}^{L} a_{k,l} \Pi \left(\frac{t - iT_b - lT_c}{T_c} \right) \right] \cos(2\pi f_c t).$$

Normally, the information bits and the PN sequence chips are completely independent. Then, from Section 3.4, the psd of the transmitted signal is

$$\Phi_1(f) = \frac{E_c}{2} \{\text{sinc}^2[(f - f_c)T_c] + \text{sinc}^2[(f + f_c)T_c]\}, \tag{6.3.1}$$

where $E_c = \int_0^{T_c} [x_k(t)]^2 dt = \frac{1}{2} A_c^2 T_c$ is the chip energy. For comparison, without spread spectrum, the transmitted signal would be $A_c d_k(t) \cos(2\pi f_c t)$ and the corresponding psd would be

$$\Phi_2(f) = \frac{E_b}{2} \{\text{sinc}^2[(f - f_c)T_b] + \text{sinc}^2[(f + f_c)T_b]\}, \tag{6.3.2}$$

where $E_b = \int_0^{T_c} [A_c a_k(t) \cos(2\pi f_c t)]^2 dt = \frac{1}{2} A_c^2 T_b = LE_c$ is the bit energy. Figure 6.8 illustrates the normalized psd without spreading, $\Phi_2(f)/(E_b/2)$, and with spreading, $\Phi_1(f)/(E_c/2)$, for $f > 0$.

The output emerging from the channel, which is also the input to the receiver, is a superposition of the spread signals from all users in the same radio cell, plus background noise and interference from neighboring cells. Let $r(t)$ denote the received signal. Under the assumption that all the K users in the cell are synchronized in time and have the same received signal power

[1]For binary signaling, the symbol rate equals the bit rate; the terms symbol and bit are synonymous.

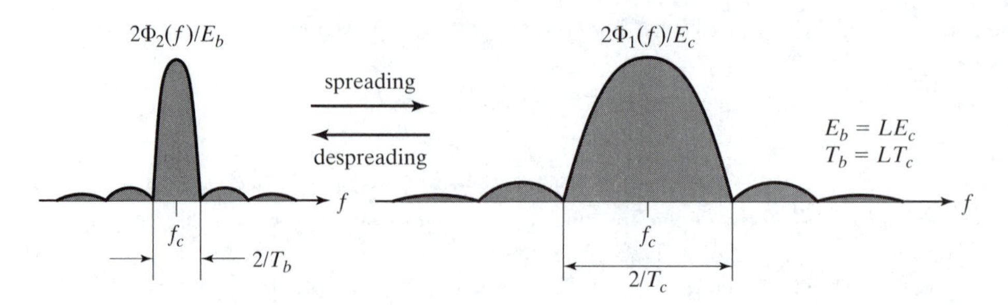

Figure 6.8 Illustration of signal power spectral density without and with spreading.

$(A_c^2/2)$, $r(t)$ is given by

$$r(t) = \sum_{k=1}^{K} x_k(t) + I(t) + w(t), \tag{6.3.3}$$

where $I(t)$ represents intercell interference and $w(t)$ represents background white Gaussian noise with zero mean and two-sided psd $N_0/2$. Both intracell interference and intercell interference are due to multiple access and, therefore, are called multiple access interference (MAI). Suppose we tag the signal transmitted by user 1 as the desired signal and the transmitted signals from all other users as interference. For the purpose of detecting user 1's signal at the receiver, it is convenient to isolate the desired signal component on the right-hand side of Eq. (6.3.3) and express $r(t)$ as

$$\begin{aligned} r(t) &= A_c a_1(t) d_1(t) \cos(2\pi f_c t) \\ &+ \sum_{k=2}^{K} A_c a_k(t) d_k(t) \cos(2\pi f_c t) + I(t) + w(t). \end{aligned} \tag{6.3.4}$$

The first block in the receiver of Figure 6.7 is the demodulator, which translates the received signal centered at frequency f_c to baseband centered at frequency zero. This is done by a correlator or a matched filter in coherent demodulation. The output of the demodulator at the end of the lth chip interval is

$$\frac{1}{T_c} \int_{lT_c}^{(l+1)T_c} r(t) \cos(2\pi f_c t) dt.$$

Over each chip interval, $a_k(t)d_k(t)$ ($k = 1, 2, \ldots, K$) is a constant. The demodulator output as a function of time, $y(t)$, can be written as

$$y(t) = \frac{A_c}{2} a_1(t) d_1(t) + \sum_{k=2}^{K} \frac{A_c}{2} a_k(t) d_k(t) + n(t),$$

where

$$n(t) = \frac{1}{T_c} \sum_l \left\{ \int_{lT_c}^{(l+1)T_c} [I(t) + w(t)] \cos(2\pi f_c t) dt \right\} \Pi \left(\frac{t - lT_c}{T_c} \right)$$

is due to the intercell interference and additive background noise. In general, the integral can be approximated by a zero-mean Gaussian random variable and is independent from chip to chip. As a result, $n(t)$ is baseband Gaussian noise with bandwidth approximately equal to $1/T_c$. To extract user 1's transmitted signal from the demodulator output in the receiver shown in Figure 6.7, $a_1(t)$ should be used as the despreading signal. Despreading is achieved by first multiplying the demodulator output, $y(t)$, with $a_1(t)$ and then integrating the product over each symbol interval. Since $a_1^2(t) = 1$ at any t, we have

$$a_1(t)y(t) = \frac{A_c}{2} d_1(t) + \sum_{k=2}^{K} \frac{A_c}{2} a_1(t) a_k(t) d_k(t) + a_1(t) n(t),$$

where the first term represents the desired signal component and is a constant over each symbol interval. It is clearly observed that the despreading process indeed recovers the original baseband signal $d_1(t)$ from the spread signal $d_1(t)a_1(t)$. The second term represents the effect of the intracell interference, where $a_1(t)a_k(t)$ can be viewed as a new spreading signal with the same chip rate. With a very high chip rate, the power of the interference is approximately uniformly distributed over the frequency band $[0, \frac{1}{T_c}]$. For the intercell interference plus noise term, since (*a*) $a_{1,l}$ takes on the values of -1 and $+1$ with equal likelihood, (*b*) $n(t)$ in each chip interval is a zero-mean Gaussian random variable, and (*c*) $a_1(t)$ and $n(t)$ have the same bandwidth, the statistical behavior of $a_1(t)n(t)$ is the same as that of $n(t)$. Therefore, the power of the intercell interference and noise is approximately uniformly distributed over the frequency band $[0, \frac{1}{T_c}]$.

To suppress the interference and noise, the next step in the receiver is to integrate the despread signal over each information symbol interval over which the desired signal component is a constant. The output of the integrator at the end of the ith symbol is

$$\frac{1}{T_b} \int_{iT_b}^{(i+1)T_b} a_1(t) y(t) dt = \frac{A_c T_b}{2} \alpha_1 d_{1,i} + \frac{A_c T_b}{2} \left[\sum_{k=2}^{K} \alpha_k d_{k,i} \right] + n_i,$$

where

$$\alpha_1 = \frac{1}{T_b} \int_{iT_b}^{(i+1)T_b} [a_1(t)]^2 dt = 1$$

is the autocorrelation of the spreading signal $a_1(t)$ over the symbol interval and

$$\alpha_k = \frac{1}{T_b} \int_{iT_b}^{(i+1)T_b} a_1(t) a_k(t) dt, \quad k = 2, 3, \ldots, K$$

is the crosscorrelation between the spreading signals $a_1(t)$ and $a_k(t)$ over the symbol interval. If all the spreading signals $a_k(t)$, $k = 1, 2, \ldots, K$, are orthogonal in the symbol interval, then there is no intracell interference in the recovered baseband signal. The effect of the intercell interference and background noise on the signal detection is given by the last term n_i, which is

$$n_i = \frac{1}{T_b} \int_{iT_b}^{(i+1)T_b} a_1(t) n(t) dt.$$

In the frequency domain, the integrator is a low-pass filter with bandwidth approximately equal to $1/T_b$. The LPF lets the desired signal component $\frac{A_c}{2} d_1(t)$ go through without distortion and greatly reduces the interference and noise power. The despreading process significantly improves the signal-to-interference plus noise ratio (SINR).

The spread spectrum system performance is measured by the processing gain, G_p, defined as the SINR improvement achieved by despreading. That is

$$G_p \triangleq \frac{\text{SINR after despreading}}{\text{SINR before despreading}}.$$

Figure 6.9 plots the psd of the signals before and after despreading to illustrate the SINR improvement achieved by despreading, where the sinc function shape of the psd of the signals, as given in Eqs. (6.3.1) and (6.3.2), is approximated by a uniform psd with cut-off frequency at the first null point of the sinc function, $W = 1/T_c$. Note that W is the spreading bandwidth and $R_s = 1/T_s$ is the baseband data symbol rate. From the simplified illustration, it is observed that (*a*) the desired signal power (the area in the figure for user 1) remains unchanged in the despreading process, and (*b*) the power of the interference and noise is reduced by W/R_s times. As a result, the processing gain is given by

$$G_p = \frac{W}{R_s} = \frac{T_s}{T_c} = L.$$

For binary signaling, R_s $(= R_b)$ is the bit rate, and T_s $(= T_b)$ is the bit interval.

Transmission Performance. In the absence of path loss, the received signal energy per bit is E_b. For transmission in an AWGN channel without MAI (by using truly orthogonal spreading sequences), the BER for the DS-CDMA user in additive white Gaussian noise of zero mean and two-sided psd $N_0/2$ is the same as that without spread spectrum modulation, due to the fact that (*a*) for the desired signal, the functions of spreading at the transmitter and despreading at the receiver cancel each other, and (*b*) the despreading process does not change the statistics of the noise component at the decision device input. For example, if BPSK is used, the BER for the DS-CDMA user with coherent detection is

$$P_b = Q\left(\sqrt{\frac{2E_b}{N_0}}\right).$$

If the spreading sequences are not orthogonal, the MAI from all other mobile users in the system will increase the transmission error rate. When the number of mobile users in the system is large and the interferences from all other users are independent and have similar stochastic behavior,

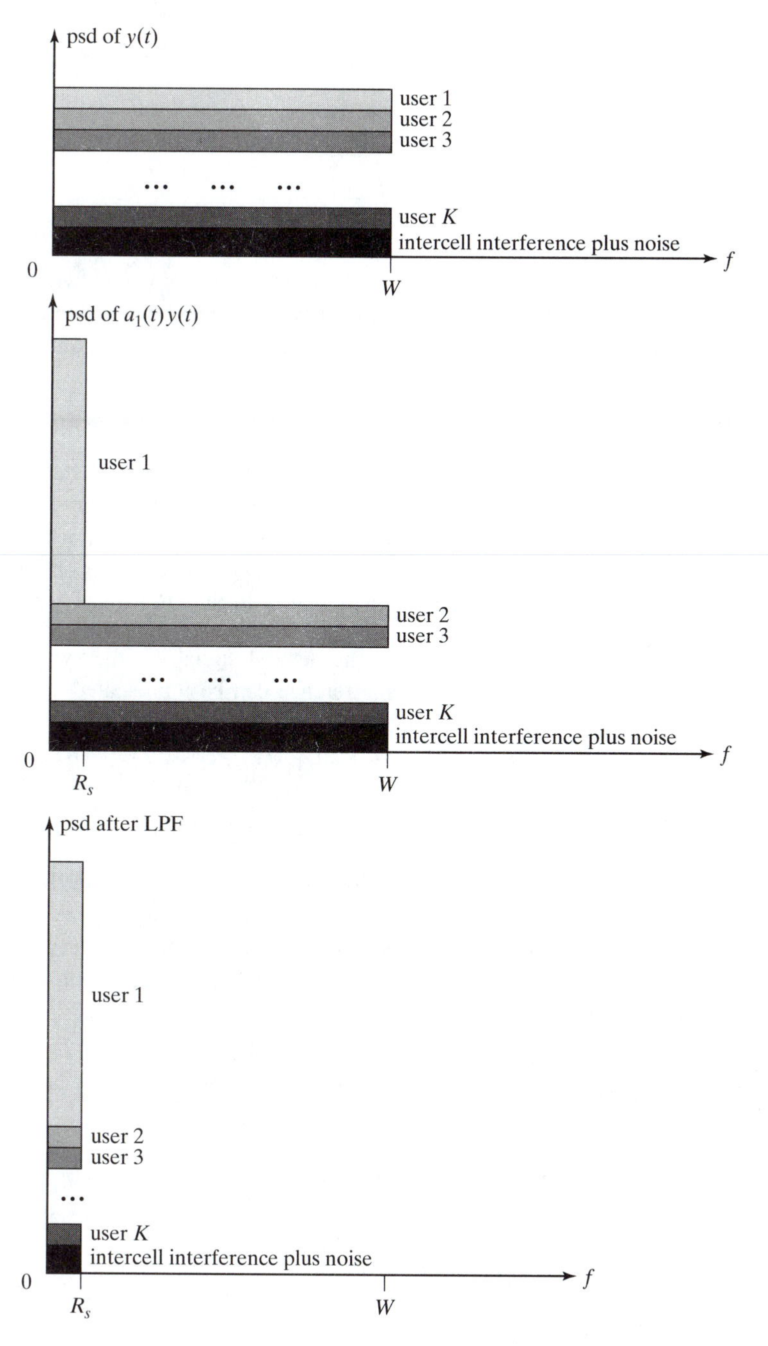

Figure 6.9 Illustration of signal power spectral density before and after despreading.

from the central limit theorem, the MAI can be approximated as a Gaussian process. Furthermore, with a large spread spectrum bandwidth W, the psd of the interference is approximately uniform over the bandwidth. As a result, the effect of the MAI on the transmission performance can be treated in the same way as the additive white Gaussian noise, and the BER is

$$P_b = Q\left(\sqrt{\frac{2E_b}{I_0 + N_0}}\right),$$

where I_0 is the two-sided psd of the MAI over the spread spectrum bandwidth. Even though spread spectrum does not provide performance gain over an AWGN channel where the interference (noise) is wideband with an infinite power, it does achieve a processing gain over narrowband interference (jamming) as shown in Example 6.2.

Example 6.2 Suppression of Narrowband Jamming in DS-CDMA

Consider a DS spread spectrum system using BPSK. The channel introduces a single-tone jamming signal with power P_J. The jamming signal is synchronized with the desired signal both in frequency and in the initial phase. Under the assumptions of (a) accurate PN code synchronization and carrier phase synchronization at the receiver and (b) rectangular chip pulses, find the signal-to-interference ratio (SIR) at the demodulator output in the absence of background noise.

Solution Let P_d denote the received power of the desired signal, T_b the symbol interval, and T_c the chip interval. The received signal plus interference can be represented as

$$r(t) = \sqrt{2P_d}d(t)a(t)\cos(2\pi f_c t + \theta) + \sqrt{2P_J}\cos(2\pi f_c t + \theta),$$

where $d(t) = \sum_i s_i \Pi\left(\frac{t-iT_b}{T_b}\right)$, s_i $(\in \{-1, +1\})$ represents the ith binary information digit, $a(t) = \sum_l a_l \Pi\left(\frac{t-lT_c}{T_c}\right)$, a_l $(\in \{-1, +1\})$ represents the lth binary chip, f_c is the carrier frequency, and θ is the carrier phase at $t = 0$. For simplicity, assume that a_l takes on the values of $+1$ and -1 with equal likelihood and all the chips are independent of each other. Without loss of generality, consider the detection of symbol s_1 over $t \in [0, T_b]$. With accurate PN code and carrier phase synchronization, the demodulator output at $t = T_b$ is

$$\begin{aligned}
&\frac{1}{T_b}\int_0^{T_b} r(t)[a(t)\cos(2\pi f_c t + \theta)]dt \\
&= \frac{1}{T_b}[\sqrt{2P_d}\int_0^{T_b} s_1 a^2(t)\cos^2(2\pi f_c t + \theta)dt + \sqrt{2P_J}\int_0^{T_b} a(t)\cos^2(2\pi f_c t + \theta)dt] \\
&\approx \frac{1}{2}\sqrt{2P_d}s_1 + \frac{1}{2T_b}\sqrt{2P_J}\int_0^{T_b} a(t)dt \\
&= (\sqrt{P_d/2})s_1 + \frac{1}{2G_p}\sqrt{2P_J}\sum_{l=1}^{G_p} a_l,
\end{aligned}$$

where $G_p = T_b/T_c$ (assumed to be an integer) is the processing gain. The desired signal component, $(\sqrt{P_d/2})s_1$, has a power of $P_d/2$. The interference component due to the jamming signal, $\frac{1}{2G_p}\sqrt{2P_J}\sum_{l=1}^{G_p} a_l$, has a power of

$$E\left\{\left[\frac{1}{2G_p}\sqrt{2P_J}\sum_{l=1}^{G_p} a_l\right]^2\right\} = \frac{P_J}{2}\left(\frac{1}{G_p}\right)^2 E\left[\left(\sum_{l=1}^{G_p} a_l\right)^2\right]$$

$$= \frac{1}{2G_p}P_J.$$

The SIR at the demodulator output is

$$\frac{P_d/2}{P_J/(2G_p)} = G_p(P_d/P_J).$$

That is, the SIR is increased by the processing gain G_p, as compared with the SIR at the demodulator input. Note that, without spread spectrum, the SIR at the demodulator output would remain the same as that at the demodulator input.

Example 6.3 Transmission Performance in Multiple Access Interference

Consider the forward link transmission of a single-cell DS-CDMA system using BPSK, where the multiple access interference due to other users in the cell is synchronized with the desired signal both in chip timing and in carrier phase. It is assumed that (a) all the PN code sequences are independent of each other, and chip values "+1" and "−1" in each sequence are equally likely and are independent of each other; (b) rectangular pulses are used for the spreading waveforms, (c) the receiver uses coherent detection, (d) the number of users in the cell is large, and (e) the received signal power levels from all the mobiles are the same at the base station receiver. Derive the probability of bit error in an AWGN channel in terms of the processing gain G_p, the number K of users in the cell, the bit energy E_b of the received signal, and the one-sided noise psd N_0.

Solution Let K denote the number of mobile users in the system. Consider the detection process at mobile user 1's receiver. The received signal $r(t)$ is

$$r(t) = \sqrt{2P_d}\left[a_1(t)d_1(t) + \sum_{k=2}^{K} a_k(t)d_k(t)\right]\cos(2\pi f_c t + \theta_0) + w(t),$$

where P_d is the desired signal power of the kth user and is independent of k, $d_k(t) = \sum_i s_{k,i}\Pi\left(\frac{t-iT_b}{T_b}\right)$, $a_k(t) = \sum_l a_{k,l}\Pi\left(\frac{t-lT_c}{T_c}\right)$, $s_{k,i}$ $(\in\{-1,+1\})$ is the ith information binary digit and $a_{k,l}$ $(\in\{-1,+1\})$ is the lth chip of the PN code sequence for the kth user, $k = 1, 2, \ldots, K$, T_b and T_c are the bit and chip intervals respectively, f_c is the carrier frequency, θ_0 is the carrier phase at $t = 0$, $w(t)$ is the additive white Gaussian noise with zero mean and two-sided psd $N_0/2$ (i.e., $E[w(t)w(s)] = (N_0/2)\delta(t-s)$). Consider the signal detection over $t \in [0, T_b]$. The output of the coherent demodulator is

$$\frac{1}{T_b}\int_0^{T_b} r(t)[a_1(t)\cos(2\pi f_c t+\theta_0)]dt.$$

The output consists of three components:

(1) the desired signal component, $(\sqrt{P_d/2})s_{1,1}$, which is a constant given $s_{1,1}$;
(2) the noise component

$$(1/T_b)\int_0^{T_b} w(t)a_1(t)\cos(2\pi f_c t+\theta_0)dt,$$

which is a Gaussian random variable with zero mean and variance

$$E\left[\frac{1}{T_b^2}\int_0^{T_b}\int_0^{T_b} w(t)w(s)a_1(t)a_1(s)\cos(2\pi f_c t+\theta_0)\cos(2\pi f_c s+\theta_0)dtds\right]$$

$$=\frac{1}{T_b^2}\int_0^{T_b}\int_0^{T_b} E[w(t)w(s)]E[a_1(t)a_1(s)]\cos(2\pi f_c t+\theta_0)\cos(2\pi f_c s+\theta_0)dtds$$

$$=\frac{1}{T_b^2}\int_0^{T_b}\int_0^{T_b} \frac{N_0}{2}\delta(t-s)E[a_1(t)a_1(s)]\cos(2\pi f_c t+\theta_0)\cos(2\pi f_c s+\theta_0)dtds$$

$$=\frac{N_0}{2T_b^2}\int_0^{T_b} E[a_1^2(t)]\cos^2(2\pi f_c t+\theta_0)dt$$

$$\approx\frac{N_0}{4T_b};\text{ and}$$

(3) the intracell interference

$$(\sqrt{P_d/2})(1/G_p)\sum_{k=2}^{K} s_{k,1}\sum_{l=1}^{G_p} a_{1,l}a_{k,l},$$

where $G_p = T_b/T_c$ (assumed to be an integer) is the processing gain. If $s_{k,1}$ takes on the values of -1 and $+1$ with equal likelihood, and is independent for different k values, $s_{k,1}\sum_{l=1}^{G_p} a_{1,l}a_{k,l}$ are iid random variables for different k values. If $K \gg 1$, then based on the central limit theorem, the interference component can be modeled as a Gaussian random variable with zero mean and variance

$$\frac{P_d}{2G_p^2}E\left\{\left[\sum_{k=2}^{K} s_{k,1}\sum_{l=1}^{G_p} a_{1,l}a_{k,l}\right]\left[\sum_{k'=2}^{K} s_{k',1}\sum_{l'=1}^{G_p} a_{1,l'}a_{k',l'}\right]\right\}$$

$$=\frac{P_d}{2G_p^2}\sum_{k=2}^{K} E[s_{k,1}^2]\sum_{l=1}^{G_p} E[a_{1,l}^2]E[a_{k,l}^2]$$

$$=\frac{(K-1)P_d}{2G_p}.$$

In summary, the demodulator output is a Gaussian random variable with mean $(\sqrt{P_d/2})s_{1,1}$ and variance $\frac{N_0}{4T_b} + \frac{(K-1)P_d}{2G_p}$, given that $s_{1,1}$ was sent. Similar to the derivation of the BER for coherent BPSK in an AWGN channel in Section 3.5, the probability of bit error is

$$P_b = Q\left(\frac{\sqrt{P_d/2}}{\sqrt{\frac{N_0}{4T_b} + \frac{(K-1)P_d}{2G_p}}}\right)$$

$$= Q\left(\sqrt{\frac{2E_b}{N_0 + 2(K-1)E_b/G_p}}\right),$$

where E_b $(= P_d T_b)$ is the received signal bit energy. If there is no multiple access interference (i.e., $K = 1$), the probability of error is the same as BPSK in AWGN with coherent detection as given in Section 3.5. On the other hand, if there is no background noise (i.e., $N_0 = 0$), then the probability of error is $Q(\sqrt{G_p/(K-1)})$, where the effective signal-to-interference ratio is the product of the processing gain G_p and the ratio of the desired signal power P_d to the total interference power $(K-1)P_d$. Given the background noise, the maximum number of users allowed in the system, K, can be determined from the transmission accuracy requirement.

Characteristics of CDMA. Based on the preceding analysis of the DS-CDMA system, we can summarize some properties of DS-CDMA as follows:

a. *Universal frequency reuse* – As CDMA achieves the orthogonality among the transmitted signals from the mobile users by using the orthogonal, or approximately orthogonal, PN sequences in spreading the signals, the total frequency bandwidth allocated to the system can be reused from cell to cell. As a result, we achieve the minimum cell cluster size ($N = 1$) and maximum frequency reuse. This significantly reduces the complexity of frequency planning in cellular system design;
b. *Soft handoff* – Because of the universal frequency reuse, a mobile user can simultaneously communicate with several nearby base stations using the same frequency band and the same spreading signal in each link. When the mobile user is at the cell boundary, it can establish a connection with the new base station before terminating the connection with the old base station. This will improve handoff performance;
c. *High transmission accuracy* – With spread spectrum, as discussed in Chapter 4, we can use Rake receivers to mitigate the fading dispersive channel impairments and, therefore, improve transmission accuracy, especially during soft handoff;
d. *Soft capacity* – As in practice, the PN sequences are not truly orthogonal, MAI will degrade the transmission BER performance. The maximum number of users that can be supported in each cell depends on the required quality of service (QoS) and is limited by MAI, to be discussed in Subsection 6.4.3. As a result, unlike TDMA and FDMA, there is no hard limit on the number of users in each cell. During peak traffic hours, if the users can tolerate a lower QoS to a certain degree, the system can accommodate more users to satisfy the high service demands in that period;

e. *Flexibility* – As CDMA is interference limited, if a user does not transmit, it does not generate any interference with other active users and, therefore, does not use the system resources. This feature translates to a high resource utilization via statistical multiplexing for on-off voice traffic and bursty data traffic. Even though TDMA can make use of the traffic activity factor to increase resource utilization, with CDMA it is easier to implement the statistical multiplexing. In addition, CDMA has more flexibility than TDMA in supporting multimedia services (with various time-varying traffic rates).

The advantages of the CDMA systems are not achieved without paying a price. First, CDMA requires stringent power control to achieve high capacity. For example, with voice services, the cell capacity is maximized when the signals received at the base station from all the mobiles in the cell have the same minimum power level. Second, with a large processing gain, the maximum transmission rate in each code channel (using a unique PN sequence in the signal spreading) is limited as compared with TDMA. This limitation can be overcome by parallel transmissions of information from/to one mobile user, with each transmission using a unique PN sequence. Third, the CDMA systems operate at a high chip rate and require accurate PN synchronization at the receiver. The complexity of the transmitter and (Rake) receiver is higher than that of TDMA and FDMA systems.

6.4 SPECTRAL EFFICIENCY

Because of the severe channel impairments, the spectrum of the wireless channel is interference limited. Techniques commonly used to enhance the spectrum utilization in a mobile communication system include

a. data compression to reduce the transmission rate,
b. bandwidth reduction,
c. channel assignment, and
d. choice of multiple access method.

The overall spectral efficiency of a mobile communication system can be estimated based on a knowledge of

a. channel spacing in kHz,
b. cell area in km^2,
c. frequency reuse factor, and
d. multiple access scheme used.

Factors (*a*)–(*c*) are attributed to the system parameters, including the modulation method used. Note that the modulation scheme used does not depend on the choice of multiple access technology.

The spectral efficiency of a mobile communication system can be represented as a combination of two independent components: one component that depends on the system parameters, and the other component that depends on the multiple access method used. Let η_{sys} denote the spectral efficiency component that depends on the system parameters, and η_{access} denote the spectral efficiency component that depends on the multiple access scheme used. The overall system spectral efficiency for a mobile communications system, η, can then be expressed as

$$\eta = \eta_{sys} \times \eta_{access}.$$

Depending on the units used, we can define η in two different ways. The unit commonly used for η is channels/MHz/km^2 or Erlangs/MHz/km^2. Here, Erlang is a measure of the traffic load, as discussed in Section 5.5. η is represented in terms of channels/MHz per km^2 or Erlangs/MHz per km^2 in order to capture the frequency reuse in the service coverage area of the system. Accordingly, η may be defined in the following ways:

Definition 6.1

$$\eta \stackrel{\Delta}{=} \frac{\text{Total number of channels available for data in system}}{\text{(system bandwidth)(total coverage area)}} \quad \text{Channels/MHz/km}^2.$$

Definition 6.2

$$\eta \stackrel{\Delta}{=} \frac{\text{Total traffic carried by the system}}{\text{(system bandwidth)(total coverage area)}} \quad \text{Erlangs/MHz/km}^2.$$

6.4.1 FDMA Systems

Spectral Efficiency of FDMA (η_{FDMA}). The AMPS system (in America and Australia) is based on FDMA/FDD. Here, a single user occupies a single channel while the call is in progress. When a call is finished or handed off to another base station, the channel is vacated so that another subscriber may use it. The single channel is actually two simplex channels which are frequency duplexed, as shown in Figure 6.6(a), with a 45 MHz split. A guard band of width B_g is used in each of the edges, as shown in Figure 6.10.

The number of channels, N_s, that can be simultaneously supported is

$$N_s = \frac{B_s - 2B_g}{B_c},$$

where B_s is the total frequency spectrum bandwidth for transmissions in one direction (uplink or downlink). The above can be rearranged as

$$B_s = N_s B_c + 2B_g. \tag{6.4.1}$$

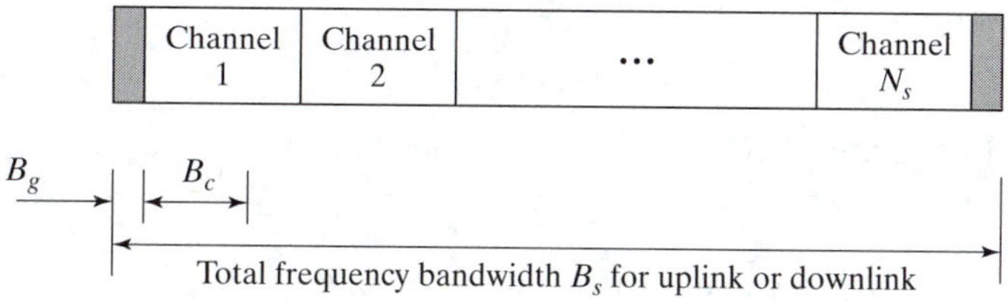

Channel 1	Channel 2	...	Channel N_s

B_g B_c

Total frequency bandwidth B_s for uplink or downlink

Figure 6.10 Channel spacing and guard bands in FDMA uplink or downlink.

Let N_{ctl} be the number of allocated control channels and N_{data} be the number of data channels in the system. Then the total number of available channels is

$$N_s = N_{data} + N_{ctl}. \tag{6.4.2}$$

Since each user in service is assigned a data channel, N_{data} is also the maximum number of simultaneous users in each cell cluster. Substituting Eq. (6.4.2) into Eq. (6.4.1), we get

$$B_s = N_{data}B_c + N_{ctl}B_c + 2B_g,$$

from which we have the inequality

$$N_{data}B_c < B_s,$$

where $N_{data}B_c$ is the total bandwidth available for data transmission.

As discussed in Chapter 5, to increase the system capacity, the entire cellular array of cells is partitioned into clusters. All available frequencies, or radio channels, are allocated to the cells in a single cluster, and the same frequencies are then reused in each and every cluster. For the purpose of discussing spectral efficiency, we will consider one cluster as the system. The spectral efficiency of FDMA is defined as

$$\eta_{FDMA} = \frac{\text{bandwidth available for data transmission}}{\text{system bandwidth}} = \frac{N_{data}B_c}{B_s} < 1. \tag{6.4.3}$$

Example 6.4 Spectral Efficiency of FDMA

In the AMPS system, the system bandwidth is 12.5 MHz, the channel spacing is 30 kHz, and the edge guard spacing is 10 kHz. The number of channels allocated for control signaling is 21. Find

a. the number of channels available for message transmission, and
b. the spectral efficiency of FDMA.

Solution We have $B_s = 12.5$ MHz, $B_c = 30$ kHz, and $B_g = 10$ kHz. Therefore,

a. the number of available channels is

$$N_s = \frac{B_s - 2B_g}{B_c} = \frac{12.5 \times 1000 - 20}{30} = 416 \text{ channels, and}$$

b. the spectral efficiency of this FDMA system is

$$\eta_{FDMA} = \frac{30 \times (416 - 21)}{12.5 \times 1000} = 0.948.$$

System Spectral Efficiency. To express η_{sys} in terms of mathematical symbols, we will use the following notation:

$B_s =$ system bandwidth in MHz

$B_c =$ channel spacing in MHz

$$
\begin{aligned}
B_g &= \text{the guard-band bandwidth in MHz} \\
N &= \text{cluster size (i.e., the number of cells in a cluster)} \\
&= \text{the frequency reuse factor} \\
N_{ch/cell} &= \text{number of available channels per cell} \\
N_{data/cell} &= \text{number of available data channels per cell} \\
N_{ctl/cell} &= \text{number of control channels per cell} \\
N_{ch/cluster} &= \text{number of available channels per cluster} \\
N_{data/cluster} &= \text{number of available data channels per cluster} \\
N_{ctl/cluster} &= \text{number of control channels per cluster} \\
A_{cell} &= \text{cell area in km}^2
\end{aligned}
$$

As discussed in detail in Chapter 5, frequency reuse is employed to increase the capacity of the entire cellular system. This is achieved by allocating the entire set of available frequencies to a single cluster. Therefore, the number of available channels per cluster is given by

$$N_{ch/cluster} = \frac{B_s - 2B_g}{B_c}.$$

Of the $N_{ch/cluster}$ channels in the cluster, $N_{ctl/cluster}$ are allocated as control channels. The total number of channels available for data traffic per cluster is thus given by

$$N_{data/cluster} = N_{ch/cluster} - N_{ctl/cluster} = \frac{B_s - 2B_g}{B_c} - N_{ctl/cluster}. \qquad (6.4.4)$$

The total number of channels available for data traffic per cell is given by

$$N_{data/cell} = \frac{N_{data/cluster}}{N} = \frac{\dfrac{B_s - 2B_g}{B_c} - N_{ctl/cell}}{N}, \qquad (6.4.5)$$

which is also the cell capacity, N_c, defined as the maximum number of mobile stations that can be served at one time in each cell.

Therefore, from Definition 6.1, we can express η as

$$\eta = \frac{\text{number of data channels per cluster}}{\text{system bandwidth times area of the cluster}} = \frac{N_{data/cluster}}{B_s \times (N \times A_{cell})}. \qquad (6.4.6)$$

Substituting Eq. (6.4.4) into Eq. (6.4.6), we get

$$\eta = \frac{1 - \dfrac{B_c}{B_s} \times \left(N_{ctl/cluster} + \dfrac{2B_g}{B_c}\right)}{B_c \times N \times A_{cell}} \text{ channels/MHz/km}^2. \qquad (6.4.7)$$

Equation (6.4.7) can be rearranged to yield the following form

$$\eta = \frac{1}{B_c \times N \times A_{cell}} - \frac{N_{ctl/cluster} + \dfrac{2B_g}{B_c}}{B_s \times N \times A_{cell}} \text{ channels/MHz/km}^2. \quad (6.4.8)$$

In Eq. (6.4.8), the second term on the right-hand side accounts for the overhead in FDMA, due to the guard bands and control channels.

In any multiple access system there is a finite probability that some of the access traffic is blocked. Let η_t be the trunk efficiency in each cell, which is a function of the blocking probability and the total number of available channels per cell, $N_{data/cell}$, as discussed in Section 5.5. Hence, the total traffic carried in a cluster, in Erlangs, is $\eta_t \times N_{data/cluster}$. Using Definition 6.2, we can express η as

$$\eta = \frac{\eta_t \times N_{data/cluster}}{B_s \times N \times A_{cell}} \text{ Erlangs/MHz/km}^2.$$

Substituting Eq. (6.4.4) for $N_{data/cluster}$ in the above equation, we have

$$\eta = \frac{\eta_t}{B_c \times N \times A_{cell}} - \frac{\eta_t \times \left(N_{ctl/cluster} + \dfrac{2B_g}{B_c}\right)}{B_s \times N \times A_{cell}} \text{ Erlangs/MHz/km}^2, \quad (6.4.9)$$

where the second term on the right-hand side is due to the overhead in FDMA.

Example 6.5 System Spectral Efficiency in Channels/MHz/km^2

Suppose a cellular system in which the one-way bandwidth of the system is 12.5 MHz, the channel spacing is 30 kHz, and the guard band at each boundary of the spectrum is 10 kHz. If (1) the cell area is 6 km^2, (2) the frequency reuse factor (cluster size) is 7, and (3) 21 of the available channels are used to handle control signaling, calculate

a. the total number of available channels per cluster,
b. the number of available data channels per cluster,
c. the number of available data channels per cell, and
d. the system spectral efficiency in units of channels/MHz/km^2.

Solution We allocate all of the available frequencies to one cluster and these frequencies, or channels, are distributed evenly among the N cells in the cluster.

a. The total number of available channels in the cluster is

$$N_{ch/cluster} = \frac{B_s - 2B_g}{B_c} = \frac{12.5 - 2 \times 0.01}{0.03} = 416.$$

b. The number of available data channels per cluster is

$$N_{data/cluster} = N_{ch/cluster} - N_{ctl/cluster} = 416 - 21 = 395.$$

c. The number of available data channels per cell is

$$N_{data/cell} = N_{data/cluster}/N = 395/7 \approx 56.$$

d. The overall spectral efficiency of the system is

$$\eta = \frac{N_{data/cell}}{B_s \times A_{cell}} = \frac{56}{12.5 \times 6} = 0.747 \text{ channels/MHz/km}^2.$$

Example 6.6 System Spectral Efficiency in Erlangs/MHz/km^2

Suppose that the system parameter values are the same as those in Example 6.5. In addition, there are the following other specifications:

(4) The area of the entire cellular system is 3024 km^2
(5) The average number of calls per user during a busy hour is 1.5
(6) The average channel holding time of a call is 180 s
(7) The trunk efficiency, η_t, is 0.95

Calculate the following parameter values:

a. the number of cells in the system
b. the number of calls per hour per cell
c. the average number of users per hour per cell
d. the system spectral efficiency in Erlangs/MHz/km^2

Solution

a. The number of cells in the system = $\frac{3024}{6} = 504$.
b. The number of calls per hour per cell, N_{call}, is

$$N_{call} = \frac{N_{data/cluster}}{N} \times \eta_t \times \text{ number of calls per hour}$$
$$= 56 \times 0.95 \times \frac{3600}{180}$$
$$= 1064 \text{ calls/hour/cell.}$$

c. The average number of users per hour per cell, N_{users}, is

$$N_{users} = \frac{\text{number of calls per hour per cell}}{\text{average number of calls per user per hour}}$$
$$= 1064/1.5$$
$$\approx 709 \text{ users/hour/cell.}$$

d. The system spectral efficiency is, from Example 6.5,

$$\eta = \eta_t \times 0.747$$
$$= 0.710 \text{ Erlangs/MHz/km}^2.$$

6.4.2 TDMA Systems

TDMA can operate as wideband or narrowband. In wideband TDMA (W-TDMA), the entire frequency spectrum is available to any individual user. In narrowband TDMA (N-TDMA), the total available frequency spectrum is divided into a number of subbands, with each subband operating as a TDMA system. An individual user only uses the allocated subband. Thus, in narrowband TDMA, both frequency and time are partitioned.

Spectral Efficiency of Wideband TDMA (η_{W-TDMA}). The spectral efficiency of wideband TDMA, η_{W-TDMA}, is defined as the percentage of the time duration used for transmitting information data symbols in each frame. For the frame structure shown in Figure 6.5, let

$$\begin{aligned}
\tau_p &= \text{the time duration for the preamble} \\
\tau_t &= \text{the time duration for the trailer} \\
T_f &= \text{the frame duration} \\
L_d &= \text{the number of information data symbols in each slot} \\
L_s &= \text{the total number of symbols in each slot.}
\end{aligned}$$

Then, we have

$$\eta_{W-TDMA} = \frac{T_f - \tau_p - \tau_t}{T_f} \times \frac{L_d}{L_s}. \tag{6.4.10}$$

In the preceding equation, the first term on the right-hand side takes into account the overhead at the frame level (due to the frame header/trailer), and the second term takes into account overhead at the slot level (due to the trailer bits, synchronization bits, and guard bits, as shown in Figure 6.5). The overhead corresponds to the overhead due to control channels in FDMA, which are necessary to coordinate the multiple access.

Spectral Efficiency of Narrowband TDMA (η_{N-TDMA}). As mentioned earlier, in narrowband TDMA, the system bandwidth is divided into a number of subbands, and each subband is partitioned into time slots. Let

$$\begin{aligned}
B_c &= \text{the bandwidth of an individual user} \\
N_u &= \text{the number of subbands} \\
B_g &= \text{the guard spacing.}
\end{aligned}$$

Then, the number of subbands is

$$N_u = \frac{B_s - 2B_g}{B_c}.$$

The channel time in each subband is divided into time slots, numbered $1, 2, \cdots, N_{\text{slot}}$, as shown in Figure 6.11. In this way, any one of the numbered time slots, say slot 1, can be used by N_u users. Therefore, the actual usable bandwidth for information transmission is $N_u B_c$ $(= B_s - 2B_g)$.

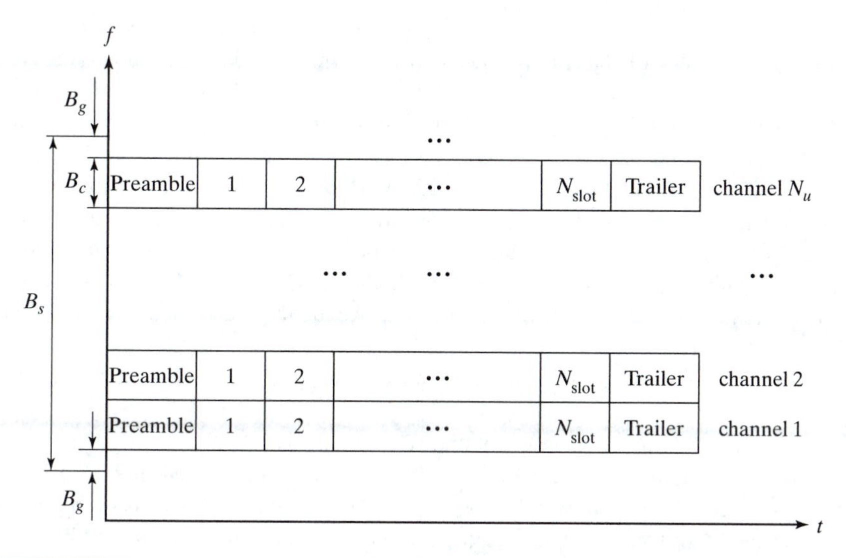

Figure 6.11 Narrowband TDMA format.

The spectral efficiency of narrowband TDMA is proportional to that of wideband TDMA. The proportionality constant is the ratio of the available information transmission bandwidth to the system bandwidth. Thus, the proportionality constant is

$$\alpha = \frac{N_u B_c}{B_s} \leq 1,$$

since $N_u B_c \leq B_s$. The spectral efficiency of narrowband TDMA is then given by

$$\begin{aligned} \eta_{N-TDMA} &= \eta_{W-TDMA} \times \frac{B_c N_u}{B_s} \\ &= \frac{T_f - \tau_p - \tau_t}{T_f} \times \frac{L_d}{L_s} \times \frac{B_s - 2B_g}{B_s}. \end{aligned} \tag{6.4.11}$$

Cell Capacity of TDMA Systems. The cell capacity is defined as the maximum number of mobile users that can be supported simultaneously in each cell. With TDMA, the maximum number of simultaneous users that can be accommodated during one use of the available frequency spectrum is

$$N_s = N_u \times N_{\text{slot}}, \tag{6.4.12}$$

where

$$N_u = \begin{cases} 1, & \text{for W-TDMA} \\ \dfrac{B_s - 2B_g}{B_c} > 1, & \text{for N-TDMA} \end{cases}.$$

With the same total available bandwidth, B_s, the bandwidth of an individual user, B_c, in W-TDMA is approximately N_u times of that in N-TDMA. If the same modulation scheme is used, then the transmission bit rate, R_b, in W-TDMA is N_u times that in N-TDMA. Therefore, with the same source traffic characteristics from mobile users, the number of time slots, N_{slot}, in each frame in W-TDMA is actually N_u times that in N-TDMA. As a result, the maximum number of simultaneous users in W-TDMA is the same as that in N-TDMA. However, in practical implementation, due to the difference in transmission rate, W-TDMA and N-TDMA have different pros and cons.

N_s is the total number of TDMA channels available for the entire cellular system without frequency reuse. With frequency reuse, all the N_s channels can be allocated to a single cell cluster. Then, the cell capacity, N_c, in a TDMA system with frequency reuse factor N is

$$N_c = \frac{N_u \times N_{\text{slot}}}{N}.$$

In general, the source stream of a user is unlikely to be always in an *on* state. Let s_f denote the source activity factor, defined as the percentage of time that a connected mobile user is actually generating information data for transmission. For example, in voice transmission the voice stream alternates between a talk spurt (*on*) state and a silence (*off*) state. If, on average, 40% of the time it stays in the *on* state and 60% of the time in the *off* state, then $s_f = 0.4$. During the silence interval, the capacity can be used by other active users (in ATDMA). Therefore, the cell capacity of the system, N_c, should be modified by the source activity factor to represent the number of connected users per cell. Assuming perfect statistical multiplexing, the cell capacity of a TDMA system is given by

$$N_c = \frac{N_u \times N_{\text{slot}}}{s_f \times N} \text{ users/cell.} \tag{6.4.13}$$

In a synchronous TDMA system where the resources corresponding to the *off* state of a source stream are not allocated for use by other users, the source activity factor equals 1 (i.e., $s_f = 1$).

System Spectral Efficiency. The cell capacity depends on the number of slots, N_{slot}, which is a function of factors that govern the efficiency of the system usage, including the modulation scheme used. The overall frequency efficiency of a TDMA system in bits/s/Hz/cell is derived in the following.

The maximum number of radio channels in an N-TDMA system is given by the system bandwidth divided by the bit rate of each user. If each user transmits at a basic bit rate R_b and the system bandwidth is B_s, then the maximum number of users that can be supported during one use of the system bandwidth is

$$N_s = \frac{B_s}{R_b},$$

under the assumption that the required bandwidth for transmitting 1 bps is 1 Hz. For example, if we use QPSK and the transmission bandwidth is equal to the width of the main psd lobe as shown in Figure 3.23 (bandpass filtering may be necessary to suppress out-of-band radiation), with a frequency reuse factor of N, the cell capacity is

$$N_c = \frac{N_s}{N}.$$

This value is reduced because of inefficiency in both bandwidth usage due to the guard bands and modulation scheme. Let ϵ_{bw} denote the modulation efficiency, defined as the maximum transmission rate in bits per second that the modulation can accommodate over one hertz bandwidth. For example, if BPSK is used and the transmission bandwidth is equal to the width of the main psd lobe as shown in Figure 3.23 (bandpass filtering may be necessary to suppress out-of-band radiation), then $\epsilon_{bw} = 0.5$. The effective number of users, $N_{effective}$, that can be supported per cell is N_c, modified by the modulation efficiency factor. Therefore,

$$N_{effective} = \epsilon_{bw} \times \frac{B_s}{R_b \times N}.$$

If we consider the overhead necessary for TDMA, from Eq. (6.4.11), the effective number should be further modified to

$$N_{effective} = \epsilon_{bw} \times \frac{T_f - \tau_p - \tau_t}{T_f} \times \frac{L_d}{L_s} \times \frac{B_s - 2B_g}{R_b \times N},$$

where, for a W-TDMA system, $B_g = 0$. In fact, there may exist guard bands in a W-TDMA system. The overhead due to the guard bands can be captured in the modulation efficiency ϵ_{bw}. Under the assumption of perfect statistical multiplexing, a channel corresponds to a constant bit rate of R_b. The overall spectral efficiency of the system in bits/unit time/unit bandwidth/cell, η, can be expressed as

$$\eta = N_{effective} \times \frac{R_b}{B_s}$$

$$= \epsilon_{bw} \times \frac{T_f - \tau_p - \tau_t}{T_f} \times \frac{L_d}{L_s} \times \frac{B_s - 2B_g}{B_s} \times \frac{1}{N} \text{ bits/s/Hz/cell.} \qquad (6.4.14)$$

Example 6.7 Spectral Efficiency of the IS-54 System

Consider IS-54 (updated as IS-136), which is a synchronous N-TDMA/FDD system that uses a one way bandwidth of 25 MHz for the forward (or reverse) channel. The system bandwidth is divided into radio channels of 30 kHz, each supporting transmission at a rate of 16.2 kbps. Guard bands with $B_g = 20$ kHz are used. The frame duration is 40 ms, consisting of 6 time slots. A single radio channel supports 3 full-rate speech channels, each using 2 slots in a frame. Each slot consists of 324 bits, among which 260 bits are for information data and the remaining 64 bits are overhead for access control. The speech codec rate is 7.95 kbps, which corresponds to a gross bit rate of 13.0 kbps with channel encoding. If the frequency reuse factor is 7, find

a. the number of simultaneous users that can be accommodated in each cell cluster,
b. the cell capacity,
c. the spectral efficiency η_{N-TDMA} of TDMA, and
d. the overall spectral efficiency.

Solution Given:

$$B_s = 25 \text{ MHz} = 25000 \text{ kHz}$$

$$B_c = 30 \text{ kHz}$$

$$B_g = 20 \text{ kHz}$$
$$R_b = 7.95 \text{ kbps}$$
$$N_{\text{slot}} = 6/2 = 3$$
$$T_f = 40 \text{ ms}$$
$$\tau_s = 0 \text{ ms}$$
$$\tau_t = 0 \text{ ms}$$
$$L_d = 260 \text{ bits}$$
$$L_s = 324 \text{ bits}$$
$$\epsilon_{bw} = 7.95 \text{ kbps}/30 \text{ kHz} = 0.265$$
$$s_f = 1$$
$$N = 7$$

a. The number of simultaneously transmitting users that can be accommodated in each cell cluster is

$$N_u = \frac{B_s - 2B_g}{B_c} = \frac{25 \times 1000 - 2 \times 20}{30} = 832 \text{ users/cell cluster.}$$

b. The cell capacity of the synchronous TDMA system is

$$N_c = \frac{N_u N_{\text{slot}}}{s_f N} = \frac{832 \times 3}{1 \times 7} \approx 356 \text{ users/cell.}$$

c. The spectral efficiency of the N-TDMA is

$$\eta_{N-TDMA} = \eta_{W-TDMA} \times \frac{B_c N_u}{B_s} = \frac{T_f - \tau_p - \tau_t}{T_f} \times \frac{L_d}{L_s} \times \frac{B_s - 2B_g}{B_s} \approx 0.8.$$

d. The overall spectral efficiency is

$$\begin{aligned}\eta &= \epsilon_{bw} \times \frac{T_f - \tau_p - \tau_t}{T_f} \times \frac{L_d}{L_s} \times \frac{B_s - 2B_g}{B_s} \times \frac{1}{N} \\ &= 0.265 \times \frac{40 - 0 - 0}{40} \times \frac{260}{324} \times \frac{25000 - 2 \times 20}{25000} \times \frac{1}{7} \\ &\approx 0.0303 \text{ bits/s/Hz/cell.}\end{aligned}$$

6.4.3 DS-CDMA Systems

Before attempting to define the spectral efficiency of a DS-CDMA system, let us first consider the cell capacity.

Cell Capacity of DS-CDMA Systems

Definition 6.3 The cell capacity, N_c, of a DS-CDMA system is defined as the maximum number of mobile stations that can be supported during one use of the wireless channel in a single cell, under the constraint that quality of service requirements are met.

The cell capacity of a DS-CDMA system is a function of many system-related factors, as follows:

E_b = energy of transmitted signal per information bit
I_0 = one-sided interference-plus-noise power spectral density
P_n = background noise power
S = signal power received at the cell-site receiver
G_p = signal processing gain
η_f = frequency reuse efficiency (to be defined)
c_d = capacity degradation factor due to imperfect power control
Q = number of cell sectors
s_f = source activity factor

As a function of the preceding parameters, the number of mobile stations, N_{MS}, that can be supported by a DS-CDMA system can be expressed as

$$N_{MS} = 1 + \frac{c_d \eta_f}{s_f}\left[\frac{QG_p}{E_b/I_0} - \frac{P_n}{S}\right]. \tag{6.4.15}$$

Derivation of Eq. (6.4.15). Let us now derive the expression for N_{MS} as given by Eq. (6.4.15). Consider a single base station supporting multiple mobile stations, as shown in Figure 6.12. We will first obtain an expression for the number of mobile stations as a function of E_b/I_0, on the basis that (*a*) each base station uses an omnidirectional antenna, (*b*) there is perfect power control, and (*c*) sources are persistently transmitting information.

Under the assumption that the uplink (from mobile users to the base station) has accurate automatic power control, the signal power received by the base station from each and every

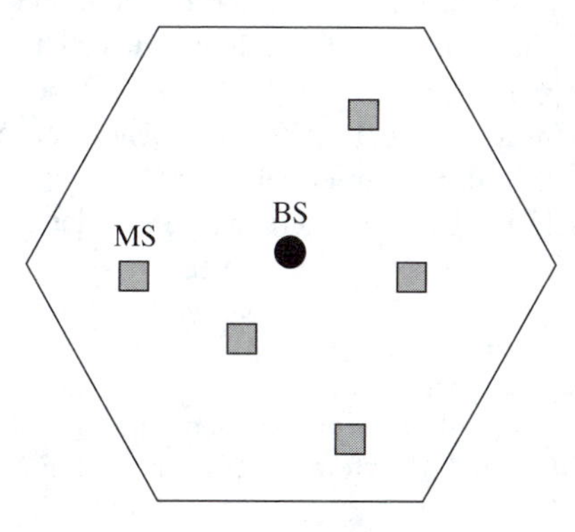

Figure 6.12 Single hexagonal cell.

mobile in the cell is the same, independent of their location within the cell. This power control is designed to circumvent the near–far problem.

The signal power received from the candidate mobile station is

$$S = R_b E_b,$$

where R_b is the bit rate. The interference power is

$$I = B_s I_0,$$

where B_s is the one-way system bandwidth. Therefore, the signal-to-interference ratio, S/I, is given by

$$\frac{S}{I} = \frac{R_b E_b}{B_s I_0}. \tag{6.4.16}$$

In the preceding equation, I_0 is the only variable. Recall that the processing gain is $G_p = B_s/R_b$. Then we can write Eq. (6.4.16) as

$$\frac{S}{I} = \frac{1}{G_p}\frac{E_b}{I_0}$$

or

$$\frac{E_b}{I_0} = G_p \frac{S}{I}. \tag{6.4.17}$$

Both S/I and E_b/I_0 represent the quality of service factor at the cell-site receiver.

Interference is a combination of intracell interference and intercell interference. In addition, there is the background noise. Consider the signal reception from a desired user, say user 1. Transmissions from the other $N_{MS}-1$ users represent intracell interference. Under the assumption of perfect power control, the signal power received from each and every user in the cell is S. Then, the intracell interference power from the $(N_{MS} - 1)$ other users is $(N_{MS} - 1)S$.

Consider now the interference from neighboring cells due to frequency reuse. In CDMA, the frequency reuse factor is 1, so that intercell interference comes from all neighbors. We can consider interference from different tiers of the cellular array. The first two tiers of interference cells are illustrated in Figure 6.13. Let $k_i, i = 1, 2, 3, \cdots$, denote the normalized interference contribution from all mobiles in a tier i cell due to frequency reuse, normalized with respect to the intracell interference from within the target cell.

The number of neighbors in the different tiers are, six in tier 1, 12 in tier 2, 18 in tier 3, etc. In general, tier i has $6i$ cells. Define the interference factor as

$$\kappa_f = 1 + 6k_1 + 12k_2 + 18k_3 + \cdots.$$

κ_f is the total intracell and intercell interference (on average) normalized to the total intracell interference. Therefore, the total interference power, including both intracell and intercell interference, is

$$I = [(N_{MS} - 1)S] \times \kappa_f.$$

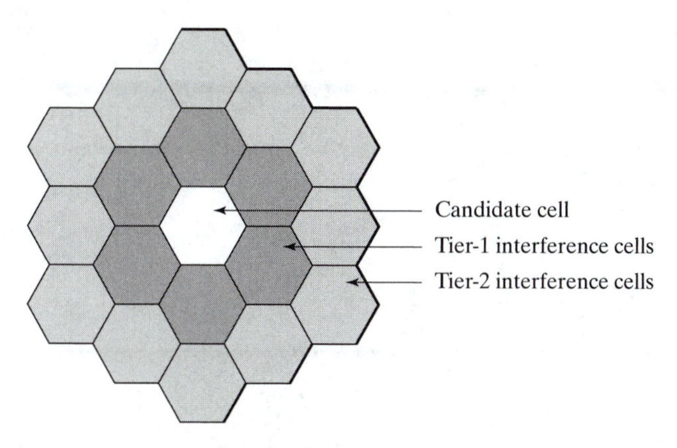

Figure 6.13 Interfering neighbors in hexagonal array.

In addition, there is always background and spurious noise. Let P_n denote the noise power. Then, the total power of interference plus noise is $[(N_{MS} - 1)S \times \kappa_f] + P_n$. The SINR can then be expressed as

$$\frac{S}{I} = \frac{1}{(N_{MS} - 1)\kappa_f + P_n/S}. \tag{6.4.18}$$

Define $\eta_f = 1/\kappa_f$ as the frequency reuse efficiency factor. Note that $\eta_f = 1$ for a single-cell system and $\eta_f < 1$ for a multiple-cell system due to intercell interference. Substituting Eq. (6.4.18) into Eq. (6.4.17), the energy-to-interference spectral density ratio can be written as

$$\frac{E_b}{I_0} = G_p \times \frac{S}{I} = \frac{G_p \times \eta_f}{(N_{MS} - 1) + \eta_f P_n/S}. \tag{6.4.19}$$

Equation (6.4.19) can be rearranged to yield an expression for N_{MS} as

$$N_{MS} = 1 + \eta_f \left[\frac{G_p}{E_b/I_0} - \frac{P_n}{S} \right]. \tag{6.4.20}$$

The derivation of Eq. (6.4.20) assumes

a. perfect automatic power control,
b. unity source activity factor (i.e., persistent transmissions),
c. the cell-site antenna has only one sector (i.e., omnidirectional antenna), and
d. every cell handles the same type of traffic.

If items (*a*) through (*c*) were to be relaxed, the number of mobile stations given in Eq. (6.4.20) would have to be modified by the capacity degradation factor c_d due to imperfect power control, source activity factor s_f, and the number, Q, of cell sectors resulting from use of directive antennas in the cell-site.

Taking into account the source activity factor, s_f, both intracell interference and intercell interference are reduced, but the background noise power remains the same. The average power of total interference and noise is now

$$I' = s_f[(N_{MS} - 1)S \times \kappa_f] + P_n.$$

With Q cell sectors resulting from use of directive antennas, the average power of interference and noise seen at the base station receiver for each sector is reduced to

$$I'' = \{s_f[(N_{MS} - 1)S \times \kappa_f] + P_n\}/Q.$$

The SINR is now S/I'', from which we get

$$N_{MS} = 1 + \frac{\eta_f}{s_f}\left[\frac{QG_p}{E_b/I_0} - \frac{P_n}{S}\right]. \tag{6.4.21}$$

With the capacity degradation factor c_d due to imperfect power control, Eq. (6.4.21) is modified to become

$$N_{MS} = 1 + \frac{c_d\eta_f}{s_f}\left[\frac{QG_p}{E_b/I_0} - \frac{P_n}{S}\right],$$

which is Eq. (6.4.15).

From Eq. (6.4.15), we see that the number of mobile stations that can be supported decreases as the E_b/I_0 value increases. The BER of any modulation system is a function of E_b/I_0. Smaller values of E_b/I_0 correspond to larger values of BER. To achieve an acceptable BER, the E_b/I_0 has to exceed a prescribed threshold. For example, with an efficient modem, powerful channel coding (using convolutional code) and two-branch diversity, to achieve satisfactory transmission quality specified as a BER of 10^{-3} on the uplink, the required E_b/I_0 is 7 dB. The cell capacity, N_c, should therefore be defined in terms of an E_b/I_0 requirement. Let $(E_b/I_0)^*$ be the QoS specification. Acceptable BER performance requires that $E_b/I_0 \geq (E_b/I_0)^*$. The cell capacity, from Definition 6.3, is then

$$N_c = 1 + \frac{c_d\eta_f}{s_f}\left[\frac{QG_p}{(E_b/I_0)^*} - \frac{P_n}{S}\right]. \tag{6.4.22}$$

N_c, as given by Eq. (6.4.22), is known as the Erlang capacity. The right-hand side of Eq. (6.4.22) is derived on the basis that each user requires the same E_b/I_0^* to satisfy the same target BER requirement under all propagation conditions, where I_0^* is the maximum total acceptable interference power spectral density and E_b is a constant. In addition, the number of users in the cell is assumed fixed. In a cellular system in which the users are in motion, the number of users in a cell at a given epoch is a random variable. The actual number of active users that can be supported in a cell will be smaller than the Erlang capacity given by Eq. (6.4.22).

Definition 6.4 The system utilization is defined as the number of users, U_{MS}, that can be supported under the constraint that $E_b/I_0 \geq (E_b/I_0)^*$.

The system utilization is then given by

$$U_{MS} = 1 + \frac{c_d\eta_f}{s_f}\left[\frac{QG_p}{E_b/I_0} - \frac{P_n}{S}\right], \quad \text{with } E_b/I_0 \geq (E_b/I_0)^*. \tag{6.4.23}$$

As CDMA systems are highly complex, the overall capacity depends on many factors, such as power control error, soft handoff, etc., whose effects on the capacity are difficult to model. As a result, the preceding derivation of the cell capacity and system utilization of the CDMA system is a first approximation.

Example 6.8 Capacity and Utilization of CDMA System

Consider a DS-CDMA system whose one-way system bandwidth is 25 MHz. Suppose the data rate per user is 8 kbps. Assume perfect power control, one sector antenna, and persistent transmissions.

a. Calculate the E_b/I_0 specification required to support a maximum number of 250 users/cell if the signal-to-background noise ratio is 26 dB and the frequency reuse efficiency is 0.9.
b. If the actual operating value of E_b/I_0 is 12 dB, calculate the system utilization for the parameter values specified in part (a).

Solution

a. With the assumptions given in the problem, the E_b/I_0 specification can be written as

$$(E_b/I_0)^* = \frac{\eta_f G_p}{N_c - 1 + \eta_f P_n/S}.$$

It is given that $S/P_n = 26$ dB or 400, $\eta_f = 0.9$, $G_p = B_s/R_b = 25 \times 1000/8 = 3125$, and $N_c = 250$. Substituting these values in the preceding equation for $(E_b/I_0)^*$ yields

$$(E_b/I_0)^* = \frac{0.9 \times 3125}{250 - 1 + 0.9/400} \approx 11.25 \text{ or } 10.5 \text{ dB}.$$

b. The utilization can be expressed as

$$U_{MS} = 1 + \eta_f \times \left[\frac{G_p}{E_b/I_0} - \frac{P_n}{S}\right], \quad \text{with } E_b/I_0 \geq 10.5 \text{ dB}.$$

If the actual operating value of E_b/I_0 is 12 dB, or 15.85, then the system utilization is

$$U_{MS} = 1 + 0.9 \times \left[\frac{3125}{15.85} - 1/400\right] \approx 178.$$

Spectral Efficiency of DS-CDMA Systems. The spectral efficiency, η, of a DS-CDMA system can be represented as a unitless quantity or in bits/unit time/unit bandwidth. It is defined as

$$\eta = \frac{U_{MS}}{N_c} = \frac{(E_b/I_0)^*}{E_b/I_0}, \qquad E_b/I_0 \geq (E_b/I_0)^* \tag{6.4.24}$$

or

$$\eta = U_{MS} \times \frac{s_f R_b}{B_s} \text{ bits/s/Hz}, \tag{6.4.25}$$

where R_b is the constant bit rate (in units of bits/s/user) when a user is in an *on* state, and B_s is the one-way system bandwidth in Hz. Note that, unlike the FDMA and TDMA systems, it is not necessary to represent the efficiency η in terms of bits/s/Hz per km^2, as in DS-CDMA the frequency spectrum is reused from cell to cell.

Example 6.9 Spectral Efficiency of DS-CDMA

Consider the CDMA system in Example 6.8, with the same assumptions about parameter values.

a. If the actual operating value of E_b/I_0 is 12 dB, find the spectral efficiency for this CDMA system.
b. Suppose the actual operating value of E_b/I_0 is now 11 dB. Find the spectral efficiency.
c. Discuss the impact of operating at a lower value of E_b/I_0 in terms of spectral efficiency and system utilization. What would be the benefit, if any, to the service provider?

Solution

a. For $E_b/I_0 = 12$ dB, the number of users that can be supported is 178 (Example 6.8), and the system capacity is 250. Therefore, the spectral efficiency is

$$\eta = 178/250 = 71.2\%$$

or

$$\eta = 178 \times \frac{8}{25 \times 1000} = 0.005696 \text{ bits/s/Hz}.$$

b. If the value of E_b/I_0 is now 11 dB or 12.59, the system utilization is

$$U_{MS} = 1 + 0.9 \times \left[\frac{3125}{12.59} - \frac{1}{400}\right] \approx 224.$$

Therefore, the spectral efficiency is

$$\eta = 224/250 = 89.6\%$$

or

$$\eta = 224 \times \frac{8}{25 \times 1000} = 0.007168 \text{ bits/s/Hz}.$$

c. A reduction in the operating value of E_b/I_0 from 12 dB to 11 dB is accompanied by an increase in spectral efficiency from 71.2% to 89.6%. With a 1 dB decrease in E_b/I_0, the system utilization is increased from 178 to 224 users. This is significant from the revenue point of view for the service provider.
With the system operating at $E_b/I_0 = 11$ dB, which is 0.5 dB above the QoS requirement, the resultant BER would be acceptable to the mobile users.

SUMMARY

Wireless communication systems are characterized by multiple access in the uplink (reverse channel) and broadcast in the downlink (forward channel). The capacity of the system, in terms of the number of users that can be supported during any one use of the wireless channel, depends on the multiple access technology used. The commonly used multiple access technologies are of the conflict-free type or, at worst, exhibit minimal interference. These technologies include FDMA, TDMA and CDMA. FDMA is narrowband and has been used as the multiple access technology in the first generation wireless standards. TDMA, which can be narrowband or wideband, is the multiple access technology of the second generation wireless standards. CDMA, an interference-limited multiple access technology, has been used as a second generation wireless standard (IS-95). This chapter has discussed the implementation issues and the spectral efficiency of each of these three multiple access techniques. As random access is often used for users to gain initial access to conflict-free multiple access systems, two random access schemes (i.e., Aloha and CSMA) have also been studied.

The first and second generation wireless systems mainly support voice communications. The information rate of these systems, as tabulated in Tables 1.1 and 1.2, are relatively low. Third generation wireless systems are expected to support multimedia services, with transmission rates up to 2 Mbps. CDMA has been targeted as the multiple access technology for the third generation wireless standards.

ENDNOTES

1. For key developments in the theory and practice of multiple user communication channels, see the book edited by Abramson [3]. In particular, for Aloha random access, see the papers by Abramson [1, 2] and by Namislo [104]; for CSMA, see the paper by Kleinrock and Tobagi [73]; for TDMA, see the book chapter by Campanella and Schaefer [21] and papers by Dill [38] and by Falconer, Adachi and Gudmundson [49].
2. For principles of spread spectrum systems, see the papers by Scholtz [135] and by Pickholtz, Schilling, and Milstein [116], and textbooks by Dixon [40] and by Peterson, Ziemer, and Borth [114].
3. As the capacity of CDMA systems is interference limited, an essential approach to achieve maximal capacity is multiuser detection, where information about multiple users is used to improve detection of each individual user [100]. The main challenge in multiuser detection is the complexity of the optimal receiver, which increases exponentially with the number of active users and with the delay spread of the channel. Various techniques have been proposed to reduce the receiver complexity while achieving near-optimum performance. For details, see the book by Verdú [154], the special issue of the *IEEE Journal on Selected Areas in Communications* [28], and papers by Divsalar, Simon and Raphaeli [39], Honig and Xiao [64], Müller [96], and Tse and Hanly [152].
4. In Subsection 6.4.3 on the cell capacity of DS-CDMA systems, the goal of accurate automatic power control is to equalize the received signal power levels from all the mobile users in the cell in order to increase the cell capacity. Power equalization (also referred to as channel

inversion) is common in spread spectrum systems to overcome the near–far problem and is simple to implement. However, the approach may not achieve the maximum capacity in extreme fading (such as Rayleigh fading) environments. When side information about the current channel state is available to both the transmitter and receiver, the optimal adaptive transmission scheme should use water-pouring in time for power adaptation. For details, see the papers by Goldsmith and Varaiya [55] and by Shamai and Verdú [138].

5. In DS-CDMA systems, spread spectrum is achieved by modulation using PN sequences. Prior to data transmission, the receiver needs to accurately synchronize the locally generated PN sequence waveform with the incoming PN sequence waveform. As a result, PN sequence synchronization is an essential part of CDMA systems. The synchronization is usually achieved in two steps (i.e., acquisition and tracking). The acquisition process is to achieve coarse alignment within some fraction of one chip interval between the two PN waveforms, and the tracking process is to achieve fine alignment between the PN waveforms which further reduces synchronization error to an allowed limit. For details, see the papers by Polydoros and Weber [118, 119], Sourour and Gupta [144, 145], Spilker, Jr. and Magill [67], Meyr [97], and Sheen and Stuber [137].
6. For CDMA in wireless communications, see the book by Viterbi [155] and papers by Kohno, Meidan and Milstein [75] and Lee [80]. For further details on Erlang capacity in CDMA systems, see the papers by Gilhousen *et al.* [54] and the book by Viterbi [155]; for spectrum efficiency, see the paper by Lee [79]; and for power control and power allocation, see the papers by Ariyavistakul and Chang [8], by Sampath, Kumar and Holtzman [134], and by Mark and Zhu [94].
7. The three pure multiple access schemes can be combined for hybrid multiple access, such as hybrid TDMA/CDMA and multi-carrier (MC)-CDMA. In hybrid TDMA/CDMA, statistical multiplexing can be achieved in both time and code domains [65, 66, 6]. Time is partitioned into frames and further into slots as in TDMA. Within each time slot, multiple packet transmissions are possible by using orthogonal PN codes as in CDMA. There are two types of MC-CDMA [58, 74]. One is a combination of orthogonal frequency division multiplexing (OFDM) and CDMA, where the spread sequence is serial-to-parallel converted and each chip modulates a different carrier frequency. The other is a parallel transmission scheme of narrowband DS waveforms in the frequency domain, where the available frequency spectrum is divided into a number (typically much smaller than the processing gain) of equal-width frequency bands and each frequency band is used to transmit narrowband DS-CDMA signals.
8. For the evolution toward wideband CDMA in third generation wireless communication systems, see the paper by Prasad and Ojanperä [121].
9. For in-depth analysis of CDMA system performance, see the textbook by Stüber [147].

PROBLEMS

P6-1 Verify that, with orthogonal spreading sequences, transmission accuracy in AWGN with spread spectrum modulation is the same as that without spread spectrum, assuming that coherent BPSK is used.

P6-2 Random access is a contention medium access control protocol used in the Aloha systems.

a. Show that the maximum throughput of pure Aloha and slotted Aloha are given respectively by Eqs. (6.2.6) and (6.2.7).
b. Plot the throughput S versus offered load G graphs for both pure Aloha and slotted Aloha, and comment on the properties characterized by the $S \sim G$ curves.
c. Plot the delay versus throughput curves for pure Aloha and slotted Aloha, and describe the behavior displayed by these curves.

P6-3 The tree search for the collision resolution algorithm described in Example 6.1 can be implemented by a software program to facilitate the handling of large user populations.

a. Write a program to implement the collision resolution algorithm.
b. Test your program for correctness by calculating the length of the collision resolution interval in Example 6.1, and then do the same for a larger population, say 100 users, for the ith collision resolution interval by assuming that a subset of the users have packets ready for transmission in the ith collision resolution interval.

P6-4 CSMA is a contention access protocol with partial coordination among the multiple users.

a. Explain why and how CSMA should provide better throughput performance than the Aloha systems. Plot the throughput curves and derive your explanation from a comparison of the characteristics displayed by the curves.
b. Compare the delay-throughput characteristics of Aloha and CSMA. The implementation of CSMA is slightly more complex compared to Aloha. At the same throughput, say 0.3, which protocol (slotted Aloha or unslotted CSMA) incurs a larger delay?

P6-5 Consider digital transmission using spread spectrum with BPSK in an AWGN channel. There is a pulse noise jammer in the system, which transmits pulses of white Gaussian noise within the system frequency spectrum (W Hz) with an average power of P_J. The ratio of the average *on* period to the average *off* period of the jamming signal is α. The one-sided psd of the background noise is N_0. The received signal is detected coherently.

a. Derive the BER.
b. Determine the α value which results in the maximal BER.
c. Determine the maximal BER using the α value obtained in (*b*).
d. Does the spread spectrum offer any advantage in suppressing the jamming signal?

P6-6 Consider the forward link transmission of a single-cell DS-CDMA system using BPSK, where the processing gain is 256. The transmitted signals for all the users are synchronized both in chip and in carrier phase with the same power level. Assume that background noise is negligible when compared with the multiple access interference. Determine the maximum number of users allowed in the system in order to achieve a BER of 10^{-3}, 10^{-4}, and 10^{-5}, respectively, with coherent detection.

P6-7 Consider a metropolitan city with a total area of 1500 km^2 to be covered by a hexagonal cellular system with 7-cell reuse pattern. Suppose each cell has a radius (R) of 5 km and the city is allocated 25 MHz of spectrum, with a full-duplex channel bandwidth of 30 kHz and total 40 kHz guard bands. The system uses FDMA with 14 control channels. Determine

a. the number of cells in the service area,
b. the number of channels without frequency reuse,
c. the cell capacity, and
d. the system spectral efficiency in channels/MHz/km^2.

P6-8 In the system of Problem 6-7, it is known that, on average, each user makes two calls in a busy hour and each call lasts 2 minutes. Given that the trunking efficiency is 0.9, calculate

a. the number of calls per hour per cell,
b. the average number of users served per hour per cell, and
c. the system spectral efficiency in Erlangs/MHz/km^2.

P6-9 The Personal Digital Cellular (PDC) TDMA system uses a 42 kbps data rate to support 3 users per frame. Each user occupies 2 of the 6 time slots per frame.

a. What is the raw data rate provided for each user?
b. If the frame efficiency is 80% and the frame duration is 6.667 ms, determine the number of information bits sent to each user per frame. Assume no overhead at the slot level.
c. If speech coding that reduces the data rate by half is used, then 6 users per frame can be accommodated. Determine the number of information bits provided for each user per frame.
d. What is the information rate per user in this half-rate compressed PDC system?

P6-10 In GSM, the total system bandwidth (50 MHz) is divided into 25 MHz for uplink and 25 MHz for downlink. There are a total of 125 duplex RF carriers, each having a bandwidth of 2 × 200 kHz with a frequency separation of 45 MHz between the uplink and downlink channels. Each carrier supports 8 users using TDMA with a frame duration of 4.615 ms. Using GMSK, the bandwidth efficiency is 1.35 bits/s/Hz. The speech codec rate is 13 kbps. Channel coding results in a coded bit rate of 22.8 kbps. Consider only the normal speech frames. Each frame consists of 8 slots, and each time slot has 156.25 bits–3 start bits, 116 coded speech bits, 26 training bits, 3 stop bits, and 8.25 bits for guard period. Assume a frequency reuse factor of 7. Determine

a. the number of simultaneous users that can be accommodated in each cell cluster,
b. the cell capacity,
c. the spectral efficiency η_{N-TDMA} of TDMA, and
d. the overall spectral efficiency in bits/s/Hz/cell.

P6-11 A GSM system has a service area of 4500 km^2. Each cell has a coverage area of 7.5 km^2. During a busy hour, on average, each user makes 2 calls and each call lasts 5 minutes. If the trunking efficiency is 0.85, using the GSM system parameters described in Problem 6-10, determine

a. the number of cells in the system,
b. the number of calls per hour per cell,
c. the average number of users served per hour per cell, and
d. the overall system spectral efficiency in Erlangs/MHz/km^2.

P6-12 In an omnidirectional CDMA cellular system, the required E_b/I_0 is 10 dB. If 250 users, each with a baseband data rate of 13 kbps, are to be accommodated, determine the minimum channel chip rate of the spread spectrum sequence for the following two cases:

a. ignoring voice activity considerations,
b. the voice activity factor is 0.4.

P6-13 Consider a DS-CDMA cellular system.

a. Discuss why power control is needed in a CDMA system.
b. Discuss how each of the following factors influence the capacity of the DS-CDMA system: (1) the processing gain G_p, (2) antenna sectorization, (3) the required E_b/N_o value, (4) imperfection in automatic power control.
c. If the following factors are specified, determine the system capacity and the spectral efficiency of the CDMA system: (1) frequency reuse efficiency = 0.55, (2) capacity degradation due to imperfect power control = 0.9, (3) E_b/N_o = 10 dB, (4) information bit rate = 16.2 kbps, (5) system bandwidth = 12.5 MHz, (6) neglecting all other sources of interference.

P6-14 Consider a CDMA cellular system using hexagonal cells with BPSK modulation for voice services. The allocated radio spectrum for reverse link has a bandwidth of W Hz. The data transmission rate for each user is R_b bps. The additive background noise component is negligible. To provide satisfactory service quality, it is required that the received signal energy per bit to interference density ratio should be at least γ_r. The source activity factor is s_f. Assuming perfect power control, derive the cell capacity of the CDMA system.

P6-15 Suppose 61 users share a CDMA system, and each user has a processing gain of 480. Assume that (1) the power control on the uplink is perfect, (2) BPSK modulation is used, (3) the signal transmitted by each user is a Gaussian process, and (4) the system is interference limited so that the background noise can be neglected. Determine the probability of bit error for each user.

P6-16 On the basis that the total interference seen at the cell-site receiver is given by the sum of intracell interference, intercell interference, and background noise, show that the Erlang capacity of a CDMA-based radio cell is given by Eq. (6.4.22). To facilitate your derivation, make the same assumptions as those used in the alternate derivation of the Erlang capacity as given by Eq. (6.4.22). (Hint: Let I_0 be the power spectral density of the total interference and W be the spread spectrum bandwidth. Then the total interference is $I_0 W$. Let I_0^* be the power spectral density of the acceptable total interference. Then, to satisfy the quality of service requirement, the total interference $I_0 W$ at anytime must satisfy the inequality $I_0 W \leq I_0^* W$.)

P6-17 Consider the uplink transmission of voice signals in a single-cell DS-CDMA system with a processing gain of 256. Here, we are interested in studying the effect of power control error on the cell capacity by computer simulation. For simplicity, the following assumptions are made:

(1) The required transmission accuracy is specified by a required E_b/I_0 of 7 dB.
(2) The background additive noise is negligible when compared with the multiple access interference.

(3) The transmission is frame based with a frame duration of 5 ms. All the transmissions from mobile users are synchronized in frame.
(4) Different users have independent power control errors. For each user, the power control error is independent from frame to frame. Over each frame duration, the power control error for each user remains constant.
(5) Each power control error (in dB) can be modeled as a Gaussian random variable with zero mean and standard deviation σ (also in dB).

Due to the power control error, there are chances that the required E_b/I_0 cannot be guaranteed from time to time. It is required that the outage probability, defined as $P(E_b/I_0 < 7\text{dB})$, should be kept below α.

a. Given $\alpha = 5\%$, determine the cell capacity for $\sigma = 1$ dB, 2 dB, and 3 dB, respectively.
b. Given $\sigma = 1$ dB, determine the cell capacity for $\alpha = 1\%$, 2%, and 5%, respectively.
c. Compute the cell capacity under the assumption of perfect power control.
d. Comment on the effects of the power control error and the required outage probability bound on the cell capacity.

7

Mobility Management in Wireless Networks

One of the salient features of wireless communications is the flexibility to support user roaming. With the desire to increase system capacity by using small radio cells in an array structure, users tend to move in and out of cells frequently. The frequency of cell boundary crossing increases with the speed of the user. To maintain continuity for a connection, when the mobile crosses the cell boundary, it is necessary to hand the connection off from the old base station to the new base station. Also, the new location of the mobile must be made known to its home location register where the mobile's permanent address resides in order to deliver messages destined for the mobile. Mobility management is thus a two-step process: handoff management and location management. This chapter is concerned with issues in, and methods for, handoff management and location management, and calculation of handoff traffic.

7.1 INTRODUCTION

A wireless network has the flexibility to support user roaming. However, the geographical coverage of a wireless network is somewhat limited. For wide area coverage, a backbone network, such as a wireline network or a satellite network, is needed to extend the coverage to provide personal communications globally. Such a network with wide area coverage is called a personal communications network (PCN). If the backbone network is a wireline network, then in general a PCN comprises both wireline and wireless links interconnecting traffic handling devices referred to as routers. The user terminals can be mobile (e.g., handsets) or fixed. A base station (BS) is employed as a hub to handle the information transfer between a source and a destination user. In this regard, the base station provides the resource for a mobile station to access the network. Hence the base station is also known as the access point (AP).

As stated in Chapter 1, the geographical area served by a single BS is called a radio cell. Depending on the cell size, a radio cell may be a picocell, a microcell, or a macrocell. With small cell sizes (e.g., microcells), a mobile can migrate out of the current cell and into a neighboring cell quite frequently. In this case, handoff can be too frequent. A cluster of neighboring cells can be grouped together and overseen by a common controller. The controller can collect and accumulate information from all cells in the cluster. In this way, user migration from one cell to another cell within the cluster does not involve the transfer of connection information. A mobile switching center (MSC), which has much more computing power and functionality than any individual base station, is the appropriate common controller to handle the handoff functions.

At a higher level of the network hierarchy, MSC's can also be interconnected by wirelines. A pictorial view of a mobile communications network is shown in Figure 7.1.

To facilitate user roaming, there must be an effective and efficient *handoff* mechanism so that the mobile's connection with its current serving base station is handed off to its target base station to maintain service continuity without the need to disconnect and reconnect. The handoff completion time must be within a prescribed tolerance. The need to perform handoff is induced by user mobility so that handoff management represents one component of mobility management.

Users subscribe to a regional subnetwork for communications services. This regional subnetwork is called the subscriber's home network. For identification purposes, the subscriber must register with its home network. The subscriber's identity is a permanent address that resides in the database of a location register, called the subscriber's home location register (HLR). When the subscriber moves away from its home network, the new network it enters is a foreign or visitor network. It must update its registration with the home location register through its visitor location register (VLR) in order to facilitate message delivery to the mobile in its new location. The procedure of maintaining an association between the mobile and its HLR when it is away from its home network is referred to as location management.

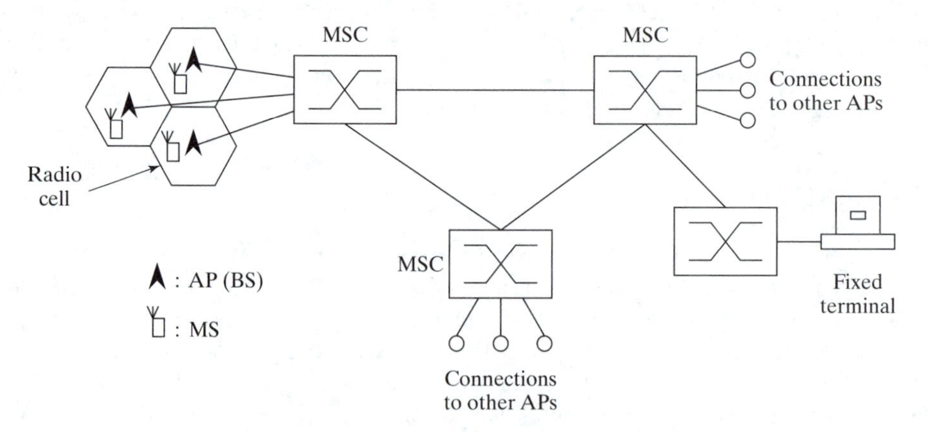

Figure 7.1 Wireless/wireline network.

Mobility management thus consists of two components:

a. *handoff management*, in which a connection is transferred from one access point to another access point, and
b. *location management*, in which the association between the mobile and its HLR is maintained as the mobile moves from one network to another.

As discussed in Chapter 6, the capacity of a cell is a hard number in FDMA and TDMA. Although the capacity of a CDMA system is soft, it is nevertheless limited by the E_b/I_0 specification that defines the quality of service (QoS) required by the participating users. For discussion purposes, suppose the capacity of the cell-site (base station) is defined as the number of basic channels, each of which is capable of handling the information transmission in one connection. Let N_c be the capacity of the cell under consideration. This total capacity is used to handle ongoing connections and to admit new and handoff requests. The mechanism that oversees the admission of new and handoff calls should protect the integrity of the network and satisfy user QoS requirements. This mechanism is referred to as call admission control.

The remainder of this chapter is organized as follows. Section 7.2 treats the issues that govern call admission control. The rationale behind handoff management and techniques for the development of handoff strategies are discussed in Section 7.3. The complexity of location management is a function of the population size. With higher transmission rates in third generation wireless networks, location management for third generation systems is expected to be relatively more involved than that for second generation systems. Location management issues are treated in Section 7.4 and Section 7.5. Traffic and handoff rate calculations are discussed in Section 7.6.

7.2 CALL ADMISSION CONTROL (CAC)

The performance requirements of users are measured in terms of quality of service (QoS) and grade of service (GoS). QoS is a packet-level factor which includes packet loss rate, packet delay, packet delay variation, and throughput rate. GoS is a call-level factor, which includes new call blocking probability (NCBP), handoff call dropping probability (HCDP), and connection forced termination probability (CFTP). CFTP is a measure of a call (connection) being forced to terminate at some point during the lifetime of the connection. In terms of CAC at the call-level, we are more concerned with NCBP and HCDP as GoS measures.

CAC ensures network integrity by restricting access to the network so as to avoid overload and congestion, and to ensure that QoS requirements of all ongoing connections are satisfied. The CAC problem can be phrased as follows. Suppose there are $(N - 1)$ ongoing calls. When the Nth request arrives, the network calculates the amount of available resources. If there are enough resources to admit the Nth call, such that its QoS requirement and those of the ongoing calls are satisfied, then the new request is admitted. Otherwise, the call request is denied. This is the case whether the new request is initiated by a new call or a handoff call.

Before making admission decisions, the CAC algorithm needs to determine the amount of unallocated capacity (i.e., the number of basic channels available for accepting new and handoff requests). Also, a handoff request tries to maintain service continuity of an ongoing session. If the available capacity for accepting connection requests is limited, handoff call requests are admitted in preference to new call requests. For example, if only one basic channel is available and there are two competing requests, one new and one handoff, the decision should be to accept the handoff request and reject the new request. Thus, handoff requests should be offered a higher admission priority than new requests.

Let P_n be the new call blocking probability and P_h be the handoff call dropping probability. The GoS can be defined by jointly taking both P_n and P_h into account. That is

$$\text{GoS} = P_n + \alpha P_h, \tag{7.2.1}$$

where $\alpha > 1$ is a balancing factor between new call and handoff call GoS. A CAC algorithm may be designed subject to satisfaction of the GoS defined by Eq. (7.2.1).

7.2.1 Prioritized Call Admission

Handoff calls can be admitted at a higher priority than new calls. To manage the admission of requests based on priority, it is necessary to reserve capacity for admitting handoff requests. Let N_g be the number of basic channels reserved for admitting handoff requests. A common technique to reserve capacity for handling handoff requests is the *guard channel* method. This is the reason that we use the subscript g to denote the amount of reserved capacity for accepting handoff requests. Let N_{ua} denote the number of unallocated channels. With the guard channel method, the admission rule is the following:

a. If $N_{ua} > N_g$, admit a new or handoff request;
b. If $N_{ua} \leq N_g$, admit a handoff request only.

The above guard channel method is a fixed reservation strategy. One can also introduce dynamic reservation methods to reserve the right amount of capacity to satisfy the demand-supply problem. The dynamic approach would offer a higher resource utilization efficiency, but may entail fairly complex parameter estimation issues. For example, if the controller knows exactly the number (or the rate) of handoff requests during any epoch, it can determine the value of N_g exactly. This information is never available; at best, the number or the rate of handoff may be estimated.

The focus of this chapter is more on handoff management and location management issues. The remainder of this chapter will be devoted to these two aspects of mobility management, rather than dwelling further on call admission control.

7.3 HANDOFF MANAGEMENT

When a mobile moves outside the footprint of its current serving base station, its connection must be transferred from the current serving base station to the target new base station. The handoff procedure consists of an *initiation phase* and an *execution phase*.

The initiation phase may employ a decision making strategy based on the measured received signal level, with or without hysteresis. Without hysteresis, a handoff is initiated as soon as the average signal level from the new base station exceeds that from the current base station. With hysteresis, a handoff is initiated when the average signal level from the new base station exceeds that from the current base station by a threshold amount specified by the hysteresis level. Also, the decision at any instant should be a function of previous decisions.

The execution phase will include the allocation of new radio resources (e.g., channel assignment) and the exchange of control messages.

7.3.1 Handoff Strategies

As mentioned earlier, an MSC is the appropriate device to oversee the handoff operation. An MSC is connected to a cluster of base stations through wirelines and obtains status information for all the base stations in the cluster on an ongoing basis.[1] In this way, the MSC has information concerning the channel occupancy status of all the BSs in the cluster. When the mobile moves between cells in the cluster overseen by the same MSC, handoff initiation and execution are handled by the MSC, and there is no need to copy connection identification states. This mode of operation is referred to as *intraswitch handoff*. When the mobile moves from one cluster of cells into another cluster of cells, overseen by another MSC, connection status information has to be transferred from the current serving MSC to the new MSC. The copying of state information from one MSC to another MSC takes time. This mode of operation is referred to as *interswitch handoff*. Because of the need to transfer state information, interswitch handoff tends to take longer than intraswitch handoff.

Depending on the information used and the action taken to initiate the handoff, the methods for handoff can be

a. mobile controlled handoff (MCHO),
b. network controlled handoff (NCHO), or
c. mobile assisted handoff (MAHO).

These handoff methods have the following significance:

a. MCHO is a desirable method because it reduces the burden on the network. However, this will increase the complexity of the mobile terminal.
b. In NCHO, the BSs or APs monitor the signal quality from the mobile and report the measurements to the MSC. The MSC is responsible for choosing the candidate AP and initiating the handoff. The mobile, on the other hand, plays a passive role in the handoff process.
c. MAHO is a variant of NCHO and is employed by GSM. In MAHO, the mobile measures the signal levels from the various APs using a periodic beacon generated by the APs (to keep track of the locations of the mobiles). The mobile collects a set of power levels from different APs and feeds it back to the MSC, via the serving AP, for handoff decision making.

7.3.2 Types of Handoff

The handoff procedure can be hard, soft, backward, or forward, as described below.

Hard Handoff This handoff mode is characterized by a mobile having a radio link with only one AP at any time. Thus, the old connection is terminated before a new connection is activated. This mode of operation is referred to as *break before make*.

[1]The size of a cell cluster in handoff management can be much larger than the size of a cell cluster in frequency reuse.

Soft Handoff In soft handoff, the mobile can simultaneously communicate with more than one AP during the handoff. That is, a new connection is made before breaking the old connection, and is referred to as *make before break*. CDMA systems use soft handoff techniques.

Backward Handoff In backward handoff, the handoff is predicted ahead of time and initiated via the existing radio link. A sudden loss or rapid deterioration of the radio link poses a major problem in backward handoff.

Forward Handoff In forward handoff, the handoff is initiated via the new radio link associated with the candidate AP. This type of handoff alleviates the problem of signaling on deteriorating old links. However, it can result in large delays.

7.3.3 Design Issues

The simplest scenario for investigating the handoff issue is a configuration involving a mobile station handing off from one access point, say AP_0, to another access point, AP_1, where the actual handoff decisions are made by the MSC. This scenario is illustrated in Figure 7.2.

Communication is a two way process since the mobile sends as well as receives information. During the dwelling time of the mobile within a cell, the mobile sends and receives information through the same access point. In the scenario shown in Figure 7.2, the serving access point is AP_0. When the mobile approaches the boundary of its current cell and a handoff is initiated, all conditions associated with sending and receiving information by the mobile must be taken care of during the *execution* phase. For example, during the transition time required to terminate transmission via AP_0 and start transmission via AP_1, it is possible that AP_0 has some leftover packets destined for the mobile after the mobile has moved into the footprint of AP_1. Therefore, the handoff strategy must ensure that packet delivery to the mobile after its connection has been handed off from AP_0 to AP_1 (see Figure 7.2) does not suffer undue loss. Also, the order of packets delivered to the mobile after handoff should be preserved.

An effective handoff scheme should have the following features:

a. fast and lossless
b. minimal number of control signal exchanges

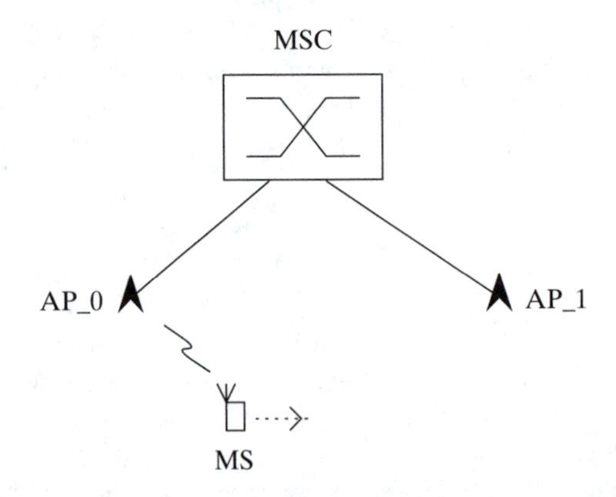

Figure 7.2 Scenario of MS moving from AP_0 toward AP_1.

c. scalable with network size
d. capable of recovering from link failures, such as abrupt loss of radio link (e.g., due to shadowing).

7.3.4 Feedback-Based MAHO Strategy

Rather than considering handoff management issues in a very general way, we shall focus attention on the movement of a tagged mobile, which is currently located in its serving cell and moving toward an adjacent cell. On the downlink from the BS to the MS, the APs broadcast pilot (beacon) signals to the mobile. With MAHO, the mobile sends the pilot signal strengths received from the APs as feedback information via the serving AP to the MSC for handoff decision making. The AP also computes and supplies the distance information to the MSC. This distance information is used by the MSC to determine the direction of movement of the mobile. To facilitate the construction of a feedback-based handoff algorithm, we assume that the mobile can always measure the strength of the pilot signals broadcast by the serving and adjacent APs.

General Design Goals. The design goals of an effective handoff scheme include

a. low handoff delay,
b. low cell loss,
c. small buffer required, and
d. efficient use of resources.

Based on the received signal strengths from the surrounding APs, the mobile creates a profile of signal strengths and sends the profile as feedback information to the MSC, via the serving AP.

Let the cell topology be a regular hexagon. Since each hexagon has six neighbors, for simplicity we consider a tagged cell and its six nearest neighbors as a cluster of seven cells. The base stations of the seven cells are connected to the MSC by wirelines. In practice, the cluster size may be much larger than seven.

The profiles to be fed back to the MSC through the serving AP contain control signals to aid decision making. Thus, feedback profiles represent an overhead on system performance. It is, therefore, desirable that the feedback profiles be as small as possible. To make the feedback profile sufficiently small, we further make the following assumptions, taking the location of the mobile within the cellular array into consideration.

Assumptions

a. A mobile receives equal signal strengths from at most three APs. This occurs when the mobile is located at the intersection point of three neighboring cells, as shown in Figure 7.3.
b. The feedback profile contains the signal strengths from the three APs with the highest signal strength being sufficient for decision making.
c. If the mobile is located in the middle of the boundary between two cells, as shown in Figure 7.4, the signals from A and B are stronger than those from C and D.

The profile needs to be formatted to provide identification of the MSCs.

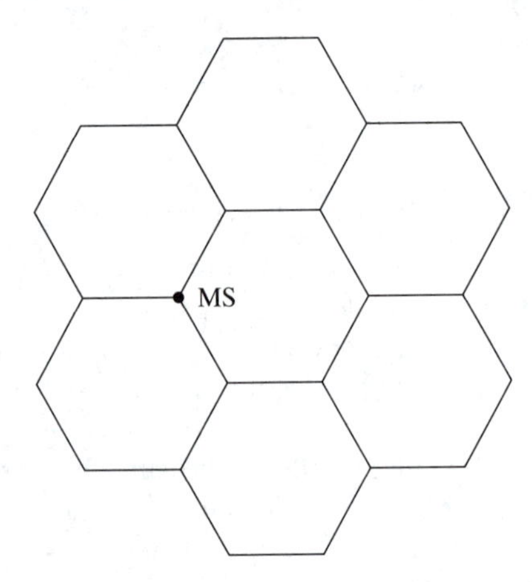

Figure 7.3 Mobile located at the vertex of three cells.

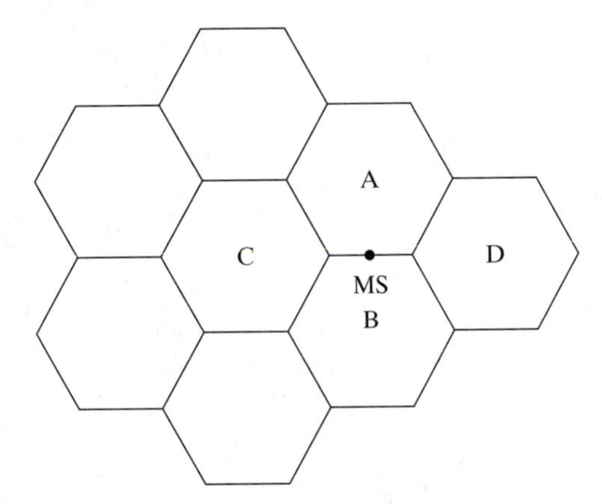

Figure 7.4 Mobile located at the boundary of two cells.

7.3.5 AP/MSC Identification

Consider the situations shown in Figures 7.3 and 7.4. For convenience, we will use the term *cluster* here to denote a group of APs served by the same MSC. There are three scenarios in which the three APs are located closest to the mobile:

a. The three APs are located within the same cluster.
b. The three APs are divided between two clusters. This happens when the mobile is located at the boundary shared by two clusters.
c. The APs are divided among three clusters (i.e., one from each cluster). This occurs when the mobile is located at the boundary shared by three clusters.

Therefore, it is only necessary to identify a single cluster or at most three adjacent clusters uniquely to the current serving MSC. Each MSC (or cluster) has six adjacent MSCs. Let MSC_{curr}

denote the MSC that is currently serving the tagged mobile. To uniquely identify the adjacent MSCs, their identities have to be different. Since there are seven MSCs altogether, three bits are sufficient to uniquely identify all adjacent MSCs. The 3-bit codeword shall be referred to as a *3-bit identity* [101].

Example 7.1 3-bit AP/MSC Identification

Describe how a 3-bit codeword can be used to uniquely identity a cluster of seven MSCs.

Solution The clustering of MSCs is shown in Figure 7.5. Here, a 3-bit codeword is used to uniquely identify a supercluster of 7 MSCs. The *3-bit identity* (000) is used to identify MSC_{curr}, and the 6 adjacent MSCs are identified by the code vector {(001),(010),(011),(100),(101),(110)}. Therefore, if MSC (000) maintains a list of all adjacent MSCs, it can identify an adjacent MSC based on a *3-bit identity*.

To see that a 3-bit identity is necessary and sufficient, consider the group of tier 2 clusters. There are a total of 12 clusters. Based on the identification system indicated above, MSC (001) will be surrounded by 6 adjacent MSCs represented by the code vector {(000),(010),(011),(100),(101),(110)}. It is important to note that there are two occurrences of

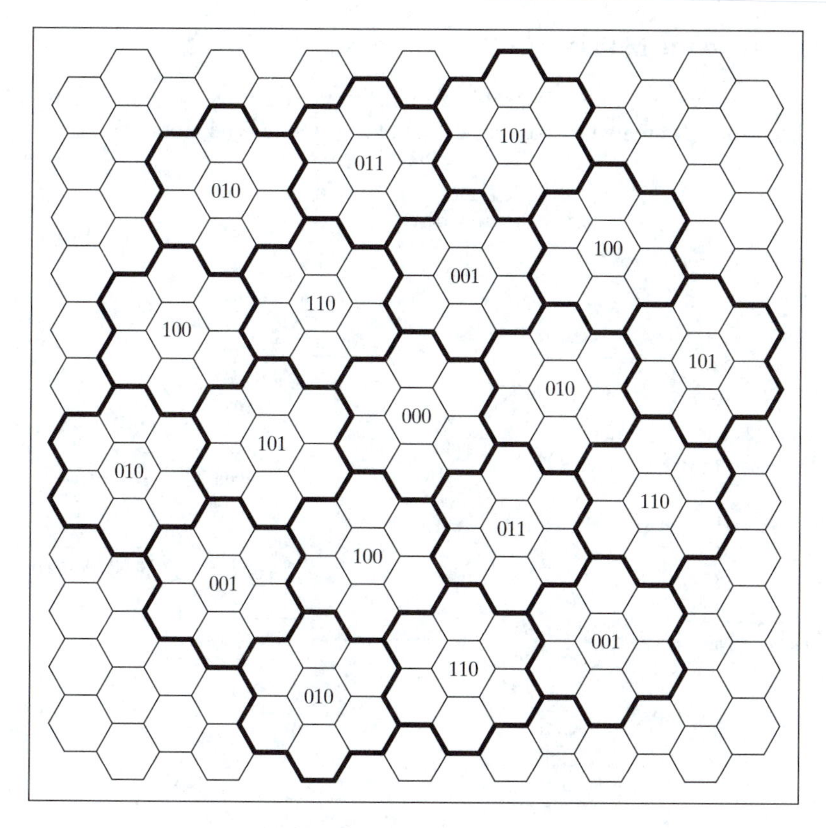

Figure 7.5 MSC identification.

each of {(001),(010),(011),(100),(101),(110)} in tier 2. To uniquely identify MSCs in tier 2 to the network, an additional bit is needed for tier 2 identification.

This identification process can continue until all the tiers are exhausted. In order to have equal size identities, if MSCs in tier L (the highest tier) have a K-bit identity (3 of the K bits identifying the MSCs and the other $(K - 3)$ bits for tier identification), with 7 APs per MSC, an additional 3 bits are needed for AP identification, so that the profile would contain $(K + 3)$ bits.

7.3.6 Profile

In addition to the AP/MSC identification, the profile has to contain bits that represent the signal strengths received by the mobile from the nearest APs. With the possibility of the strongest signals from three APs (see Figure 7.3), only three power fields are needed to carry the information about the signal strengths. Suppose eight bits are allocated to digitally represent each power level. Then, the profile should allocate 24 bits (three bytes) to represent the power levels. One of the three APs should be the serving AP, residing in the serving MSC. The profile does not need fields to identify the serving AP or MSC, but must contain fields to identify the other two APs and their corresponding MSCs.

7.3.7 Capability of the Mobile

The flowchart for implementing the profile format discussed in Subsection 7.3.6 is shown in Figure 7.6. The profile is created by the transceiver at the mobile. The mobile transceiver processes

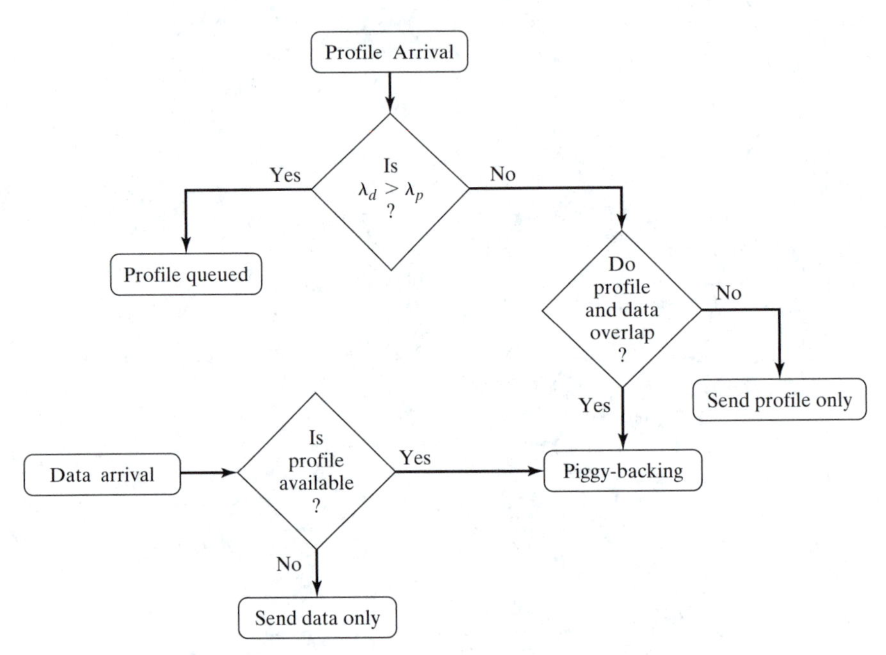

Figure 7.6 Implementation of piggy-backed profile.

the beacon signals received from the APs to create a profile, and then transmits the profile and the generated data packet according to the implementation shown in Figure 7.6. The procedure works as follows.

Let λ_d and λ_p be, respectively, the data generation rate and the profile generation rate. The transceiver compares the data rate, λ_d, with the profile rate, λ_p. If $\lambda_d > \lambda_p$, the profile is queued in the profile buffer. Upon arrival of the next data packet, the profile is dequeued and piggy-backed onto the data before transmission. Since the data packets will be generated at a rate faster than that of the profiles, there is only one profile in the queue at any time. On the other hand, if $\lambda_d < \lambda_p$, the profile is transmitted separately if there is no overlap with data transmission; otherwise it is piggy-backed and transmitted with the data packet.

Based on the preceding description of the generation of profiles, it is clear that, when a data packet is generated, the profile queue and also any overlaps must be checked for piggy-backing. Thus, if there is a profile in the queue, or if there is an overlap, the profile is piggy-backed and transmitted with the data. On the other hand, if none of these two possibilities exist, the data packet is sent separately.

It is important to note that the utilization of this scheme is significantly affected by the relationship between λ_d and λ_p. If $\lambda_d > \lambda_p$, the utilization is governed by the data rate. However, for $\lambda_d < \lambda_p$, the profile rate heavily influences the utilization and can result in significant degradation if the data rate is small. This is because most of the profiles will be sent separately.

Assuming that the profile is properly formatted, the performance measures include

a. buffer requirement,
b. mean handoff delay,
c. feedback interval, and
d. system resource utilization efficiency.

The listed performance measures are influenced by the movement patterns of the mobile. To facilitate performance assessment, a suitable mobility model is needed.

7.3.8 Mobility Model

Knowledge of mobility information is critical to the design of handoff algorithms. The mobility pattern of a mobile is defined by the speed and the direction of the mobile's movement. Accurate information about speed and direction of movement can only be obtained through monitoring and measurement. An alternative is the construction of a mobility model, based on assumed statistics.

In general, the mobility pattern is a function of the mobile's speed and the direction of movement (i.e., the mobility pattern is a function of two variables). The change in direction of movement relative to the mobile's current direction of movement can be any value in the $[0, 2\pi]$ range. Also, the speed of travel can be variable. Therefore, in general, mobility patterns can be quite complex. Of particular interest is the expected time interval between the time a mobile initiates a call in a cell and the time the mobile reaches the cell boundary. We refer to this as the sojourn time[2] and denote it by t_s. For simplicity, assume that the mobile travels in a straight line. This reduces the mobility pattern to be a function of the mobile's speed only. This assumption is

[2]If the mobile immediately hands off to the neighboring cell when it reaches a cell boundary, the sojourn time is the same as the channel holding time discussed in Section 5.5.

appropriate for motor vehicles that travel at relatively high speeds. For a low speed mobile (e.g., a pedestrian), direction changes can be made easily. In this case, the mobility patterns will be a function of both speed and direction.

Note that mobility models are not unique. We will consider the following mobility model and determine the expected sojourn time of the mobile's movement.

Mobility Model 1. Let X be the random variable representing the distance traveled by the mobile before reaching the cell boundary. Let Y be the random variable representing the velocity of the mobile's movement. The sojourn time t_s depends on both X and Y. To simplify the modeling problem, assume further that

a. X is uniformly distributed between 0 and $D_{\max}$, and
b. Y is uniformly distributed between $V_{\min}$ and $V_{\max}$,

where $D_{\max}$ is the maximum distance traveled by the mobile, and $V_{\min}$ and $V_{\max}$ are, respectively, the minimum and maximum mobile speeds.

The sojourn time t_s is related to X and Y by $t_s = X/Y$. From the performance consideration point of view, the expected sojourn time is related to the rate at which the mobile needs to send profile information to the MSC in the MAHO strategy. To determine the expected sojourn time, we need to first find the probability density function (pdf) of t_s. The parameters in this model are $D_{\max}$, $V_{\max}$ and $V_{\min}$. For notational convenience, let

$$\beta = \frac{D_{\max}}{V_{\max}}, \quad \alpha = \frac{D_{\max}}{V_{\min}}, \quad \frac{\beta}{\alpha} = \frac{V_{\min}}{V_{\max}}$$

and

$$u(t) = \begin{cases} 1 & t \geq 0 \\ 0 & t < 0 \end{cases}.$$

It can be derived that the pdf of t_s is

$$f_{t_s}(t) = \frac{(\alpha+\beta)}{2\alpha\beta}[u(t) - u(t-\beta)] + \left[\frac{\alpha\beta}{2t^2(\alpha-\beta)} - \frac{\beta}{2\alpha(\alpha-\beta)}\right][u(t-\beta) - u(t-\alpha)]. \quad (7.3.1)$$

Using the preceding pdf, the expected value of the sojourn time is given by

$$E[t_s] = \frac{\alpha\beta}{2(\alpha-\beta)} \ln\left(\frac{\alpha}{\beta}\right). \quad (7.3.2)$$

This mean has the following asymptotic results:

$$\lim_{\frac{\alpha}{\beta}\to 1} E[t_s] = \alpha/2 \quad (7.3.3)$$

and

$$\lim_{\frac{\alpha}{\beta}\to \infty} E[t_s] = \beta/2. \quad (7.3.4)$$

The performance of a feedback-based handoff scheme, in which the mobile sends profile information to the MSC, depends on the frequency with which the mobile feeds information back to the

MSC. Profiles contain control signals, but do not carry information in a messaging sense. Sending feedback information too frequently reduces efficiency; on the other hand, sending feedback information too infrequently degrades effectiveness. Thus, the selection of the feedback interval is important. Let I_f denote the average feedback interval. The rate of transmitting profiles is $r_p = 1/I_f$. Let N_p denote the mean number of profiles sent by the mobile before the mobile moves to the cell boundary. N_p is approximately given by $E[t_s] \times r_p$. To make a good compromise between resource utilization efficiency and handoff performance, it may be appropriate to decide on the mean number of profiles to be sent and determine the rate of sending profiles from $r_p \approx N_p/E[t_s]$, from which we can get a measure of the feedback interval, I_f.

7.3.9 Intraswitch Handoff Algorithm

Handoff from one AP to another AP under the control of the same MSC is referred to as intraswitch handoff; handoffs between AP's under the control of different MSC's is referred to as interswitch handoff. Figure 7.7 depicts a situation of intraswitch handoff where all handoff decisions are under the control of the MSC. In the situation shown in Figure 7.7, the fixed terminal and AP_0 and AP_1 are each connected to the MSC by wirelines. As shown, the fixed terminal communicates with the mobile via the MSC.

As mentioned earlier, the MSC collects data sent to it by the APs within the cluster. These data include profile information sent by the mobile and measurements taken by the candidate AP to facilitate estimation of the current location of the mobile within the current cell. These data, together with the accumulated information about the channel occupancy of each of the cells in the cluster, enable the MSC to take the following items into consideration in making handoff decisions:

a. Based on the profile information from the mobile and measurements taken by the candidate AP, the MSC makes handoff decisions;
b. The MSC determines the readiness of the new AP to accommodate the handoff; and
c. The MSC ensures that the handoff algorithm maintains packet sequencing after handoff.

A general design consideration is that the mobile must hand off upon receipt of a command from the MSC. Depending on the operational procedure associated with a handoff scheme, there

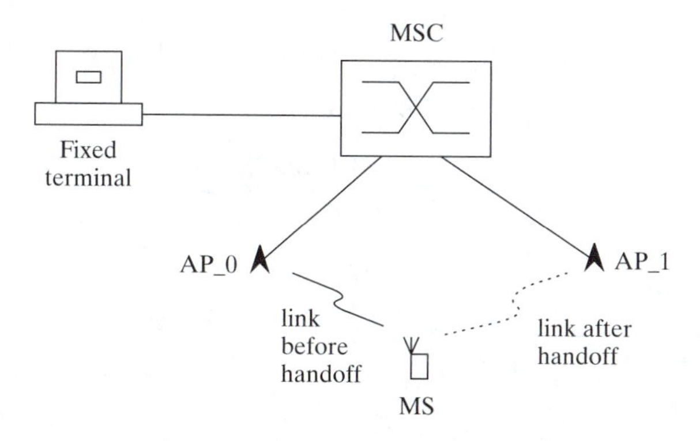

Figure 7.7 Intraswitch handoff scenario.

may be different approaches to constructing the handoff scheme. Here, we consider one design example [101].

Example 7.2 Intraswitch Handoff Design

Consider Mobile Assisted Handoff for intraswitch operation. The system shown in Figure 7.7 consists of five components: MSC, AP_0, AP_1, fixed terminal, and mobile station. In this scenario, the fixed terminal is used to generate data packets destined for the mobile, and acts as a sink for data packets arriving from the mobile via the MSC. The other components perform the following functions:

a. The APs provide access points for the mobile via a radio link;
b. All packets to/from the mobile must transit through the serving AP;
c. The MSC ensures packet ordering.

Consider the situation in which the tagged mobile being served by access point AP_0 is moving toward the neighboring cell, with access point AP_1. Both AP_0 and AP_1 are connected to an MSC, which acts as the central controller to initiate and execute handoff algorithms. Using an MAHO approach, design an intraswitch handoff scheme to hand the ongoing connection of the tagged mobile off from AP_0 to AP_1.

Solution The handoff scenario is shown in Figure 7.8. The mobile is currently located in the cell being served by AP_0, and is moving toward the cell being served by AP_1. When the

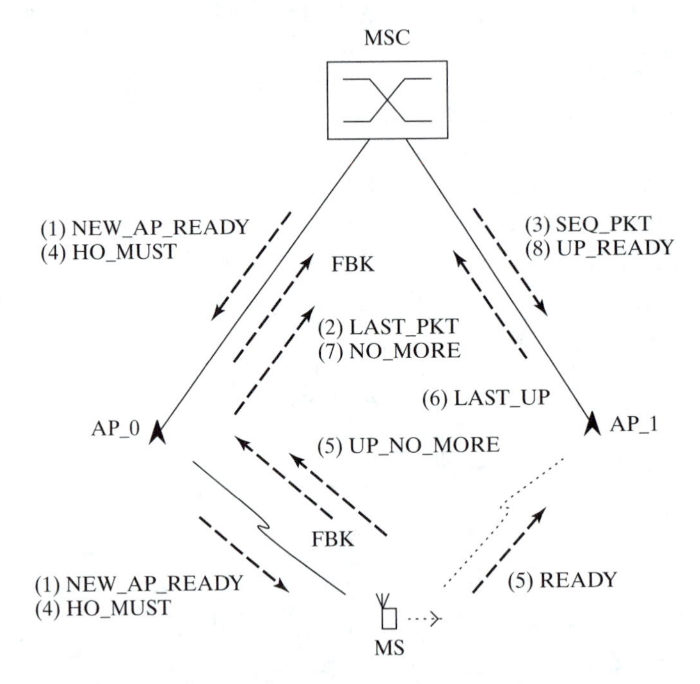

Figure 7.8 Signaling sequence for intraswitch handoff.

mobile reaches the cell boundary of AP_0, the MSC initiates and executes the handoff algorithm to be designed.

The sample handoff design is portrayed in Figure 7.8. This is an MAHO strategy. Germane to the intraswitch mobile assisted handoff scheme shown in Figure 7.8 is a set of control signals. The numbered control messages indicate the order and actions of the signal flows, under the control of the MSC.

Suppose the MSC, based on information collected, knows that the mobile (currently served by AP_0) is now at the cell boundary between AP_0 and AP_1 (Figure 7.8) and that AP_1 has a channel available to accept the handoff. The MSC now directs the mobile to handoff to AP_1. The handoff procedure, as indicated in Figure 7.8, is as follows.

Steps of handoff procedure for the scheme in Figure 7.8:

1. Send the message

   (1) NEW_AP_READY

   to the mobile via AP_0. This message indicates to the mobile that the candidate AP_1 is ready and it can handoff whenever it desires to do so. The identity of the candidate AP_1 is contained in this message.
2. Upon receipt of the *NEW_AP_READY* message, AP_0 responds with the message

   (2) LAST_PKT

   which contains the sequence number of the packet sent (or waiting to be sent) to the mobile prior to receiving the *NEW_AP_READY* message. When the MSC receives the sequence number, it calculates the sequence number of the last downlink packet transmitted to AP_0 by the MSC. Then the MSC sends the next number as the sequence number of the following downlink packet to AP_1 using the message

   (3) SEQ_PKT.
3. The message

   (4) HO_MUST

   implicitly indicates the last downlink packet from the MSC to AP_0. This message indicates the end of uplink packets from the mobile. When AP_0 receives this message, it sends all uplink packets to the MSC, if any, and flags the termination of the connection by sending the message

   (5) UP_NO_MORE

   to the MSC.
4. After sending the *UP_NO_MORE* message, the mobile switches its operating frequency and sends the message

   (6) READY

   to AP_1, and continues uplink transmission via the new connection. This message contains the sequence number of the last packet correctly received by the mobile. Sending *UP_NO_MORE* in parallel with the *READY* message speeds up the approval process of uplink transmission by the MSC.

5. As soon as AP_1 receives the *READY* message, it starts the downlink transmission and, at the same time, buffers all uplink packets arriving from the mobile. It also sends the message

   (7) LAST_UP

   to the MSC requesting for the approval of uplink transmission.
6. Upon receipt of the *LAST_UP* message, the MSC waits for the message

   (8) NO_MORE

   from AP_0 (if it has not yet been received). Once this message is received, the MSC switches the uplink connection from AP_0 to AP_1 and sends the message

   (9) UP_READY

   to indicate that the uplink flow can resume. In this way, uplink packet sequence is also maintained.

The handoff procedure is complete when AP_1 receives the *UP_READY* message from the MSC.

7.4 LOCATION MANAGEMENT FOR CELLULAR NETWORKS

To provide a wide geographical coverage, a wireless mobile cellular network is normally configured as an interconnection of many regional wireless subnetworks. The interconnection mechanism may be an intelligent router or a backbone network. The intelligent router may be in the form of an MSC, while the backbone network may be a wireline network (e.g., the Internet), or a satellite network.

The cellular network is a message transport platform for point-to-point, point-to-multipoint, or multipoint-to-multipoint communications. The end user can be a mobile station or a fixed terminal. For message delivery purposes, a mobile is identified with a home network (i.e., one of the subnetworks). The home network must know the current location of the mobile at all times for message delivery purposes. The association between the mobile and its home network is made through a *registration* process. The identity of the mobile is kept in a database of a *location register*. The location register that belongs to the mobile's home network is called the *home location register* (HLR). When the mobile moves into a different regional subnetwork, it still has to maintain an association with its home network through the registration process. Thus, whenever the mobile moves away from its home network, it must update its registration with its HLR. The regional subnetwork where the mobile now resides is called the *foreign* or *visitor network*. The mobile has to register its current location with a *visitor location register* (VLR) in the visitor network. When the mobile moves away from its home network, if the mobile does not update its registration with its HLR, the HLR has no way of knowing the current location of the mobile. All messages destined for the mobile will end up at the mobile's home network. Since the HLR has no knowledge of the mobile's current location, it is unable to deliver the messages. To facilitate message delivery, the mobile must update its registration with the HLR. Before the registration update process can take place, the HLR needs to know that a mobile attempting to update registration is the rightful owner of the ID residing in the database of the HLR. This is a security measure to protect the network from fraudulent attacks. The process of ascertaining

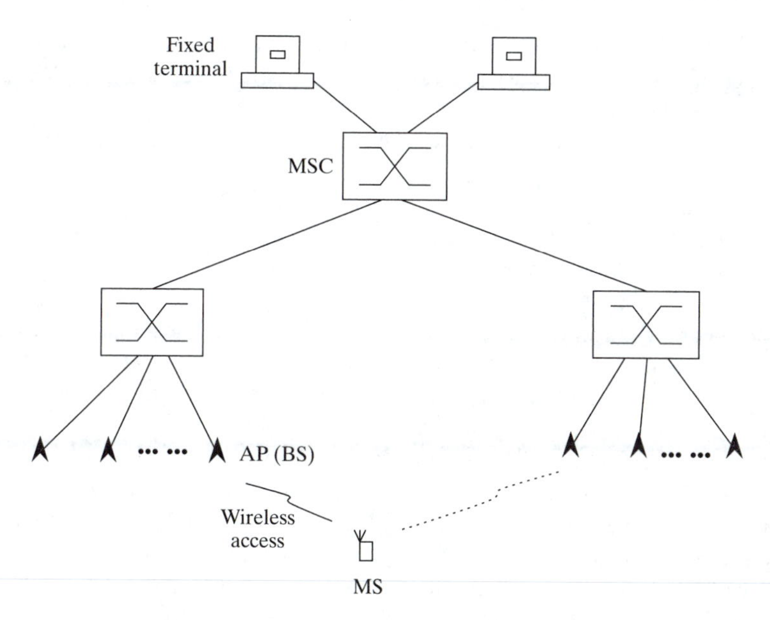

Figure 7.9 Cellular network architecture.

that the mobile attempting to update its registration is, in fact, the rightful owner of the ID in the HLR's database is referred to as an *authentication* process.

For the purpose of discussing location management issues, we will use the cellular network architecture shown in Figure 7.9. Here, the cellular network consists of BSs connected to MSCs by wirelines, MSCs connected to other MSCs by wirelines, and fixed terminals connected to MSCs by wirelines. The system shown in Figure 7.9 portrays a mobile station in motion. For the purpose of discussing location management issues, we focus attention on the processes needed for the mobile to reestablish an association with its HLR and for delivery of messages from a fixed terminal to the mobile terminal. In this regard, location management consists of three distinct components: authentication, location update, and call delivery.

Authentication. An essential function to achieve secured communication in cellular networks is reliable authentication of communicating parties and network components. Thus, authentication is a process to ensure that any party that claims to be the owner of a particular permanent address is, in fact, the rightful owner. Authentication typically depends on exchanges of cryptographic messages between the communicating parties. Encryption of a message is tantamount to putting the message in an indestructible box and locking the box. A key is needed to open the lock in order to read the message. The mobile's identity is its permanent address; only the mobile and its HLR know the mobile's permanent address (i.e., the identity of the address is locked in a box and the mobile and the HLR are the only parties who have a key to open the box).

In this text, we do not consider the design of data encryption algorithms or authentication protocols. Location management will be discussed with respect to location update and call delivery on the assumption that authentication between the mobile and the HLR has already taken place.

Location Update. For discussion purposes, we assume that the footprint covered by all the cells in a subnet of the cellular network constitutes a registration area (RA). When the mobile moves within a registration area, there is no need for the mobile to update its registration with its HLR. When the mobile moves into a new registration area, it must update its registration with the HLR.

Location update is a process in which the mobile generates periodic update messages to inform the network of its current location. In the context discussed above, the mobile informs the HLR of its current location whenever it moves into a new registration area.

Call Delivery. For the purpose of discussing message delivery to the mobile, we assume that the fixed terminal generates a message and sends it to the mobile's permanent address for delivery to the mobile in its current location. This is possible if the mobile has updated its location information with the HLR when the former moves into a new registration area.

Both location update and call delivery involve signaling exchanges between various network entities. To accomplish this, processing power is required at various network nodes to properly handle the location information. Signal flows for location update and call delivery are supported by a specific network architecture and control signals are supported by a specific control signaling network. We will base signal flow discussions on the two-tiered network architecture of IS-41 and the handling of control signals by the SS7 signaling network.

7.4.1 Two-Tiered Architecture of IS-41

The second generation systems have been operational for sometime, whereas the third generation systems are still evolving. In second generation mobile wireless networks, there are two standards currently available for location management. These are

a. IS-41, which is a companion standard to IS-54 (or the new version IS-136), and
b. Mobile Applications Part (MAP) of GSM.

Both IS-41 and GSM's MAP are two-tiered architectures, which maintain a database in the HLR and the VLR. The two-tiered architecture of IS-41 is illustrated in Figure 7.10. In this chapter, we will use the two-tiered architecture of IS-41 as the basis for discussions of registration updates.

The cluster of cells connected to an MSC, as shown in Figure 7.10, is referred to as an RA. No registration is required when the mobile roams within a given RA, but the mobile must register with the HLR when it moves to a new RA. The IS-41 architecture has the following characteristics:

a. Each RA consists of one or more radio cells.
b. Cells belonging to the same RA are connected to the same MSC.
c. The MSC is connected to both the backbone network and the signaling network, and provides typical network switching functions in addition to coordinating the location registration and call setup.
d. A VLR is associated with each RA.
e. The VLR contains an entry for all mobiles which are currently visiting its RA.
f. There is an HLR associated with each mobile in the cellular network. In IS-41, as shown in Figure 7.10, there is a central HLR associated with each cellular network.

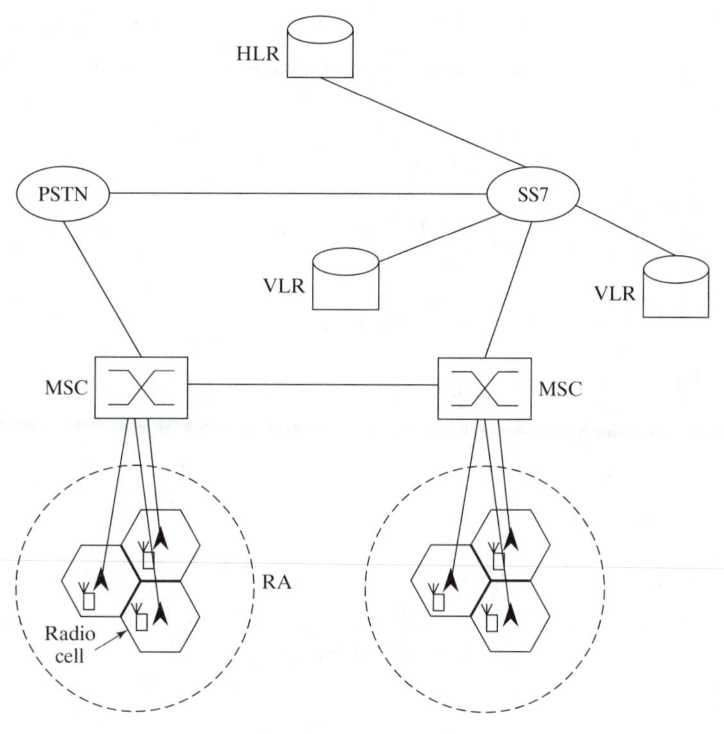

Figure 7.10 IS-41 network architecture.

g. The HLR contains an entry for each user in the network.
h. Each HLR entry indicates, for its corresponding mobile, the current VLR at which the mobile is registered.

In IS-41, network signaling exchange is carried out through the SS7 signaling network using common channel signaling.

7.4.2 SS7 Network and Common Channel Signaling

Although all signal transmissions between mobile stations and base stations propagate through wireless channels, MSCs are connected to base stations, to other MSCs, and to the backbone wireline network by wire links. All control signals in the wireline domain are handled by the SS7 signaling network. A functional block diagram of the SS7 signaling network is shown in Figure 7.11. For conciseness, acronyms are used as labels in Figure 7.11. In the following, we describe the signal flows in the signaling network using the acronyms to portray each of the service points. The acronyms are defined as follows:

STP - signaling transfer point
SCP - service control point
SSP - service switching point

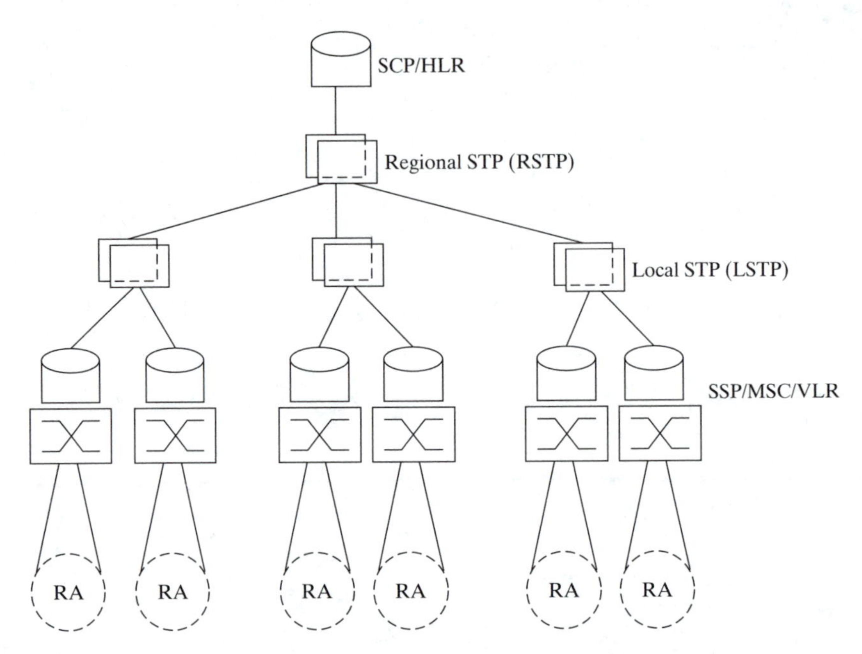

Figure 7.11 SS7 signaling network.

LSTP - local STP
RSTP - regional STP.

As shown in Figure 7.11, the SSP, MSC and VLR are all functionally colocated. Also, the SCP and the HLR are functionally colocated. From the registration update point of view, all base stations connected to the same MSC represent one registration area. Signaling between the mobile in the current RA and the HLR passes through the SSP, LSTP, RSTP and then the SCP.

The base stations of an RA are connected via a wireline network to an end-office switch, or an SSP. The SS7 signaling network has the following features:

a. Each SSP serves one RA;
b. SSPs of different RAs are, in turn, connected to a two-level hierarchy of signaling transfer points, comprising a regional STP connected to all local STPs;
c. The STPs, which are installed in pairs to ensure robustness, perform message routing and other SS7 functions;
d. The RSTP is also connected to an SCP which contains the functionality associated with an HLR.

As mentioned earlier, the SS7 network carries out network signaling exchange operations using common channel signaling (CCS). Briefly, the salient features of CCS are as follows. CCS is a digital communications technique that provides simultaneous transmission of user data, signaling data, and other related traffic throughout a network. Using out-of-band signaling channels that logically separate the network control data from the user information (e.g., voice or data) on the same channel, CCS is implemented in a time division multiplexing (TDM) format for serial data transmissions, even though the concept of CCS may imply the use of dedicated, parallel channels.

7.4.3 Location Update, Call Setup, and Paging

For second generation wireless communications systems, CCS is used to pass user data and control/supervisory signals between the mobile and the BS, between the BS and the MSC, and between MSCs.

Once the mobile has identified its current registration area, location management is mainly concerned with location update and call setup. Once the registration update is complete, call delivery is then a very simple procedure. Whenever the mobile moves into a new RA, it must update the registration with its HLR.

A mobile is identified by a mobile identification number (MIN). A registration request sent by the mobile should contain its MIN. The base stations in each RA periodically broadcast beacon signals. A mobile listens to the pilot signals from the base stations in the RA and uses these to identify its current RA. A location registration occurs when a mobile enters a new RA. The signaling flow diagram for location registration in IS-41 is shown in Figure 7.12. Registration update is a query (request) and response (ACK) process. The steps involved in the location update procedure trace out the signal flow shown in Figure 7.12.

Example 7.3 Location Update Procedure

Consider the two-tier network architecture of IS-41, shown in Figure 7.10, and the signal flow diagram shown in Figure 7.12. Describe the steps for implementing the location update procedure.

Solution The location update by the mobile with its HLR can be implemented using the following six steps:

(1) By monitoring the beacon signals from the base stations in the RA, the mobile senses that it is in a new RA and sends a registration request to the base station of the current cell. The registration request contains the mobile's MIN.

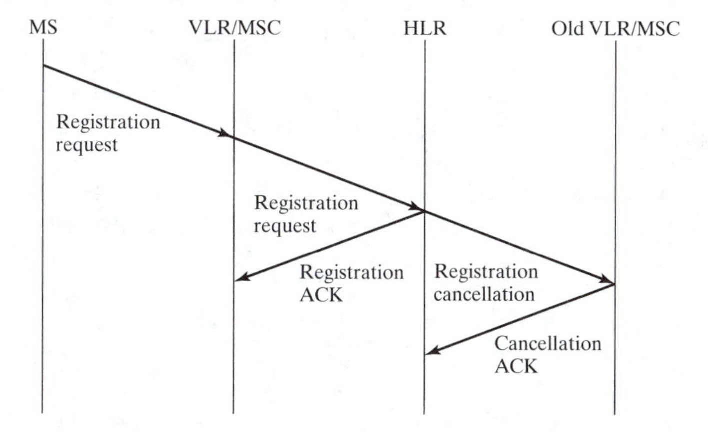

Figure 7.12 Location registration signaling flow diagram for IS-41.

(2) The base station forwards the request to the MSC which, in turn, launches a registration query to its associated VLR.
(3) The VLR adds an entry in its records on the location of the mobile and forwards the registration request to the HLR. The HLR of the mobile is determined from the MIN of the mobile through a global title translation.
(4) The HLR performs the necessary authentication routine and records the identity of the new serving VLR in its database entry for the mobile. The HLR then sends a registration acknowledgment message to the new VLR.
(5) The HLR sends a registration cancelation message to the old serving VLR of the mobile.
(6) The old VLR removes the record of the mobile from its location database and returns a cancelation acknowledgment message to the HLR.

The mobile's registration with its HLR is now complete. With the dismantling of the association between the HLR and the *old* VLR, the HLR now has an association with the *new* VLR.

When the network receives a call for a mobile, the network access point, which receives the request, must locate the called mobile and deliver the call to it. The procedure to set up the call is as follows. The calling mobile sends the call request to the MSC in the registration area via its serving base station. The MSC which handles the call request is referred to as the calling MSC. The calling MSC locates the called MSC/VLR combination through its association with the HLR. The HLR routes the call request to the called MSC which, in turn, pages the called mobile to set up the call. The call setup is illustrated by the signaling flow diagram shown in Figure 7.13, where TLDN stands for temporary local directory number.

Example 7.4 Call Setup Procedure

Consider the call setup signaling flow diagram shown in Figure 7.13. Describe the steps involved in setting up the call between the calling MSC and the called MSC.

Solution The steps involved in setting up the call should trace out the signal flows shown in Figure 7.13. The call can be set up using the following six steps:

(1) The calling BS receives a call request. The request should contain the MIN of the called mobile. The calling BS forwards the request to the VLR of its RA. The VLR then forwards the request to its corresponding MSC.
(2) The calling MSC sends a location lookup request to the HLR of the called mobile.
(3) The HLR of the called mobile determines the current VLR of the called mobile (from its database) and sends a routing request message to the VLR. The VLR then forwards the message to its corresponding MSC.
(4) The called MSC allocates a temporary local directory number to the called mobile and sends a reply to the HLR, together with the TLDN.
(5) The HLR forwards the TLDN to the MSC of the calling mobile.
(6) Using the TLDN, the MSC of the calling mobile initiates a connection request to the called MSC through the SS7 network.

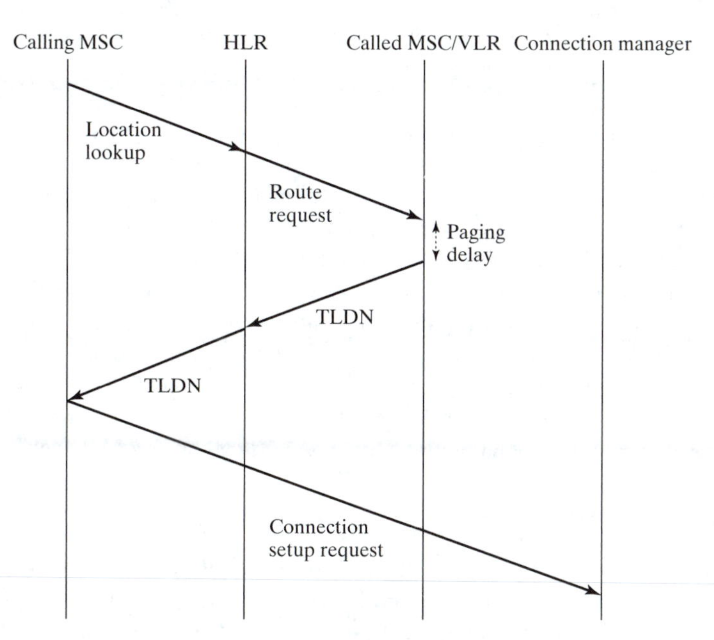

Figure 7.13 Call setup signaling flow diagram.

The above call setup procedure establishes a linkage between the calling MSC and the called MSC. In general, the called MSC only knows that the called mobile is in the RA under its jurisdiction, but does not know in which cell the called mobile is located. A *paging* procedure is needed to locate the called mobile.

Paging Procedure. Paging is a means of locating a mobile within a registration area. There may be different approaches to performing paging. A commonly used approach is *polling*. To locate the called mobile using the polling mechanism, the called MSC broadcasts a polling message, which contains the MIN (mobile identification number) of the called mobile, to all cells within the RA. The BS of each cell relays the paging message to the mobile terminals. Upon receiving the paging message, the called mobile responds to the called MSC, through its currently serving BS, with the base station ID of its current cell. The called MSC then knows where to forward the call.

The second generation wireless systems mainly support voice communications. It is anticipated that future wireless systems, called *personal communication service* (PCS) systems, will support multimedia services at transmission rates that are orders of magnitude larger than that for voice. The two-tiered architecture of IS-41 and GSM's MAP may not be adequate.

7.5 LOCATION MANAGEMENT FOR PCS NETWORKS

A PCS network is expected to allow communications by anyone, from anywhere, and at anytime. Compared with the second generation cellular networks such as IS-136 and GSM, future PCS networks (e.g., third generation wireless networks) are expected to provide higher transmission rates and operate at higher regions of the frequency spectrum (e.g., in the spectral range of 1.85–1.99 GHz as compared with the 850 MHz range in second generation cellular systems). In this case, the two-tiered location management system of IS-41 and MAP, which is suitable for second generation cellular systems, may not be adequate to support future PCS networks. Also, future PCS networks will have smaller cell sizes to enlarge the system capacity (see Chapter 5 for detailed discussions on the capacity expansion issue). This will result in smaller registration areas and an increase in the number of times that the HLR is accessed. With more mobile users in the PCS than in a second generation cellular network, the HLR can become the bottleneck of the location management network. Moreover, in a large network, the HLR may be very far away from its point of access. The large distances will introduce considerable signaling delays. It is therefore desirable to introduce methods that can reduce the number of HLR accesses. Different approaches may be used to reduce the delay in call setup and delivery. We will discuss below two such approaches: (*a*) an overlay approach and (*b*) a local anchor approach.

7.5.1 Overlay Approach

One strategy is to use an overlay approach to provide enhancement to the two-tiered architecture of IS-41 by reducing the number of HLR accesses. One such overlay approach is called pointer forwarding. A functional block diagram of the pointer forwarding scheme is illustrated in Figure 7.14. In the pointer forwarding scheme, instead of reporting to the HLR on every RA crossing, the pointer forwarding strategy sets up pointers from the old VLR to the new VLR, thus eliminating signaling exchanges between the HLR and the VLR's. To locate a mobile during call delivery, the system must traverse the link chain from the initial VLR to the mobile's current VLR. If the length of the chain is overly long, the delay in call setup and message delivery may

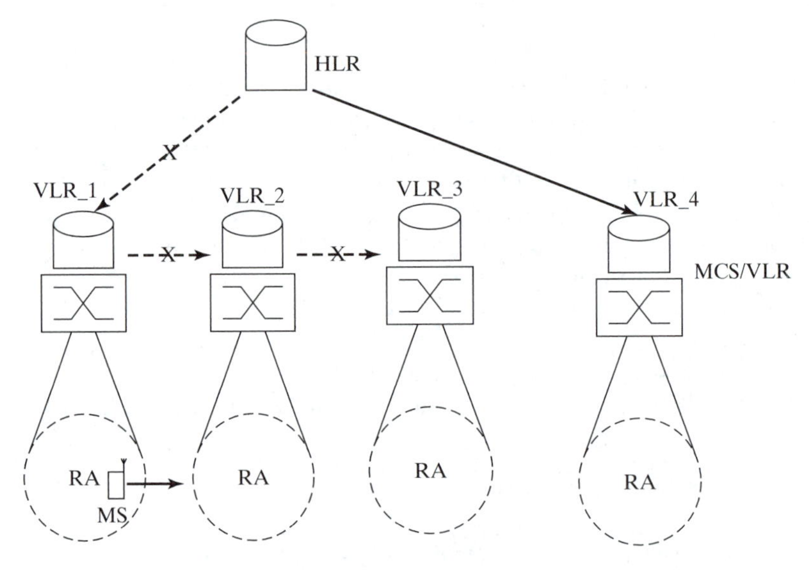

Figure 7.14 Pointer forwarding scheme.

be intolerable. Thus, in this form of overlay, an upper bound on the cost of call setup needs to be incorporated.

To place an upper bound on the cost of call setup, a limit of value K can be placed on the length of the link chain. When the length of the link chain exceeds K, an update to the HLR is forced and the link chain is reset to NULL. As an example, in Figure 7.14, consider the case of $K = 2$. When the mobile moves from VLR_3 to VLR_4, the link chain length limit is exceeded, and the HLR is notified of the location change.

The rate of call arrivals and the rate of RA crossings have an impact on overlay approaches. The call arrival rate and the RA crossing rate can be combined into a single call-to-mobility ratio (CMR), which is defined as

$$\text{Call-to-mobility ratio} \triangleq \frac{\text{rate of call arrivals}}{\text{rate of RA crossings}}.$$

The pointer forwarding scheme works well if the CMR is relatively low. Also, after K moves, the mobile is forced to update the HLR even if the mobile is still very close to its previously reported location.

7.5.2 Local Anchor Approach

Another approach to reducing the frequency of accessing the HLR is to create a virtual HLR in the form of a local anchor (LA) [5]. When the mobile crosses an RA boundary, it registers with the LA. The LA has an association with the HLR, and plays the role of the HLR as far as registration updating is concerned. This is the reason why we call the LA a virtual HLR. In the local anchor scheme, the MSC of the newly entered RA registers the mobile's location at the colocated VLR. This colocated VLR is called the LA. The associated LSTP region (see Figure 7.11) is the anchor LSTP region. Note that each mobile may have a different LA and the LA for a mobile may change from time to time. Once a VLR is selected as the LA for a particular mobile, an entry is set up in a table at this VLR indicating the current serving MSC/VLR for the mobile. A question of how to select the LA arises. One strategy is to select the serving VLR of the mobile during the last call arrival as the LA. A diagram that portrays the signal flow of reporting location change to the LA is shown in Figure 7.15. The numerals shown in Figure 7.15 indicate the steps involved in the location registration.

Example 7.5 Location Registration Procedure

Consider the numbered signal flow diagram for reporting location change shown in Figure 7.15. The steps used to implement the location registration procedure should correspond to the numbered signal flows shown in the diagram. Describe the steps needed to implement the registration procedure.

Solution Following the numbered signal flows, we can establish the following six steps to implement the location registration procedure:

(1) Start by considering a mobile moving into the new RA and sending a location update message to the nearby BS.
(2) The BS forwards this update message to the new MSC/VLR.

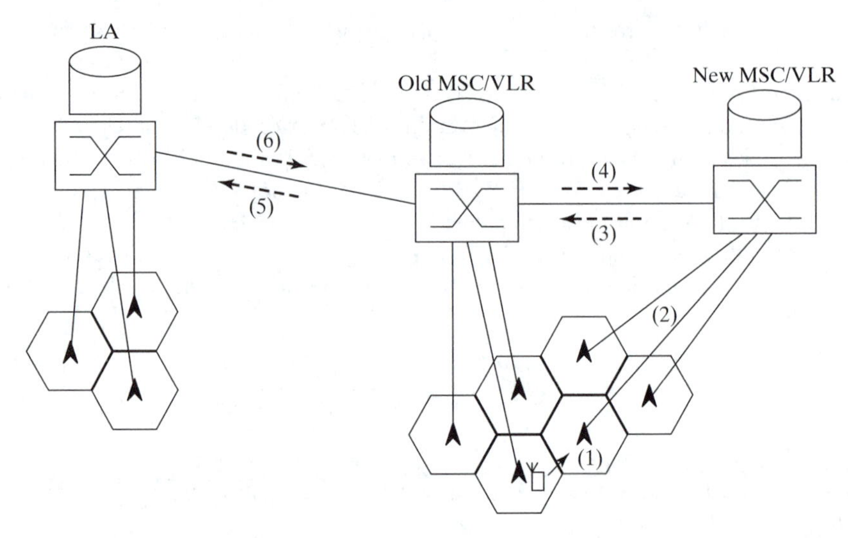

Figure 7.15 Signaling flow for reporting location change to the local anchor.

(3) The new MSC sends a message to inform the old MSC that the mobile has moved out of its associated RA.
(4) The old MSC sends an acknowledgment message to the new MSC together with a copy of the mobile's profile.
(5) The old MSC removes the record of the mobile station in its associated VLR and sends a message to inform the LA of the location change.
(6) The LA updates its record indicating the new location of the mobile and sends an acknowledgment message to the old MSC.

After executing step (6), the reporting process is complete.

Consider a mobile-to-mobile communication scenario. The calling mobile's message is to be delivered to the called mobile. The calling mobile sends the message to the calling MSC via its serving base station. Since there is no direct association between the calling MSC and the called MSC, communication between these two MSCs has to hop through the central HLR. The signal flow for call delivery is illustrated in Figure 7.16. The numerals in the figure indicate the actions to be executed in the call delivery process.

Example 7.6 Call Delivery Procedure

Consider the call delivery problem shown by the signal flows in Figure 7.16. Describe the steps needed to implement the call delivery procedure.

Solution The call delivery process involves seven steps that correspond to the numbered signal flows shown in Figure 7.16.

(1) A call is initiated by a mobile which sends a call initiation signal to its serving MSC through the nearby BS.

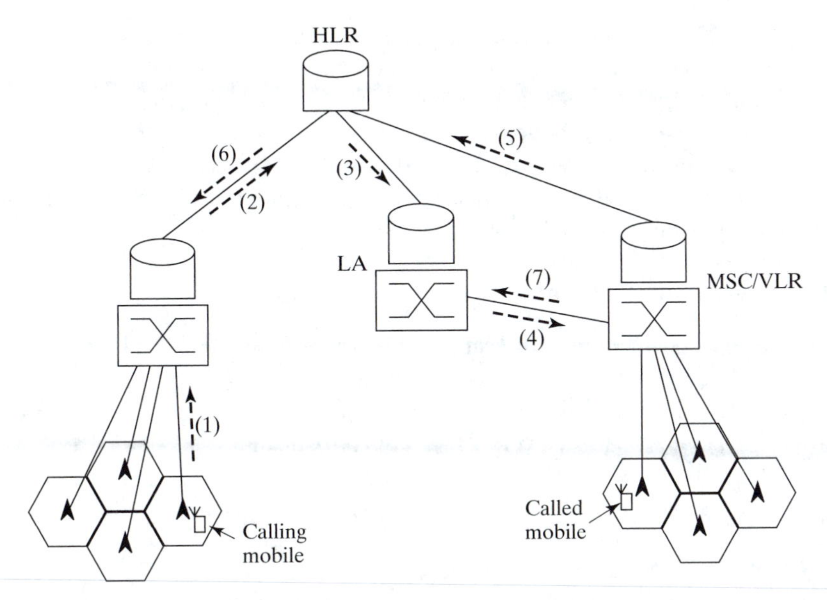

Figure 7.16 Signaling flow for call delivery in LA-based scheme.

(2) The MSC of the calling mobile sends a location request message to the HLR.
(3) The HLR sends a location request message to the LA of the called mobile.
(4) The LA forwards the location request message to the MSC serving the called mobile.
(5) The called MSC allocates a TLDN to the mobile and sends this TLDN to the HLR. The HLR updates its record such that the current serving VLR of the called mobile becomes the new LA.
(6) The HLR forwards the TLDN to the MSC of the calling mobile terminal. The calling MSC can now set up a connection to the called MSC using this TLDN.
(7) The serving MSC of the called mobile sends an acknowledgment message to the old LA. The old LA removes its record of the called mobile terminal.

In the preceding call delivery procedure, it is assumed that the LA is different from the serving VLR. If they are the same, the messages between the LA and the serving VLR are not needed. Step (7) can take place concurrently with steps (5) and (6).

7.6 TRAFFIC CALCULATION

The rate of boundary crossing by the mobile is an important parameter for performance assessment of both handoff and location management schemes. Handoff can be intraswitch or interswitch with different handoff rates. In location management, a location update is necessary whenever the mobile crosses the boundary of a registration area. Knowledge of the handoff rate and location update rate can provide a good indication of system performance and a basis for designing handoff schemes.

For a given mobility model, network architecture, and traffic parameters, it is possible to calculate the intraswitch handoff rate, the interswitch handoff rate, and the location update rate. The traffic calculation can be model-based or measurement-based. Analytical models are normally constructed based on known distributions. Therefore, an analytical model may not portray the real scenario very well. Nevertheless, results obtained using analytical models can provide guidance for the design of handoff and location management strategies. The traffic calculations in this section are based on a universal mobility model.

7.6.1 System and Traffic Parameters

Consider the cellular layout shown in Figure 7.17, where a cluster of cells is connected to an MSC. Assuming a hexagonal cell topology, a cluster comprises N hexagonal cells. Although in general the shape of a cluster is not hexagonal, from the modeling point of view, we use a hexagon (the large hexagon in Figure 7.17) as an approximation. The area of a hexagonal cell, A_{cell}, is approximately given by

$$A_{cell} = 2.598R^2,$$

where R is the radius of the small hexagon in Figure 7.17, measured from the center of the cell to the vertex (it is also the length of each side of the hexagon). The area of the cluster is approximately given by

$$A_{cluster} \approx N \times 2.598R^2 \approx 2.598R^2_{cluster}, \tag{7.6.1}$$

where $R_{cluster}$ is the radius of the large hexagon. From Eq. (7.6.1), we have

$$R_{cluster} \approx \sqrt{N}R. \tag{7.6.2}$$

To calculate the intracluster handoff rate, we need a knowledge of the number of boundary crossings per cell per unit time. To calculate the intercluster handoff rate, we need a knowledge of the number of cell crossings per cluster per unit time. Handoff rate calculation requires information on the network parameters and traffic parameters, and the mobile's movement pattern. Note that a mobile is in a dormant state unless it is powered on. Also, there is no traffic loading from a mobile unless it is active (i.e., either transmitting or receiving messages). Taking these factors into

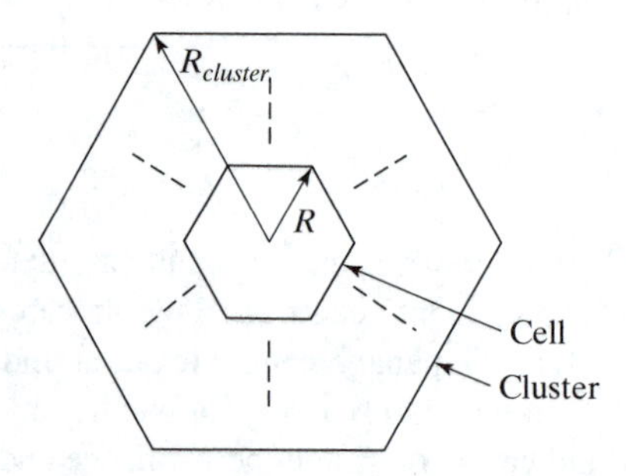

Figure 7.17 Cellular layout of clustering.

consideration, the following is a list of the relevant system and traffic parameters that contribute to the formulation of an expression for the handoff rate:

Population density of mobiles = ρ (number of mobiles/km^2)
Hexagonal cell radius = R (km)
Number of cells per cluster = N
Speed of mobile: V (km/hr)
Traffic load per mobile station = λ_{ar} (Erlangs/MS)
Percentage of powered stations = δ
Percentage of active mobiles among the powered stations = ϵ
Number of crossings per cell per unit time = μ_{cell}
Number of crossings per cluster per unit time = $\mu_{cluster}$

Note that a mobile can be in motion even if it is not powered on or not active. When a mobile terminal is powered on, it is in a listening mode. When the mobile is active, it has an ongoing connection. Therefore, a mobile is active only if it is powered on. Thus, the percentage of the total population that is active is given by $\epsilon \cdot \delta$ and the number of handoffs per unit time is considerably smaller than the number of boundary crossings per unit time.

In formulating Mobility Model 1 in Subsection 7.3.8, we focused attention on the movement of a single mobile. To consider handoff rates in general, we need to account for all active mobiles within the cluster. We call the model that characterizes the movements of all mobiles Mobility Model 2.

Mobility Model 2. Mobility Model 2 is governed by the following assumptions:

a. The mobiles are uniformly distributed in the cell;
b. The direction of travel of all mobiles relative to the cell boundary is uniformly distributed on $[0, 2\pi]$;
c. Once a mobile has been admitted into the cell, it assumes a direction to travel and then travels along that direction with constant velocity V.

7.6.2 Handoff Rate Calculation

With Mobility Model 2, the rate of boundary crossing can be determined by the rate at which a randomly chosen mobile crosses the cell boundary times the number of mobiles per cell. The rate of boundary crossing is a function of the cell size, the population density, and the speed of mobile movement. Let L_{cell} be the perimeter of the cell. For a hexagon, $L_{cell} = 6R$. The rate of mobiles departing the cell (i.e., the number of mobiles departing a cell per unit time), μ_{cell}, is given by

$$\mu_{cell} = \frac{\rho V L_{cell}}{\pi} = \frac{6\rho V R}{\pi}. \tag{7.6.3}$$

By the flow conservation law, the number departing a cell per unit time equals the number entering the cell per unit time. With the area of the cluster approximated by a hexagon of radius, $R_{cluster}$, given by Eq. (7.6.2), the rate of mobiles departing the cluster, $\mu_{cluster}$, can also be expressed as

$$\mu_{cluster} = \frac{\rho V L_{cluster}}{\pi}, \tag{7.6.4}$$

where $L_{cluster}$ is the perimeter of the hexagon used to approximate the area of the cluster. From Eq. (7.6.2), we can approximate the perimeter of this large hexagon as

$$L_{cluster} \approx 6R_{cluster} \approx 6\sqrt{N}R.$$

Hence,

$$\mu_{cluster} \approx \frac{6\rho V \sqrt{N} R}{\pi}. \tag{7.6.5}$$

Let $\lambda_{ho\_cluster}$ be the total handoff rate of a cluster (i.e., the number of handoffs in the cluster, both within the cluster and to neighboring clusters, per unit time). Then, the total handoff rate is given by

$$\begin{aligned}\lambda_{ho\_cluster} &= \text{number of cells in the cluster} \\ &\quad \times \text{number of crossings per cell per unit time} \times \%\text{ active mobiles} \\ &\quad \times\%\text{ powered stations} \times \text{traffic load per MS} \\ &= N \times \mu_{cell} \times \epsilon \times \delta \times \lambda_{ar}.\end{aligned} \tag{7.6.6}$$

The total handoff rate of a cluster includes both intracluster handoff rate and intercluster handoff rate. The intercluster handoff rate, $\lambda_{ho\_intercluster}$, is given by

$$\begin{aligned}\lambda_{ho\_intercluster} &= \text{number of crossings per cluster per unit time} \times \%\text{ active mobiles} \\ &\quad \times\%\text{ powered stations} \times \text{traffic load per MS} \\ &= \mu_{cluster} \times \epsilon \times \delta \times \lambda_{ar}.\end{aligned} \tag{7.6.7}$$

It follows that the intracluster handoff rate, $\lambda_{ho\_intracluster}$, is given by

$$\lambda_{ho\_intracluster} = \lambda_{ho\_cluster} - \lambda_{ho\_intercluster}. \tag{7.6.8}$$

Example 7.7 Intracluster Handoff Rate

Consider a cellular network with the following traffic and system parameter values:

$$\rho = 45{,}000 \text{ mobiles/km}^2$$

$$V = 6 \text{ km/hr}$$

$$R = 1.0 \text{ km}$$

$$N = 16$$

$$\epsilon = 10\%$$

$$\lambda_{ar} = 0.06 \text{ Erlangs/mobile station}$$

$$\delta = 50\%$$

Find the intracluster handoff rate, $\lambda_{ho\_intracluster}$.

Solution We first determine the average number of crossings/cell/hr, μ_{cell}, and the average number of crossings/cluster/hr, $\mu_{cluster}$, using Eqs. (7.6.3) and (7.6.4), respectively. From Eq. (7.6.3), we have

$$\mu_{cell} = \frac{6 \times 45{,}000 \times 6 \times 1}{\pi} = 515{,}595 \text{ crossings/cell/hr.}$$

From Eq. (7.6.4), we have

$$\mu_{cluster} = \frac{6 \times 45{,}000 \times 6 \times \sqrt{16} \times 1}{\pi} = 2{,}062{,}381 \text{ crossings/cluster/hr.}$$

The total handoff rate can then be obtained using Eq. (7.6.6) as

$$\lambda_{ho\_cluster} = 16 \times 515{,}595 \times 0.1 \times 0.5 \times 0.06 = 24{,}748.56/\text{hour.}$$

From Eq. (7.6.7), we get

$$\lambda_{ho\_intercluster} = 2{,}062{,}381 \times 0.1 \times 0.5 \times 0.06 = 6{,}187.14/\text{hour.}$$

Therefore, from Eq. (7.6.8), we get

$$\lambda_{ho\_intracluster} = 24{,}748.56 - 6{,}187.143 = 18{,}561.42/\text{hour.}$$

From Example 7.7, it is noted that the intracluster handoff rate, $\lambda_{ho\_intracluster}$, is much larger than the intercluster handoff rate, $\lambda_{ho\_intercluster}$. This result can be inferred from the fact that the total handoff rate in a cluster, $\lambda_{ho\_cluster}$, is proportional to the number of cells, N, in the cluster, while the intercluster handoff rate, $\lambda_{ho\_intercluster}$, is proportional to $\sqrt{N}$. The limiting case is $\lambda_{ho\_intracluster} = \lambda_{ho\_intercluster}$, which occurs when $N = 1$.

Example 7.8 Effect of Cell Splitting on Handoff Rates

Consider the effect of cell splitting on handoff rates. Suppose a cell of radius R_{cell} is split into microcells, each with radius $R_{microcell} = R_{cell}/4$. If the traffic and system parameters are those given in Example 7.7, what is the impact of cell splitting on handoff rates?

Solution Since the total handoff rate is proportional to the number of cells in the cluster, the effect of cell splitting on handoff rates can be obtained by first calculating the number of cells in the cluster after cell splitting. Let K be the number of microcells per original cell after cell splitting. The number of microcells per large cell can be determined by taking the ratio of the area of the large cell to the area of the microcell. Let A_{cell} be the area of the large cell and $A_{microcell}$ be the area of the microcell. We have

$$A_{microcell} = 2.598 R^2_{microcell}$$

and

$$A_{cell} = 2.598 R^2_{cell} = K \times 2.598 R^2_{microcell}.$$

Then

$$K = \frac{R^2_{cell}}{R^2_{microcell}} = \frac{R^2_{cell}}{(R_{cell}/4)^2} = 16.$$

The number of cells per cluster is, therefore,

$$N \times K = 16 \times 16 = 196.$$

The total handoff rate in the cluster is increased by $K = 16$-fold, while the intercluster handoff rate is increased by $\sqrt{K} = 4$-fold. This means that most of the increase in the total handoff rate is with the intracluster handoff.

SUMMARY

The salient feature of wireless communication and networking is the provision of an information transport platform that allows the mobile users to roam. To facilitate roaming, effective mobility management procedures need to be in place. Mobility management consists of two components: handoff management and location management. This chapter describes and discusses the fundamental aspects of handoff management and location management. The latter is illustrated by the two-tier architecture of the IS-41 standard.

ENDNOTES

1. For a general discussion of network issues in wireless communications, see the paper by Jabbari, Colombo and Kulkarni [68] and the book by Lin and Chlamtac [87]. For fundamentals of communication networks, see the book by Bertsekas and Gallager [13].
2. For a general discussion of handoff issues in wireless communications, see the papers by Pollini [117] and by Noerpel and Lin [105]. For feedback-based mobile assisted handoff, see the paper by Mukhi and Mark [101]. For soft handoff in CDMA systems, see the paper by Wong and Lim [159].
3. For location management, see the papers by Akyildiz and Ho [5], Tabbane [149], and Zhang and Mark [164].
4. For details of IS-41, see EIA/TIA interim standard [44]. For details of the Mobile Application Part of GSM, see ESTI technical report recommendation [47]. For an overview of signaling system no. 7, see the paper by Modarressi and Skoog [98].

PROBLEMS

P7-1 The cell capacity is defined as the maximum number of users in a cell in which each user transmits at the basic rate R_b. How much of the cell capacity can be utilized to handle services with QoS satisfaction depends on how the users' transmission rates and transmission rights are controlled. A mechanism that controls user transmission rate is known as rate allocation, and a mechanism that controls transmission rights is known as connection (or call) admission control.

a. Propose and describe a strategy to implement rate allocation. (Hint: The rate allocator located at the base station computes the available bandwidth, makes decisions on rate allocation, and sends the commands to the users via the downlink.)
b. Discuss why call admission control (CAC) is important. Propose a strategy to perform CAC on the assumption that blocking a new call and dropping a handoff call are equally important.
c. Propose a strategy to perform CAC on the basis that admitting a handoff call to maintain session continuity is more important than blocking a new call.

P7-2 In a cellular system, each mobile host monitors the signal strengths received from its serving base station and the surrounding base stations. The strengths of the signals received by the mobile hosts are used for making handoff decisions.

a. Describe three different approaches in which handoff can be performed using the knowledge of the signal strengths received by the mobile hosts.
b. In your assessment, which of the three approaches in part (*a*) is more efficient? Explain.
c. Propose and describe a mobile-initiated scheme to perform handoff.
d. Does mobile-initiated handoff involve actions by the mobile switching center? If so, what is the involvement of the mobile switching center in a mobile-initiated handoff scheme?

P7-3 By connecting an array of adjacent base stations to a single mobile switching center (MSC), handoffs between cells connected to the same MSC are referred to as intraswitch handoff, while handoffs between cells connected to different MSCs are referred to as interswitch handoffs.

a. Describe the functions needed to effect intraswitch handoff and interswitch handoff. What are the basic differences between intraswitch handoff and interswitch handoff?
b. To extend the geographical coverage of a wireless communications system, an MSC is normally connected to other MSCs in a hierarchical structure and/or to a backbone wireline network. When a mobile host has to be handed off from a base station in one MSC to a base station in another MSC, the attachment point changes. Messages destined for the mobile will continue to be sent to the former MSC for delivery. One way to maintain undisrupted message delivery is to extend the path from the former MSC to the new MSC. This approach is referred to as "path extension". Propose a procedure to perform path extension, and discuss the advantages and drawbacks in the path extension approach.

c. In interswitch handoff, the information about the mobile host that resides in the old MSC needs to be copied to the new MSC. Copying incurs delay. Explain why copying is necessary.
d. In either intraswitch or interswitch handoff, the target base station to which the mobile is to be handed off may not have resources available to accept the handoff. In this case, the handoff call will be dropped. Describe an approach to reduce the probability of handoff call dropping. If methods to reduce the probability of handoff call dropping were incorporated, what side effect would this have on the overall performance of the wireless communications system?

P7-4 The design of handoff algorithms depends to a large extent on the user mobility pattern defined by the speed and direction of movement. For a fast moving mobile, the speed is the dominant parameter in determining the cell boundary crossing.

a. Explain why, with fast moving mobiles, speed is more dominant in characterizing the use mobility pattern than direction?
b. Consider the situation in which, when the mobile initiates and establishes a connection, its current location is 10 km from the cell boundary along the direction it is currently traveling. Also, the mobile is moving at a speed uniformly distributed between 40 km/hr and 60 km/hr. Calculate the mean time for the mobile to reach the cell boundary.
c. In an MAHO scheme, the mobile sends profiles containing the signal strengths received from all neighboring base stations in the same MSC. If a profile is sent every three minutes on average, what is the mean number of profiles sent before the mobile crosses the cell boundary?

P7-5 With hexagonal cells, each cell has six nearest neighbors. Consider the situation is which an MSC oversees the operation of a cluster of seven cells. A mobile assisted handoff strategy is used to perform handoff by the MSC, if the mobile's movement is within the cluster of seven cells connected to the MSC. The MSC makes handoff decisions based on distance information it collects from the cells in the cluster and profile information sent to it by the mobile. The profile contains the signal levels received by the mobile for all neighboring base stations. The mobile can send the profile via a separate control channel. However, this would waste channel bandwidth. Also, the size of the profile and the frequency with which profiles are sent have an impact on the amount of channel bandwidth not available for transmission of information-bearing signals.

a. Based on the description of the profile given in Subsection 7.3.6, describe and draw the format you would use to transmit the profile.
b. Instead of sending the profile over a separate control channel, it may be piggy-backed with the transmission of the data packet. Describe and draw the format of a wireless packet. (Note that the MSC must be able to decipher the contents of the received packet.)
c. Suppose you decide to send the profile as a separate wireless packet. Discuss how you would format the wireless packet such that the serving MSC can distinguish whether the packet contains profile or payload.

P7-6 In mobile communications, a mobile host is associated with a home network. When the mobile moves to a foreign network, certain actions need to be performed in order to enable uninterrupted communication.

a. What are the actions that need to be performed in order to maintain continuous communication when a mobile crosses the boundary shared by two subnetworks?
b. Describe a procedure to implement each of the actions named in part (*a*).

P7-7 Derive Eqs. (7.3.1)–(7.3.4).

P7-8 Write a program to implement the intraswitch handoff algorithm described in Example 7.2, and test for consistency.

P7-9 Consider a cellular system in which the location of mobile stations within the cellular network is managed by the IS-41 two-tiered architecture.

a. With the help of diagrams, describe the sequence of actions that is required for (1) reporting location update and (2) call delivery.
b. If an overlay scheme such as local anchoring is incorporated into the two-tiered architecture of IS-41, explain and discuss the advantages and disadvantages.

P7-10 In a location management network architecture such as that of IS-41, network signaling exchange is performed using the SS7 network.

a. The service switching point (SSP), the mobile switching center (MSC), and the visitor location register (VLR) are colocated. Describe the role of each of these entities.
b. Explain why both the regional service transfer point (RSTP) and local service transfer point (LSTP) are used for SS7 signaling. How do RSTP and LSTP differ functionally?

P7-11 When a message is to be routed to a mobile host, the network (i.e., the calling MSC) needs to determine the mobile's location.

a. Is paging a necessary procedure in locating the mobile host? Explain.
b. Locating the mobile host can be done by using a three-part location mechanism. Name the three parts and discuss the action involved in each of the parts.

P7-12 Whenever a mobile moves away from its home network, it has to register with its home location register (HLR) to facilitate message delivery to the mobile in its current location. The network that the mobile moves into, that contains the HLR with which the mobile needs to update its registration, is called a registration area (RA).

a. Discuss, with diagrammatic illustration, the meaning of a registration area, the relationship between cells and their corresponding RA, and the relationship between RAs in terms of the need for registration update.
b. The frequency of registration update may further be reduced by grouping registration areas into clusters. Suppose you want to group five registration areas into one cluster for the purpose of reducing the frequency of registration. Assuming that an RA is approximated by a square, discuss how you would arrange the cluster of the five RAs, one of which is the last reported RA.

P7-13 Consider the setting up of a call between a caller and a callee.

a. A mobile wishes to initiate a call to a remotely located mobile. Using the two-tiered architecture of IS-41, describe the procedure involved in establishing the call.
b. The calling end may have an idea of which registration area the called mobile would be in, but does not know in which cell the mobile is located. In order for the communication between the caller and the callee to proceed, the called mobile needs to be assigned a temporary local directory number (TLDN) for identification purposes. The process to locate the mobile so that it can be assigned a TLDN is referred to as paging. Describe and discuss the paging procedure in locating the called mobile.
c. In part (*b*), the rate of RA crossing and the rate of call arrivals have an impact on the cost of call delivery and the cost of paging. Define CMR (call-to-mobility ratio) as the ratio of the call arrival rate to the RA crossing rate. Discuss the tradeoff between the costs of call delivery and paging with CMR as a parameter (i.e., as the value of CMR varies from low to high).

P7-14 Reducing the frequency of accessing the HLR has the effect of reducing registration cost. The use of a local anchor (LA) can achieve reduction in registration cost.

a. Explain how the use of a local anchor can reduce registration cost.
b. One rarely gets something for nothing. Describe the operational features of the LA and discuss the price entailed in the deployment of the LA approach for location management.

P7-15 Consider a PCS system where the physical size, traffic load, and speed of mobile movement are as follows:

Hexagonal cell radius = 0.2 km
Number of cells per cluster = 40
Mobile speed = 30 km/hr
Percentage of subscribers in the coverage area that are active = 10%
Percentage of mobile stations that are powered on = 40%
Traffic load per mobile station = 0.06 Erlangs

a. Assume a mobility model in which the mobiles are distributed uniformly in a cell and the direction of mobile movement relative to the boundary of a cell or a cluster is uniformly distributed on $[0, 2\pi]$. For the traffic conditions specified in the problem, the resultant rate of boundary crossing is 28.64 crossings/cell/second. Determine the population density of subscribers per cell.
b. Assuming the cluster shape is approximated by a hexagon, determine the number of boundary crossings/cluster/second.
c. Determine the number of intracluster handoffs per hour and the number of intercluster handoffs per hour.
d. Determine the number of location updates per hour in the cluster.

P7-16 In cellular communications, the system capacity increases as the cell size decreases. This increase in system capacity by using smaller cell sizes is accompanied by more complex issues for managing user mobility.

a. What is the added complexity in mobility management when the cell size decreases?
b. Suppose the rate of handoff from one cell to an adjacent cell is known. Discuss how you would use this knowledge to construct a scheme to give handoff call requests a higher admission priority than new call requests.
c. If the information on handoff rate is not readily available, how would you handle call admission such that handoff calls are given a higher admission priority than new calls?

P7-17 Statistics of user mobility patterns are important for designing efficient handoff management schemes. Due to the high degree of randomness in user movement patterns, it is in general a very complex task to obtain the statistics. Certain assumptions on user movements need to be made to simplify the task, even when computer simulation is used. Consider a hexagonal cell with cell radius $R = 5$ km. Assume that mobile users are uniformly distributed in the cell and all the mobiles are active. Over a time period of T, each mobile travels in a straight line at a constant velocity. The initial direction is uniformly distributed over $[0, 2\pi]$. The direction change in the next period is uniformly distributed over $[-\alpha, +\alpha]$, where $\alpha \leq \pi/2$. The velocity can be modeled as a Gaussian random variable with mean μ and standard deviation σ (the negative values are unlikely when $\mu \gg \sigma$ and are to be replaced by 0), and is independent from period to period. The movement pattern of each mobile is independent of those of any other mobiles. We want to obtain the following mobility statistics based on computer simulation.

a. Given $T = 1$ minute, $\alpha = 0.25\pi$, $\sigma = 5$ km/hour, find the mean sojourn time (see Subsection 7.3.8) for μ equal to 40, 45, and 50 km/hour, respectively. Comment on the effect of μ.
b. Given $T = 1$ minute, $\mu = 40$ km/hour, $\sigma = 5$ km/hour, find the mean sojourn time for α equal to $\pi/6$, $\pi/4$, and $\pi/2$, respectively. Comment on the effect of α.
c. Assume that there are 200 active mobiles in the cell all the time. When a mobile moves out of the cell, a new mobile is admitted to the cell with an initial location uniformly distributed in the cell. Given $T = 1$ minute, $\alpha = 0.25\pi$, and $\sigma = 5$ km/hour, determine the mean rate of mobile handoff to a neighboring cell for μ equal to 40, 45, and 50 km/hour, respectively. Comment on the effect of μ.

8

Wireless/Wireline Interworking

As the most pervasive wireline network, the Internet is the target wireline backbone network for the hybrid wireless/wireline network considered in this chapter. The Internet allocates a fixed address to each and every mobile user. All messages sent to a prescribed mobile user are first delivered to its permanent address located in the mobile's home network. The basic information transfer mechanism for the Internet is the Internet Protocol (IP), which resides in the network layer of the ISO (International Standards Organization) seven-layer reference model. As a protocol, IP is simple, but has no provision for traffic control. This means that IP can only offer best effort service. The current approach to forwarding messages from the mobile's fixed address to it in its new location is Mobile IP, introduced by IETF (Internet Engineering Task Force).

This chapter presents the salient features of the IP protocol, the operation of Mobile IP as an extension network to bridge information delivery to mobile users in their new locations, and TCP (transmission control protocol), in the transport layer, to oversee flow control for the IP-based network. We also present an overview of the Wireless Application Protocol (WAP), which has lighter overhead than TCP, and wireless ad hoc networks.

8.1 BACKGROUND

In wireless/wireline interworking, the front-end wireless segment provides a communications environment to support user roaming, while the backbone wireline segment extends the geographical coverage. The Internet, which can provide global information delivery, is the most pervasive wireline network for use as the backbone network in wireless/wireline interworking. In this chapter, we consider the interworking of a wireless network with an IP-based network.

The Internet is a mesh connection of routers (nodes) by wirelines. The nodes at the edge of the Internet are called edge routers, while those in the network proper are called core routers. IP is used to route traffic among the Internet nodes; it is a simple and connectionless network layer protocol that provides a unique interface to diverse upper layer protocols. As a connectionless traffic transfer mechanism, IP handles the transfer of datagrams (e.g., packets). The main responsibility of IP is to address the end-hosts and to supervise the routing of datagrams to their destinations. The IP, which has no traffic control capability, is designed to provide *best effort* services only. To ensure network integrity, the Transmission Control Protocol (TCP) is introduced as a transport layer functionality to exercise flow control. TCP is a connection-oriented point-to-point transport layer protocol to provide end-to-end reliable and in-sequence data transfer over the best effort IP-based network. TCP uses a window mechanism to throttle the traffic at the edge router by shrinking the window size when necessary. A shrinkage of the window size at the edge router restricts the amount of traffic allowed to enter the network. Besides TCP, User Datagram Protocol (UDP) is also used to transfer datagrams in the transport layer. The difference between TCP and UDP is that UDP delivers packets from the source to the destination without any reliability or in-sequence guarantee.

A hybrid wireless/IP-based network supports both mobile and fixed hosts. From a mobility management perspective, it is more efficient to have a cluster of neighboring base stations, instead of a single base station, connected to the backbone network through a single attachment point. The cluster of base stations forms a subnetwork of the wireless segment. In the sequel, we will refer to these subnetworks as wireless networks. A pictorial view of this scenario is shown in Figure 8.1, where the wireless networks (home network and foreign network) are attached to the backbone IP-based network. The home network and the foreign network have different attachment

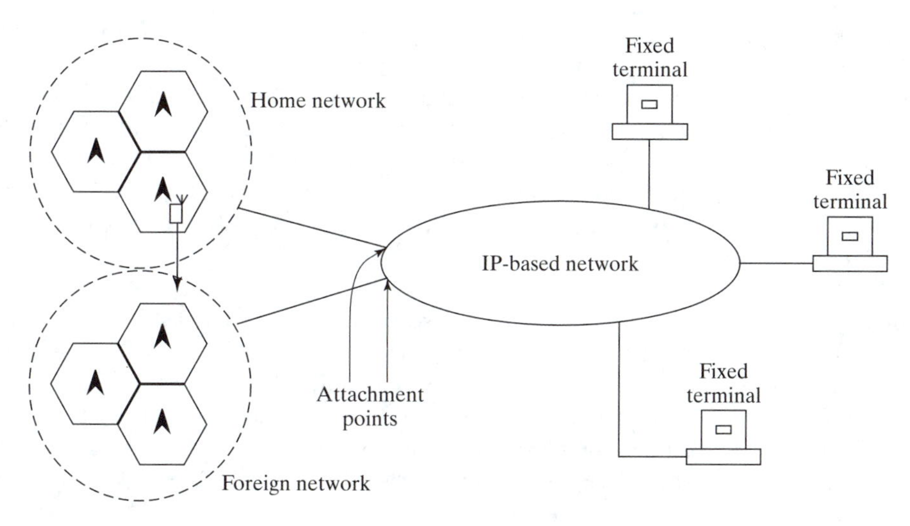

Figure 8.1 Hybrid wireless/IP-based network.

points. When a mobile signs on as a subscriber, it is allocated an IP address. The attachment point to the network where the mobile host initially resides houses the IP address. This network is the mobile's home network and the IP address is the mobile's home address. If the mobile moves to a foreign network without changing its IP address, it will be unable to receive information at the new site; if the mobile changes its IP address when it moves, it will have to terminate and restart any ongoing session.

In a wireless/Internet interworking environment, two of the most pressing problems are

a. the delivery of messages from the Internet to the mobile at its current location, and
b. traffic control to protect network integrity and to satisfy end-to-end quality of service (QoS) requirements.

The focus of this chapter is to address these two problems of wireless/Internet interworking. We first consider the problem of information delivery to the mobile in its current location in Section 8.2. We then describe the changes of IP version 6 from version 4, and the associated changes in the information delivery to mobiles in Section 8.3. The problem of traffic control to ensure end-to-end QoS satisfaction is studied in Section 8.4. End-to-end network performance is then evaluated in Section 8.5, taking into consideration the effect of traffic control. Wireless Application Protocol (WAP), with lighter overhead than TCP/IP, was introduced by the WAP Forum to facilitate Internet access from wireless terminals such as cellphones and portable terminals. The basic principle of WAP is discussed in Section 8.6. Section 8.7 gives an overview of mobile ad hoc networks.

The delivery of messages from the Internet server to the mobile user in its current location in a seamless manner is to be performed by a method referred to as Mobile IP.

8.2 MOBILE IP

When a mobile moves from its home network to a foreign network, messages from the Internet server for the mobile are still sent to the mobile's home address. A mechanism to allow the home network to forward the messages to the mobile in its new location is needed. The Internet Engineering Task Force (IETF) has proposed Mobile IP as an interface between the mobile's home network and the foreign network where the mobile currently resides [112, 15].

Mobile IP is a protocol that keeps track of the mobile's whereabouts and delivers Internet messages to the mobile at its current location. The operation of Mobile IP is enabled by the following functional entities:

Mobile node is a host or a router which can travel around the Internet while maintaining any ongoing communication session. In this text, the terms *mobile node* (MN), *mobile host* (MH) and *mobile station* (MS) are used interchangeably. The reason for this is that these terms are being used quite freely in the literature. A mobile node has a home address, which is a long-term IP address residing in its home network. When away from its home network, the mobile node is assigned a *care-of address* which reflects the mobile node's current point of attachment.

Correspondent node is a peer host with which a mobile node communicates.

Home address is an IP address that is assigned for an extended period of time to a mobile node. It remains unchanged regardless of where the node resides in the wireless segment.

Care-of address is the termination point of tunneling datagrams destined to a mobile node while it is away from home.

Collocated care-of address is an externally obtained local IP address temporarily assigned to an interface of the mobile node.

Home agent is a router with an interface on the mobile node's home network link, which the mobile node keeps informing of its current location, *care-of address*, as the mobile node moves from link to link. The home agent can intercept packets destined to the mobile node's home address and tunnel them to the mobile node's current location.

Foreign agent is a router with an interface on a mobile node's visiting network, which assists the mobile node in informing its home agent of its current care-of address.

Foreign agent care-of address is an IP address of a foreign agent, which has an interface on the foreign network being visited by a mobile node. A foreign agent care-of address can be shared by many mobile nodes simultaneously.

Home network is a network having a network prefix matching that of a mobile node's home address.

Foreign network is a network other than a mobile node's home network to which the mobile node is currently connected.

Virtual network is a network with no physical instantiation beyond its router. The router usually uses a conventional routing protocol to advertise reachability to the virtual network.

Link is a facility or medium over which nodes can communicate at the link layer.

Link-layer address is an address that identifies the physical endpoint of a link. Usually, the link-layer address is the interface's Medium Access Control (MAC) address.

Mobile node's home link is the link which has been assigned the same network-prefix as the network prefix of the mobile node's home address.

Mobile node's foreign link is the link that the mobile node is visiting, which has been assigned the same network prefix as the network prefix of the mobile node's care-of address.

Agent advertisement is the process in which foreign agents advertise their presence by using a special message.

Agent solicitation is the message sent by a mobile node to request agent advertisement.

Tunnel is the path followed by a datagram while it is encapsulated.

Binding entry is an entry in the home agent's routing table. Mobile IP maps the mobile node's home address into its current care-of address.

Messages from the Internet, destined for the mobile, are always sent to the mobile's permanent address in the mobile's home network. The Mobile IP interface is designed to deliver Internet messages from the mobile's home network to the mobile in its current location in a seamless manner. In Mobile IP, the routing of messages from the mobile's home network to the mobile in its current location is accomplished by allowing each mobile to have two IP addresses: a fixed *home address* for identification and a *care-of address* for routing. The home address remains unchanged regardless of where the mobile resides in the wireless segment. However, the care-of address changes at different access points.

In summary, Mobile IP uses an *agent* concept. The mobile has a *home agent* (HA) and a *foreign agent* (FA). The home agent maintains a database in which the mobile's home address resides. When the mobile moves to a foreign network, it establishes an association with its foreign agent which, in turn, establishes an association with the mobile's home agent. That is, the mobile updates its registration with its home agent through the foreign agent. The registration updating procedure is similar to that described in Chapter 7. The operational features of Mobile IP are described in the next subsection.

8.2.1 Operation of Mobile IP

Figure 8.2 shows the functional relationships among the different entities in Mobile IP. The operation of Mobile IP is as follows. The home and foreign agents make themselves known by sending agent advertisement messages. After receiving an agent advertisement, the mobile determines whether it is in its home network or in a foreign network. The mobile basically works like any other node in its home network when it is at home. It routes packets using traditional IP routing protocols. When the mobile moves away from its home network, it obtains a care-of address on the foreign network by soliciting or listening for agent advertisements. The mobile node registers each new care-of address with its home agent, possibly by way of a foreign agent. Datagrams sent to the mobile node's home address are intercepted by its home agent, tunneled

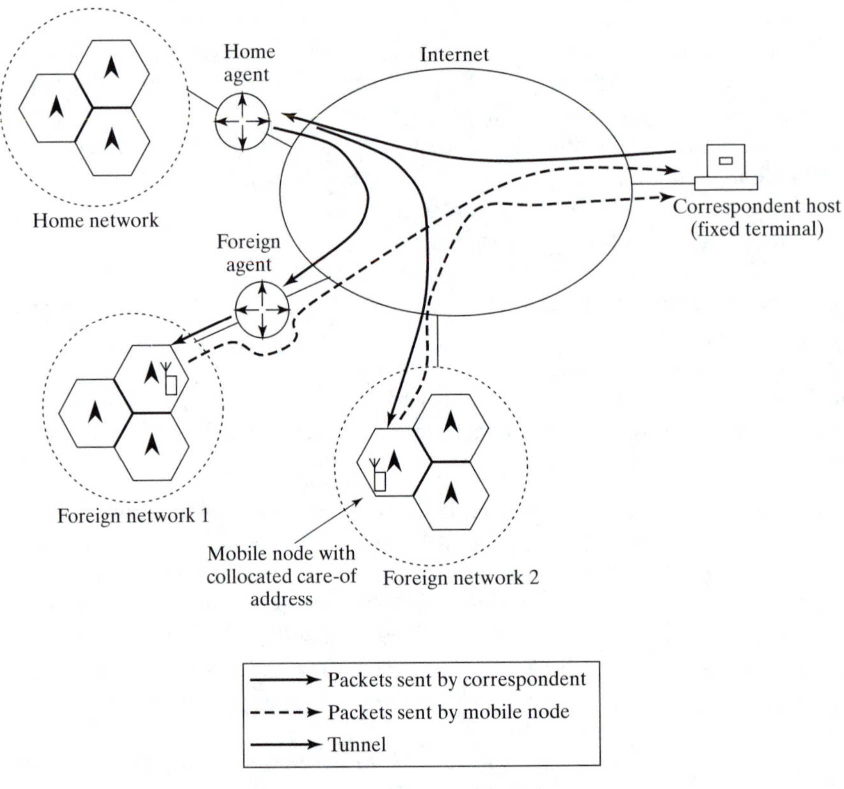

Figure 8.2 Entities in Mobile IP.

by its home agent to the care-of address, received at the tunnel endpoint (at either a foreign agent or the mobile node itself), and finally delivered to the mobile node. In the reverse direction, datagrams sent by the mobile node are generally delivered to their destination using standard IP routing mechanisms. The operation of Mobile IP is based on the cooperation of three major processes: agent discovery, registration, and tunneling (routing).

Agent Discovery is a process by which a mobile node determines its new attachment point or IP address as it moves from place to place within the wireless segment of the wireless/IP network. By agent discovery, a mobile node can (*a*) determine whether it is connected to its home link or foreign link, (*b*) detect whether it has changed its point of attachment, and (*c*) obtain a care-of address if it is connected to a foreign link. The mobile node identifies whether it is connected to the home or foreign link from agent advertisements sent periodically by agents (home, foreign or both) as multicasts or broadcasts to the link. In case a mobile node does not receive any agent advertisement, or it does not have the patience to wait for the next agent advertisement, the mobile node can send an agent solicitation to request an agent advertisement from the agent to which it is attached. When a mobile node is connected to its home link, it works exactly as a traditional node in a fixed place. When a mobile node detects that it has moved, it acquires a care-of address by reading it directly from an agent advertisement, or contacting Dynamic Host Configuration Protocol (DHCP), or using the manual configuration. Registration follows once the mobile node gets a new care-of address.

Registration is a process performed as a mobile node enters and remains in a foreign link. This process involves requesting services from a foreign agent and informing the home agent of a mobile node's new care-of address. Registration also involves reregistration upon expiration of a current registration and deregistration as the mobile node returns to its home link. Some of the characteristics of registration include having multiple, simultaneous care-of addresses and the ability to remove any number of them while retaining others. Registration consists of an exchange of two messages, a registration request and a registration reply, between the mobile node and its home agent, possibly involving a foreign agent as well, depending on the type of the mobile node's care-of address. While an agent discovery message is carried by the Internet Control Message Protocol (ICMP) payload portion, the registration message is carried by the User Datagram Protocol (UDP).

Tunneling (routing) is a process by which Mobile IP tunnels datagrams to the mobile node, whether it is or it is not away from its home network.

Example 8.1 Registration and Deregistration

When a mobile user moves to a visiting location, it has to register with its home agent. When the mobile returns to its home network, it also has to deregister with its home agent. Describe and illustrate with diagrams, using the concepts of care-of address and collocated care-of address, how registration and deregistration can be performed.

Solution

a. When the mobile host moves to a new foreign (serving) agent, it must register with its home agent using the foreign agent's care-of address. This registration process involves

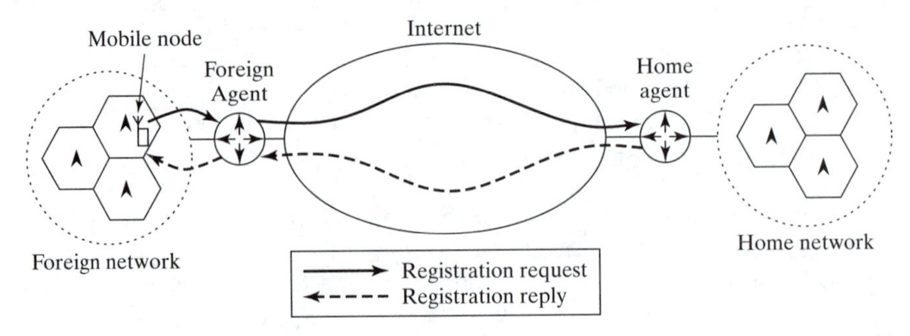

Figure 8.3 Registration with foreign agent's care-of address.

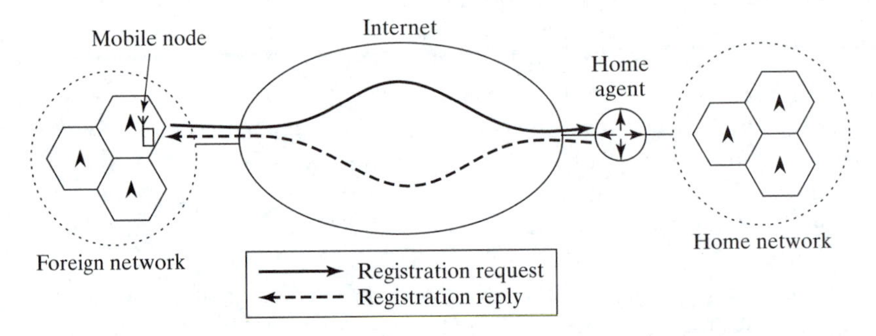

Figure 8.4 Registration with collocated care-of address.

the foreign agent. Figure 8.3, which shows both registration request and registration reply, illustrates the registration process with the foreign agent's care-of address.

b. The mobile host may register using a collocated care-of address. The collocated care-of address is an externally obtained local IP address temporarily assigned to an interface of the mobile host. As such, registration can be performed without the help of a foreign agent. Registration with a collocated care-of address is illustrated in Figure 8.4.
c. When the mobile host returns to its home network, it should deregister with its home address. This is done locally, without the involvement of the Internet. Figure 8.5 illustrates the deregistration process.

Registration can also serve as a means for a new mobile node to obtain the address of a home agent as it initially configures itself for Mobile IP. Registration in Mobile IP must be made secure so that fraudulent registrations can be detected and rejected. Otherwise, any malicious user on the Internet could disrupt communications between the home agent and the mobile node by the simple expediency of supplying a registration request containing a bogus care-of address.

Tunneling. Tunneling is a procedure in which the home agent encapsulates the message from the IP host for delivery to the mobile via its foreign agent. The encapsulation process involves

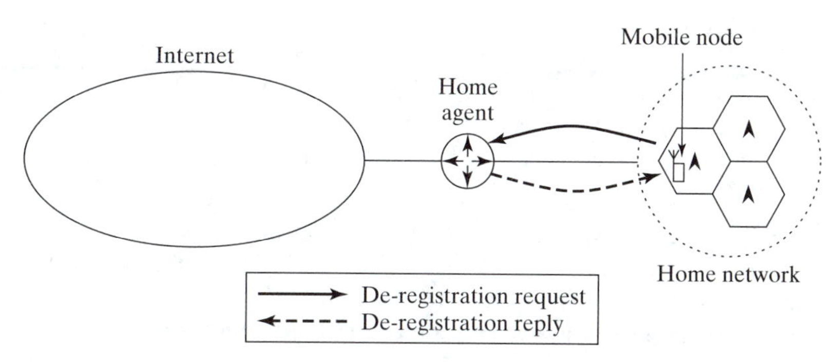

Figure 8.5 Deregistration.

shielding the inner IP header destination address (i.e., the mobile's home address) from intervening routers between the mobile's home network and its current location.

Mobile IP has proposed two routing approaches: *triangle routing* and *optimized routing*. The operational features and the merit and demerit points of triangle routing and optimized routing are as follows.

Triangle Routing A pictorial view of triangle routing is shown in Figure 8.6. The correspondent host (CH) is a fixed host connected to the Internet and communicates with the mobile host through the Internet. Datagram delivery between the correspondent host and the mobile node is performed using the following steps:

(1) A datagram from the correspondent for the mobile is sent to the mobile's home network using standard IP routing.

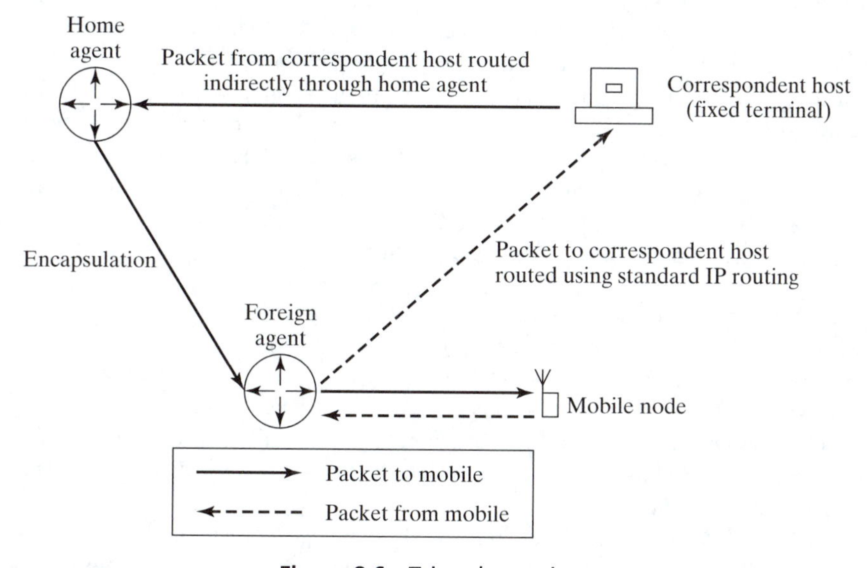

Figure 8.6 Triangle routing.

(2) Upon arrival at the home network, the datagram is intercepted by the home agent which, in turn, tunnels the datagram to the mobile's care-of address.
(3) At the foreign agent, the datagram is detunneled and delivered to the mobile.
(4) For datagrams sent by the mobile, standard IP routing is used to deliver each datagram to its destination. Note that the foreign agent is the mobile's default router.

The Mobile IP protocol with triangle routing is simple; the number of control messages to be exchanged is limited, and the address bindings are highly consistent since they are kept at one single point for a given host. One of the drawbacks of triangle routing is that the destination home agent is a fixed redirection point for exchanging every IP packet, even if a shorter route is available between source and destination. This can lead to unnecessarily large end-to-end packet delay. The other drawback is that the network links connecting a home agent to the network can easily be overloaded. Indeed, all session paths sharing the subnet field of their destination address converge into that subnet home agent, even if adjacent network links are idle.

Optimized Routing In optimized routing, the mobile host informs the correspondent host of its care-of address and has the packets tunneled directly to the mobile host, bypassing the home agent. The Mobile IP protocol with optimized routing allows every traffic source to cache and use binding copies. It supports a further update process by which a binding copy can be sent to the requiring nodes, which may keep it in their cache for immediate or future use. Local bindings enable most packets in a traffic session to be delivered by direct routing, with an apparent gain in terms of quality of service and scalability. In addition, a moving host can always inform its previous foreign agent about the new care-of address, so that packets tunneled to the old location (owing to out-of-date binding copy) can be forwarded to the current location. This should increase the overall quality of service in the case of high mobility. The disadvantage of the protocol is that it is quite complex, and the overhead incurred by message exchanges and processing (due to cache queries) can be critical. Also, cached bindings are possibly inconsistent since they are being kept in a distributed fashion. The main obstacle to implementing optimized routing resides in the security issues. The correspondent node must be informed of the mobile host's care-of address in order to tunnel data to the mobile host. In a hostile environment, an intruder can easily cut off all communications to the mobile host by sending a bogus registration if he/she knows the mobile's care-of address. Therefore, authentication/security measures have to be incorporated in the optimized routing.

Triangle routing is much simpler than optimized routing. In many cases, this is the preferred mode of routing datagrams from the correspondent host to the mobile host. However, as the mobile moves further and further away from its home network, the cost (delay) involved in the registration with the home agent can become prohibitively large. Methods to reduce registration costs are desirable. One approach is the incorporation of a *local anchor* to act as a *virtual home agent* such that the mobile host only needs to register with the close-by virtual home agent instead of the far away home agent.

The concept of local anchor as a means of reducing registration cost is introduced in Subsection 7.5.2. The same concept can be used in conjunction with the operation of Mobile IP.

8.2.2 Local Anchor for Mobile IP

In Chapter 7, we define a registration area within which the mobile does not have to register with its home location register (HLR). Here, we define an anchoring region within which the mobile only needs to register with the local anchor. The local anchor strategy can be described as follows:

a. Choose one agent as the focus of an anchoring region and name this agent as an anchor.
b. When the mobile moves within the anchoring region, it does not need to register with its home agent; instead, it registers with the anchor. That is, the local anchor acts as a virtual home agent.
c. When the mobile moves out of the anchoring region, it will register with its home agent and the new foreign agent will become the focus of the new anchoring region.

It is noted that the local anchor strategy only reduces registration costs, but has virtually no impact on delivery, since messages from the correspondent host are still sent to the mobile's home address, to be intercepted by the home agent and then tunneled to the mobile's care-of address. Thus, packets destined for the mobile will be forwarded from the home agent to the anchor agent first and, from there, to the foreign agent where the care-of address resides. This mode of packet forwarding is illustrated in Figure 8.7.

The anchor agent and the new foreign agent are two candidates that can decide whether the mobile should register with its home agent or not. The mobile does not have the knowledge of the network to make this decision. The decision making process can be based on static or dynamic information. The dynamic approach will have to use the mobile's past movement information and its current traffic information for decision making. The static approach makes use of information that is fixed for all time. For the purpose of establishing a new anchoring region, we will use the static approach, and use the distance from the old anchor agent to the new foreign agent as the criterion to decide whether or not to establish a new anchoring region.

The registration process in the local anchor scheme is shown in Figure 8.8. Depending on which agent (the home agent or the current anchor) the mobile should register with, the

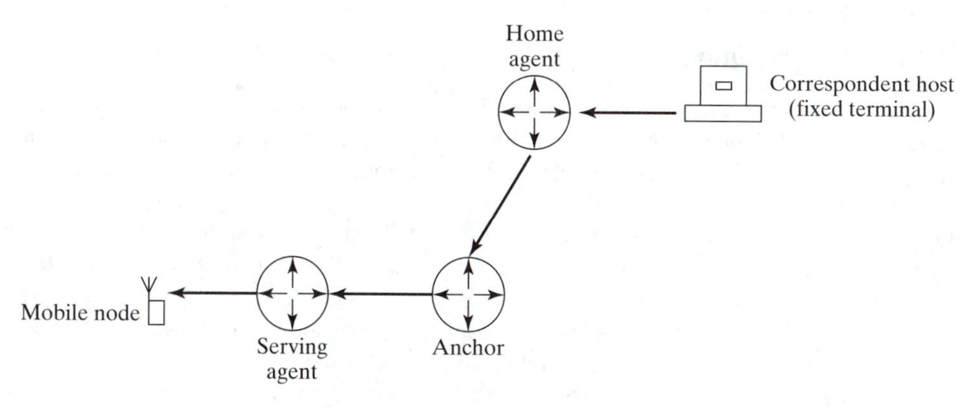

Figure 8.7 Packet forwarding in local anchor approach.

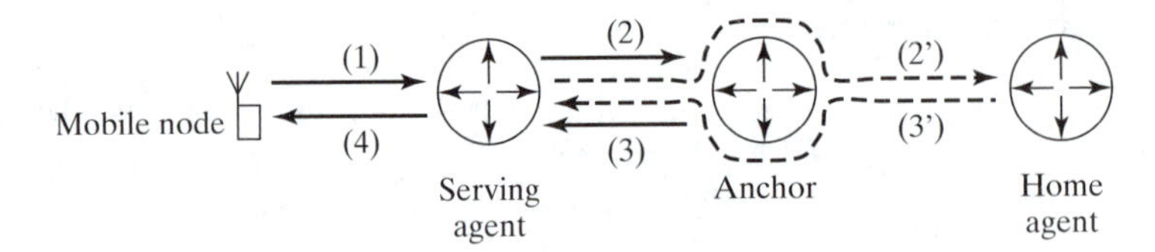

Figure 8.8 Registration in local anchor scheme.

registration process needs to consider two cases. The following example sets up the procedure for the registration process.

Example 8.2 Registration with Local Anchor

Using a static approach, construct the steps needed for the mobile to register, either with the anchor agent or the home agent. Assume that the new foreign agent takes responsibility for making the decision regarding which agent the mobile should register with.

Solution The following four steps constitute a procedure for the mobile to update its registration.

(1) The mobile sends the registration request indicating the current anchor agent and the home agent.
(2) There are two cases–
Case I: the new foreign agent decides that the mobile is still in its current anchoring region so it forwards the mobile's registration request to the anchor.
Case II: the new foreign agent decides that the mobile is out of its current anchoring region, so it forwards the mobile's registration request to the home agent.
(3) The anchor agent or the home agent sends the registration reply back to the foreign agent.
(4) The foreign agent returns an acknowledgment to the mobile and indicates who, the anchor or the home agent, sends this registration reply. In Case I, the mobile knows that it has not moved out of the current anchoring region and the anchor does not change. In Case II, the foreign agent becomes the focus of the new anchoring region and the mobile will update its anchor agent's IP address for later use.

8.2.3 Hierarchical Routing

For highly mobile users, the amount of registration traffic generated between the visited and the home networks can be quite large. In the preceding subsection, we have considered a local anchor approach to reduce the amount of registration with the home agent. The amount of registration between the home and the visited networks can also be reduced by using a hierarchical routing strategy. In hierarchical routing, a hierarchy of foreign agents is established in a tree structure, and multiple foreign agents are advertised in the agent advertisement. In this way, registrations can be localized to the foreign agent that is the lowest common ancestor of the care-of addresses at the two points of attachment of interest. To enable this, the mobile has to determine how high up the tree its new registration has to go, and then arrange for the transmission of the registration to each level of the hierarchy between itself and the closest common ancestor between its new and previous care-of addresses.

Example 8.3 Hierarchical Routing

Suppose that a mobile currently using the services of one foreign agent is migrating to use the services of a different foreign agent. If the foreign agents are hierarchically connected in a binary tree structure, a mobile moving from one foreign network to another foreign network may not involve a direct registration with its home agent.

a. Draw a binary tree, with a population of ten foreign agents, connected to the Internet as a subnetwork and, hence, to the mobile's home agent through the Internet.
b. Describe how the mobile moving from one serving foreign agent to another foreign agent will receive agent advertisements, and the situations under which the mobile does not have to register with its home agent.

Solution

a. The binary tree structure for a population of ten foreign agents is illustrated in Figure 8.9, where the mobile moves from location A to location B and then to location C. The attachment point to the Internet is FA_1. The home agent only directly "sees" foreign agent FA_1, while all the other foreign agents in the binary tree are not visible to the home agent.

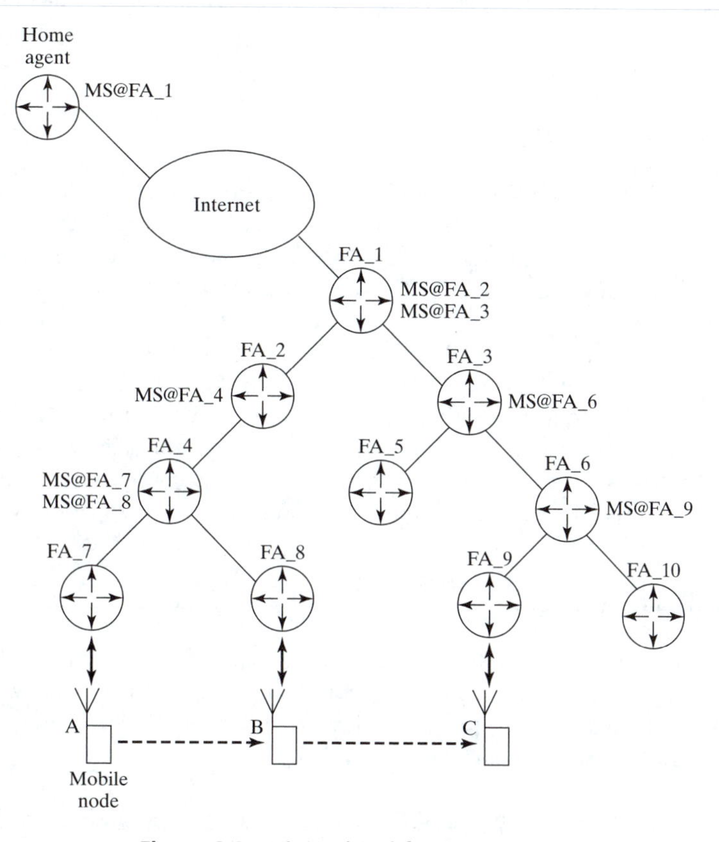

Figure 8.9 Hierarchical foreign agents.

b. As shown in Figure 8.9, the mobile is using the service of FA_7 while at location A. But it also receives agent advertisement from FA_4, FA_2 and FA_1. A registration is transmitted to each of these foreign agents as well as its home agent. Since an agent only "sees" its nearest neighbor in the hierarchy, the home agent believes that the mobile is located at the care-of address of FA_1; FA_1 believes that the mobile is located at FA_2, and so on, until foreign agent FA_7, which actually knows the whereabouts of the mobile. For illustration purposes, consider the following two scenarios.

   (1) When the mobile moves to FA_8 (at location B), it only has to cause the new registration to travel as far as FA_4.

   (2) When the mobile moves to foreign agent FA_9 (at location C), it receives advertisements which indicate the lineage of FA_9, FA_6, FA_3 and FA_1. By comparing the previous and the current lineages, the mobile determines that it has to cause the registration to travel up the hierarchical tree to foreign agent FA_1, but the registration still does not have to reach the home agent.

Note that the original datagram must be relayed to a number of intermediate nodes in the hierarchy. Each of the nodes is then charged with the responsibility of retunneling the datagram, if necessary, to the next lower level in the hierarchy.

8.3 INTERNET PROTOCOL (IP)

Until recently, Internet Protocol version 4 (IPv4) has been the protocol which supports Internet administration and operation. IPv4, with network entities such as mobile nodes, home agent, home address, foreign agent, care-of address, and the like, is the basis for the original development of Mobile IP. The basic IPv4 header, with a 32-bit source address and a 32-bit destination address, is shown in Figure 8.10. With only a 32-bit source address and a 32-bit destination address, the address allocation scheme in IPv4 is insufficient to support the rapidly growing Internet subscriber population. Since 1994, the IETF has been working on Internet Protocol version 6 (IPv6) to provide a remedy to the limitations inherent in IPv4, in terms of addressing, routing, mobility support, quality of service (QoS) provisioning, and so forth. The network entities of IPv6 are similar to those of IPv4, except that IPv6 does not have the concept of a foreign agent. With a 128-bit source address and a 128-bit destination address, IPv6 now supersedes IPv4. The basic IPv6 header is shown in Figure 8.11.

8.3.1 IPv6 versus IPv4

As the successor to IPv4, IPv6 can be installed as a normal software upgrade in Internet devices and is interoperable with IPv4. Also, IPv6 can run well on high-performance networks (e.g., ATM, fast Ethernet, and the like), and is efficient for low-bandwidth networks. The changes from IPv4 to IPv6 fall mainly into the following categories [77].

Expanded Routing Addressing Capabilities. With a 32-bit IP address in IPv4, which can hold up to $2^{32} - 1$, or over 4 billion hosts, one might think this address range is more than enough

<table>
<tr><td>Version
4 bits</td><td>Header
Length</td><td>Type of Service
8 bits</td><td colspan="2">Total Length of Datagram
16 bits</td></tr>
<tr><td colspan="3">Datagram Identification (16 bits)</td><td>Flag
3 bits</td><td>Fragment Offset (13 bits)</td></tr>
<tr><td colspan="2">Time to Live
8 bits</td><td>Protocol
8 bits</td><td colspan="2">Header Checksum
16 bits</td></tr>
<tr><td colspan="5">Source IP Address (32 bits)</td></tr>
<tr><td colspan="5">Destination IP Address (32 bits)</td></tr>
<tr><td colspan="5">IP Options</td></tr>
<tr><td colspan="5">Data Portion of Datagram</td></tr>
</table>

Figure 8.10 IPv4 header.

<table>
<tr><td>Version
4 bits</td><td>Priority
4 bits</td><td colspan="3">Flow Label (24 bits)</td></tr>
<tr><td colspan="3">Payload Length (16 bits)</td><td>Next Header (8 bits)</td><td>Hop Limit (8 bits)</td></tr>
<tr><td colspan="5">Source IP Address (128 bits)</td></tr>
<tr><td colspan="5">Destination IP Address (128 bits)</td></tr>
</table>

Figure 8.11 IPv6 header.

to support the addresses needed on the Internet. However, with this address space, one can easily run out of available addresses, as explained in the following:

a. Although the traditional two-level addressing scheme, with a network prefix and a host address, is convenient, there is a waste of address space. Once a network address is assigned to a particular network, a block of IP address is assigned to that network. Any improper assignment of network address will lead to a great waste of available IP addresses.

b. Many private networks, which are currently not connected to the Internet, are reusing IP addresses used by the public network or other private networks. They require many more IP addresses when connected to the Internet.
c. There may be devices other than the traditional hosts, possibly both wireless and wireline products, such as mobile telephones, wireless organizers, and the like, which need additional IP addresses to make themselves identifiable on the Internet.

The increase of IP address size from 32 bits to 128 bits in IPv6 allows the Internet to support more levels of addressing hierarchy, a much greater number of addressable nodes, and simpler autoconfiguration of addresses. The traditional two-level IP address structure, network address and host address, is modified. Unicast address, anycase address and multicast address are defined in IPv6. The meaning of unicast, multicast and anycast addresses are as follows:

Unicast address is simply a 128-bit network node address. Unicast address can be divided into two parts: a subnet prefix, which indicates the node's subnetwork; and an interface ID, which indicates the node's interface.

Multicast address in IPv6 replaces broadcast in IPv4. Multicast address is divided into two groups: the predefined groups, which are permanently assigned; and the transient groups, which are defined by specific organizations. The most common predefined multicast addresses are: All Nodes (all nodes connected), both routers and hosts; All Routers (not including hosts); and All Hosts (not including routers).

Anycast address is a new type of address defined to identify sets of nodes such that a packet sent to an anycast address is delivered to any one of the nodes assigned that address. Packets destined to a multicast address are sent to all nodes in that group, while packets destined to an anycast address are sent to only one node in that group. The use of anycast address in IPv6 allows nodes to control the path along which their traffic flows. The scalability of multicast routing is improved by adding a "scope" field to the multicast address.

There will be coexistence of both IPv6 addresses and IPv4 addresses and, also, it is impossible to replace all IPv4 routers with IPv6 routers. In IPv6, special addresses, called IPv4-compatible-IPv6 addresses, are introduced for assignment to those hosts and routers running IPv6, but which must route traffic across IPv4 networks. On the other hand, IPv4-mapped-IPv6 addresses are assigned to those hosts running IPv4.

Header Format Simplification. The IPv6 header has a fixed length of 40 octets. Some IPv4 header fields have been dropped, or made optional, to reduce the common-case processing cost of packet handling and to keep the bandwidth cost of the IPv6 header as low as possible, despite the increased size of the addresses. IPv6 extension headers, optional parts following the IPv6 header, are defined to carry additional information about the traffic being sent. Extension headers include

a. hop-by-hop option header, which defines special options that require hop-by-hop processing;
b. fragment header, which contains fragmentation and reassembly information;
c. destination options header, which contains optional information to be examined by the destination node;
d. routing header, which provides extended routing; and
e. authentication header, which provides packet integrity and authentication.

Improved Support for Options. Changes in the way that IP header options are encoded allow for more efficient forwarding, less stringent limits on the length of options, and greater flexibility for introducing new options in the future.

Priority. A 4-bit priority field is introduced for the source node to indicate the desired transmit and delivery priority of every packet relative to other packets from the same source. Traffic types are first classified as congestion-controlled traffic or non-congestion-controlled traffic; then one of eight levels of relative priority is assigned to each type of traffic.

Congestion-controlled traffic refers to traffic which can tolerate congestion, or delay. If network congestion happens, congestion-controlled traffic will be buffered or "backed off". A variable amount of packet delay or even out-of-order packet arrival is acceptable. IPv6 defines the following types of congestion-controlled traffic, in decreasing priority: Internet control traffic, interactive traffic, attended bulk transfer, unattended data transfer, filler traffic, and uncharacterized traffic.

Non-congestion-controlled traffic refers to that traffic which requires constant data rate, constant delivery rate (or at least relatively smooth data rate), or delivery delay. Examples of non-congestion-controlled traffic are real-time audio and real-time video.

Quality-of-Service Capabilities. A new capability is added to enable the labeling of packets belonging to particular traffic flows for which the sender requests special handling, such as voice or video. A flow is basically a series of packets originated by the source and having the same transmission requirements. No special transmission process is assigned to any particular flow label. A source must specify, or negotiate, what kind of special handling is requested before a flow is transmitted, possibly by means of other Internet control protocol. Flow labels are assigned pseudorandomly to ensure that there is no flow label reuse during the lifetime of that flow label.

Security Capabilities. IPv6 will support the five proposed security-related standards published by IETF. These security features, which are optional, are

RFC 1825 - Security Architecture for the Internet Protocol,
RFC 1826 - IP Authentication Header,
RFC 1827 - IP Encapsulating Security Payload (ESP),
RFC 1928 - IP Authentication Using Keyed MD5, and
RFC 1829 - The ESP DES-CBC Transform.

Two IP security mechanisms, security association and authentication, are combined to transmit IP packets that require both privacy and authentication. IPv6 includes the definition of extensions that provide support for authentication, data integrity, and confidentiality. This is included as a basic element of IPv6 and will be included in all implementations.

8.3.2 Mobile IPv6

Current activities by IETF on Mobile IP capture the salient features of IPv6. Mobile IPv6 uses the new and improved IPv6 *Routing Header*, along with the *Authentication Header*, and other pieces of IPv6 functionality to simplify routing to the mobile node and to perform route optimization

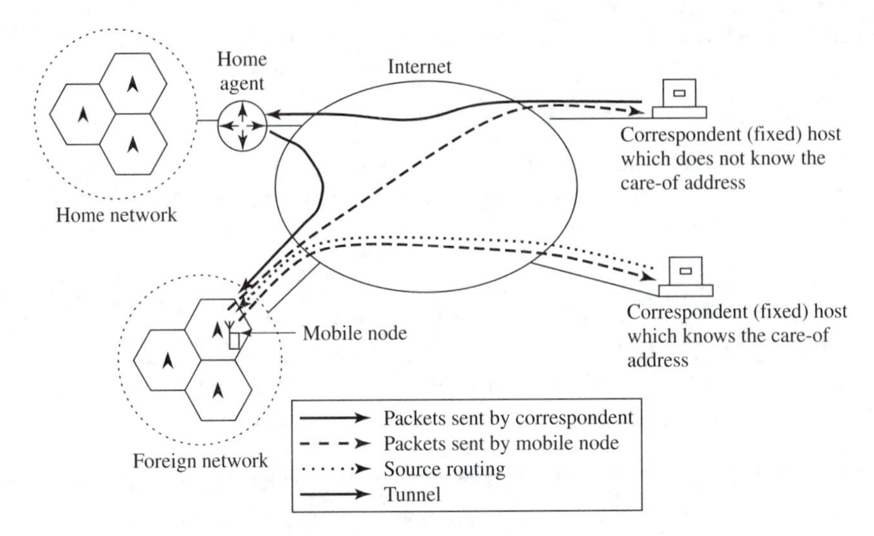

Figure 8.12 Mobile IPv6 operation.

in a secure manner. Mobile IPv6 has no foreign agent. The mobile node uses the *Address Autoconfiguration* procedure defined in IPv6 to acquire a collocated care-of address on a foreign link, and reports its care-of address to its home agent and selected correspondents. The operational procedure of Mobile IPv6 is depicted in Figure 8.12.

In Mobile IPv6, a correspondent node which knows the mobile's current care-of address can send packets directly to the mobile node by using an IPv6 Routing Header. Those correspondent nodes who do not have this information send packets without such a header. The packets are routed to the mobile node's home link, intercepted by the home agent, and tunneled to the mobile node's care-of address. When the mobile node returns to its home link, it notifies its home agent.

Mobile IPv6 has almost the same terminologies as Mobile IP (also called Mobile IPv4) discussed in Section 8.2, except for the absence of foreign agent. The concept of home agent, home link, care-of address and foreign link are roughly the same as in Mobile IPv4. Compared with Mobile IPv4, Mobile IPv6 has the following advantages:

a. The enormous address space in IPv6 allows very simple address autoconfiguration by means of Stateless Address Configuration (SAC). Because the mobile node can easily obtain a collocated care-of address by SAC, the foreign agent functionality is no longer needed. As a result, the foreign agent is eliminated from Mobile IPv6. This also implies that all Mobile IPv6 care-of addresses are collocated care-of addresses.
b. Mobile IPv6 uses the new IPv6 routing header to simplify routing to mobile nodes.
c. With the enhanced authentication header in IPv6 and the mandatory implementation of IP authentication header, Mobile IPv6 might adopt a wide scale of route optimization if a key management infrastructure becomes widely available on the Internet.
d. Mobile IPv6 uses both tunneling and source routing to deliver packets to mobile nodes. In the case of Mobile IPv4, tunneling is the only option.

8.4 TRANSMISSION CONTROL PROTOCOL (TCP)

8.4.1 Flow Control

As mentioned at the start of this chapter, the Internet Protocol only provides best effort services. Traffic control in the Internet uses the TCP in the transport layer to exercise flow control. TCP, which is designed to provide reliable end-to-end services, is a sliding window flow control mechanism. Control decisions are made based on control signals fed back from the receiver. In this sense, TCP is a reactive control method. The control signals fed back from the receiver are in the form of acknowledgment (ACK) packets. The main indicator of a problem of unsuccessful transmission through the IP network is congestion experienced in the routers along the path connecting the sender and the receiver. In conventional TCP schemes, designed primarily for exercising flow control in IP-based networks, control decisions to regulate the traffic allowed to enter the network are strictly based on network congestion. In a hybrid wireless/IP network, transmission errors incurred in the wireless propagation channel will also have to be taken into consideration in formulating the flow control policy.

As a reactive control scheme, the window size in TCP is regulated by a control signal fed back from the receiver to the transmitter. Packets arriving at the sender are stored in a buffer. The sender releases packets from the buffer and adjusts the size of the sliding window based on the acknowledgment fed back from the receiver. To facilitate flow control, the sender keeps a timer. If the sender fails to receive an acknowledgment for a packet from the receiver after a timeout interval, it retransmits that packet and exponentially backs off the timer. To determine an appropriate timeout value, TCP keeps track of the round-trip time for the data packets and the corresponding acknowledgment packets, and uses the accumulated knowledge to calculate the timeout value.

The sender transmits packets based on the size of the sliding window. At the start, the sender assumes a minimum window size and probes the network to determine the available network capacity. If the network is not in a congested state, so there is bandwidth to support an amount of traffic greater than that specified by the minimum window size, the window size is increased exponentially until the amount of traffic entering the network reaches a preset threshold. This exponential increase in the window size is referred to as a *slow-start*. When the threshold is reached, the sender probes the network continuously in a linear manner in an attempt to avoid network congestion.

If the sender fails to receive an ACK from the receiver for a particular packet after a timeout period, then the sender assumes the packet is lost inside the network, due to network congestion. The sender sets the threshold to one half of the last window size and throttles the window size to its initial value to avoid network congestion, and then retransmits the packet.

There are a number of ways the receiver can send acknowledgments. For example, the receiver may send an ACK only for each correctly received packet. In this case, the sender uses a timeout to retransmit the lost packets. An alternative is for the receiver to send duplicate acknowledgment packets for the last correctly received packet. Upon receiving a predetermined number of duplicate acknowledgment packets, the sender infers that there are missing packets in the receive buffer, and starts retransmitting the first unacknowledged packet, and then reduces the window size proportionately. A scenario in which TCP starts with slow-start and a preset

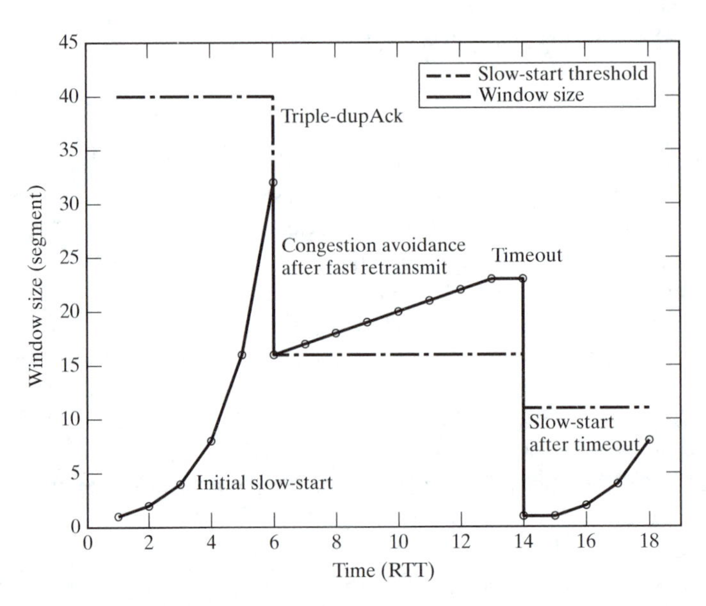

Figure 8.13 TCP window flow control strategy.

threshold value is shown in Figure 8.13, where the horizontal axis is in units of round-trip time (RTT) and the timeout interval is assumed to be one round-trip time. When the sender receives three duplicate acknowledgment packets (i.e., *Triple-DupAck* in Figure 8.13), it infers that the network is experiencing congestion. TCP then performs Fast Retransmit (see Subsection 8.4.1) and enters the congestion control phase by reducing the threshold value to one-half of the last window size.

8.4.2 Modified TCP

TCP is a connection-oriented transport layer protocol that is designed to provide reliable and in-sequence data delivery. However, if TCP is used without any modification in a hybrid wireless/IP network, a serious drop of throughput may occur. The reason is that a high bit error rate (BER), or disruption caused by poor wireless link quality, can corrupt packets, which may result in losing TCP data segments or acknowledgments. When acknowledgments do not arrive at the TCP sender within a prescribed interval of time, a timeout occurs. The sender retransmits the segment, exponentially backs off its retransmit timer for the next retransmission, and then reduces its window to one segment. Repeated errors will cause the window to remain small, resulting in a low throughput, especially on long links.

In order to ensure that the TCP connection to a mobile is efficient, it is necessary to prevent the sender from shrinking its congestion window when packets are lost, either due to bit errors or due to disconnection. When the mobile is reconnected, it should begin to receive data immediately. Several proposals have been reported in the literature for a modified or new TCP that is optimized for use over wireless links. These include Indirect Transmission Control Protocol (I-TCP), Berkeley Snoop Module, Fast Retransmit, and TCP for Mobile Cellular Network (M-TCP) [10, 20].

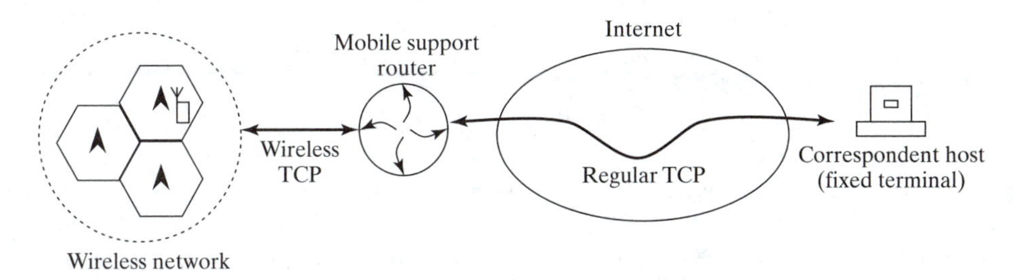

Figure 8.14 Indirect Transmission Control Protocol (I-TCP).

Indirect Transmission Control Protocol (I-TCP). In I-TCP, a connection between a mobile host and a fixed host is split into two separate connections at the base station—one between the mobile host and the base station or its mobile support router (MSR) over the wireless medium, and the other between the base station, or the MSR, and the fixed host over the wired network. The I-TCP scenario is shown in Figure 8.14. In this way, the special requirements of mobile hosts can be accommodated, which is backward compatible with the existing fixed network. All the specialized supports that are needed for mobile applications and for the low speed and unreliable wireless links can be built into the wireless side of the interaction while the fixed side is left unchanged. Data sent to the mobile host are received and acknowledged by the base station before being delivered to the mobile host. The wireless/wired link characteristics would be hidden from the transport layer, and only the wireless resources would be used for error control when the error is caused by the wireless link. With the I-TCP, the resulting benefits lie in that:

a. the flow control and congestion control functionalities on the wireless link are separated from those on the wired link;
b. a separate transport protocol for the wireless link can support notification of events such as disconnections, user movements and other features of the wireless link (e.g., changes in the available bandwidth) to the higher layers;
c. a partition of the connection into two distinct parts allows the base station to manage much of the communication overhead for a mobile host.

Throughputs are increased with I-TCP since the node, where the connection is split, may be one or two hops away from the mobile host's radio cell, and can adapt more quickly to the dynamic mobile environment because the round-trip time is shorter. However, I-TCP does not maintain end-to-end TCP semantics. This is because the TCP acknowledgments are not end-to-end; instead, there are separate acknowledgments for the wireless and the wired portions of the connection. One consequence is that the sender may believe that a segment is delivered correctly to the mobile host since the base station acknowledges it even if the mobile host is disconnected before receiving this segment. In other words, the sender does not know whether packets are actually received by the mobile host, and this may be a serious problem for many applications.

Berkeley Snoop Module. Snoop is another proposed solution for losses caused by a high BER. The Berkeley Snoop Module makes changes to the network layer software at the base station. It caches packets at the base station, inspects the TCP header of TCP data packets and acknowledgments which pass through, and buffers copies of the data packets. Using the information from the headers, the snoop module detects lost packets (a packet is assumed lost when duplicate acknowledgments are received) and performs local retransmissions across the wireless link to alleviate problems caused by a high BER. The module also implements its own retransmission timer, similar to the TCP retransmission timeout, and performs selective retransmissions when an acknowledgment is not received within this interval. Routing protocol is also modified to enable low-latency handoff to occur with negligible data losses. Experiments have shown that the Berkeley Snoop Module achieves a throughput up to 20 times that of regular TCP, and handoff latencies over 10 times shorter than those of other mobile routing protocols. A drawback of the snoop module is that it does not perform as well in either the presence of lengthy disconnections or environments in which there are frequent handoffs. If the mobile host is disconnected for a lengthy period of time, the sender will automatically invoke congestion control because it does not receive acknowledgments for some segments. The snoop module will persistently generate packets and these packets will serve no purpose since the mobile is disconnected. If the mobile host moves into a new cell, the new base station starts up a copy of the snoop module on behalf of this mobile host. This snoop module begins with an empty cache and slowly builds up the cache, and the mobile host will see a poor TCP throughput. If the radio cell sizes are small, the performance degradation can be serious.

Fast Retransmit. Fast Retransmit is proposed to combat the effects of short disconnections on TCP throughput. During a handoff, since the mobile host cannot receive packets, unmodified TCP at the sender will think that a congestion has occurred and will begin congestion control (by reducing the window size and retransmitting) after a timeout. The timeout period may be long; even though the mobile host may have completed the handoff, it will have to wait for the full timeout period before it can begin receiving packets from the sender. Fast Retransmit forces the mobile host to retransmit, in triplicate, the last old acknowledgment as soon as it finishes a handoff. This forces the sender to reduce the congestion window to one-half and retransmit one segment immediately. Fast Retransmit does not split the TCP connection. However, if the mobile host were disconnected for a long time, the sender would already have invoked congestion control and shrunk its window to one segment. Similarly, if disconnections are frequent or the wireless links are poor, Fast Retransmit will do little to improve the throughput because the sender's congestion window will repeatedly get shrunk to half of its previous size.

TCP for Mobile Cellular Network (M-TCP). M-TCP works in a three-level hierarchy: mobile hosts, supervisor host, and the Internet, from the lowest to the highest by introducing the supervisor host (SH). The three-level M-TCP hierarchy is shown in Figure 8.15, where several base stations are controlled by one SH. The SHs are connected to the Internet and handle most routing and other mobile users' requirements. The advantages of this hierarchy are:

a. When a mobile host roams from one cell to another, the two base stations do not need to transfer any state information if they are controlled by the same SH.
b. The roaming mobile host remains within the domain of the same SH for long time periods because several base stations are controlled by the SH.

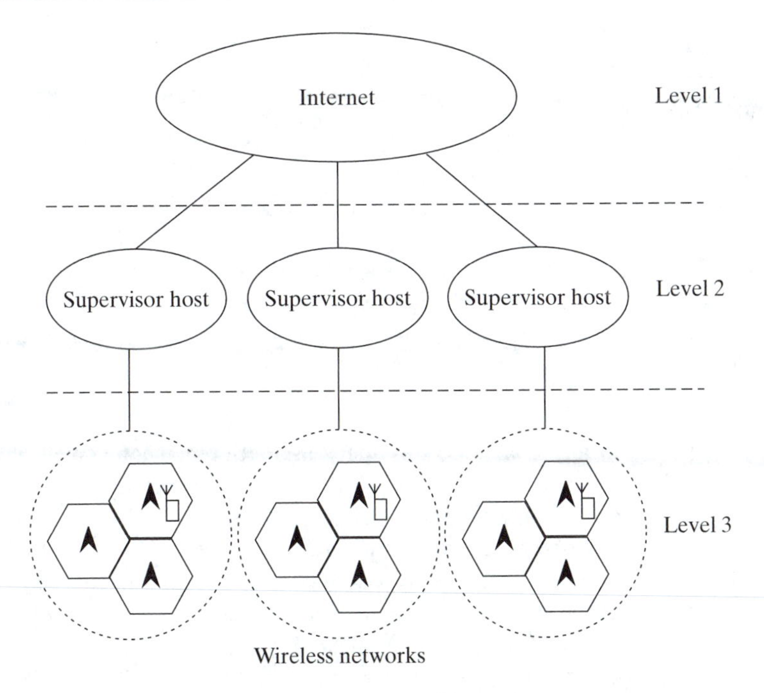

Figure 8.15 M-TCP hierarchy.

By introducing the SH, M-TCP maintains end-to-end TCP semantics while it delivers excellent performance when mobile hosts encounter disconnections. This is done by splitting the TCP connection at the SH. As packets arrive from a sender on the Internet, an acknowledgment is sent back and the SH deals with ensuring the completion of delivery. The drawback of this scheme is that it is fairly complex. Also, there will likely be a shortage of buffer space if a supervisor host services too many mobile hosts.

8.4.3 Modified UDP

User Datagram Protocol (UDP) is a datagram communication service built on top of IP. It adds multiplexing and error detection to the IP capabilities. In contrast to TCP, UDP does not use acknowledgments, and does not retransmit erroneous packets or control the flow. In wireless/IP interworking, a large percentage of packets will be lost by using UDP over wireless links. This is because UDP will continue to send packets even when transmission to a mobile host experiences signal fading. A simple concept would be to stop sending datagrams to a mobile host once it encounters fading. The goal of creating a new modified UDP (M-UDP) is to ensure that packets that have been lost are retransmitted. In the M-UDP protocol, the UDP connection is split in two at some host close to the mobile host. The host attempts to use any free bandwidth to retransmit packets lost during a fade, thus ensuring that the number of lost packets is kept small.

8.5 NETWORK PERFORMANCE

Communications networking operations are relatively complex. Exact analytical evaluation of network performance is often too difficult to tackle. Approximate analysis based on certain reasonable assumptions can yield valuable guidance for specifying system and traffic parameters. In this section, we evaluate the performance of the local anchor scheme described in Subsection 8.2.2, as an example. It is necessary to use some metrics as performance measures. Two such metrics are average handoff delay and average cost between two consecutive handoffs.

It is not easy to precisely model the Internet, since the route between any two hosts may change dynamically. But a network model is needed to facilitate performance analysis and evaluation.

8.5.1 Network Model

To simplify the analysis problem, assume that the route between the involved agents is fixed and, within the mobile's residence time, the delay on the corresponding route is also fixed. The delay of a path is proportional to the length of the path or the number of hops on the path. For simplicity, here we take the delay on a path as the cost of the path.

Consider the case in which the mobile receives information from a fixed terminal (the correspondent host). Assume that the correspondent host generates data packets destined for the mobile at a fixed rate λ. The network model for cost analysis is shown in Figure 8.16, where a, b, d, and f are the costs (delays) associated with the paths, respectively.

Assume a, b, and d are fixed, but f is a variable. In order to quantify f, assume that the relationship between the physical distance and the network distance can be modeled by a bifork tree, as shown in Figure 8.17. From this figure, it can be seen that, if any two agents are neighbors in

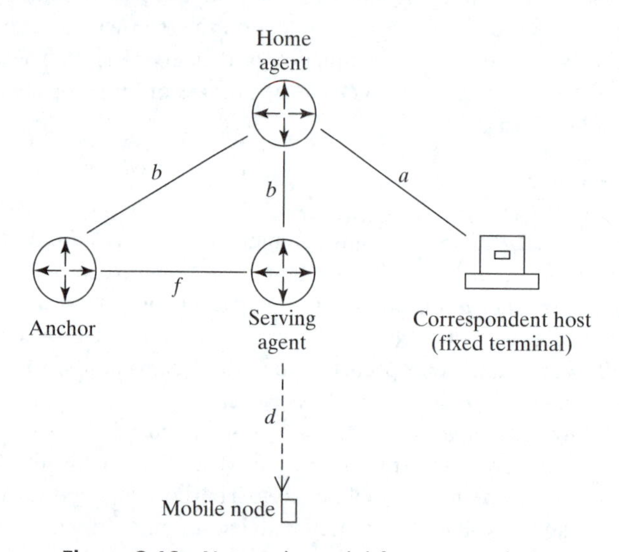

Figure 8.16 Network model for cost analysis.

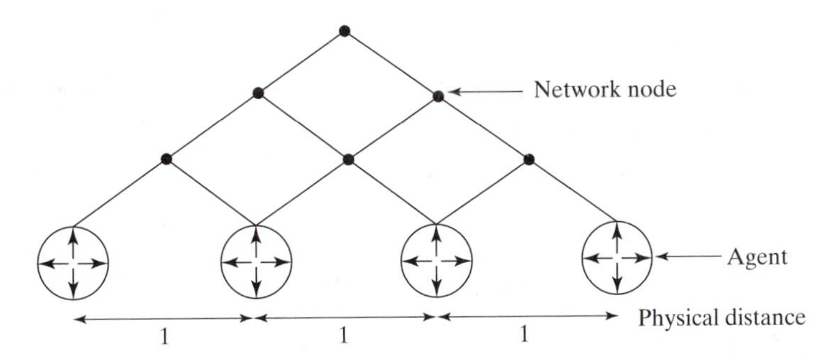

Figure 8.17 Network distance versus physical distance.

the physical location, they will have a distance (and therefore a delay) of two hops between them. Let d_p be the physical distance between two agents and assume each hop has a unity delay. Then $f = 2 \times d_p$. For example, if $d_p = 2$, there will be four hops between the two agents in the network.

8.5.2 Mobility Model 3

To capture the effect of the locality of the user movement on the average handoff delay, we introduce the following two-dimensional mobility model.

User movements in the network are modeled by boundary crossings between serving areas. The residence time of the mobile in each serving area (the interval between two consecutive handoffs) is assumed to be an exponentially distributed random variable with a constant mean value. Calls to the mobile are modeled as a Poisson arrival process with a constant rate. In addition, calls are assumed to be generated for randomly selected serving areas. The assumptions allow us to characterize the mobile's movement by a Markov state transition diagram. Furthermore, for convenience we consider rectangle radio cell clusters of equal size as the service areas. Each cluster has a unique attachment point to the Internet. As each cell cluster has four adjacent neighboring clusters, on each crossing of a cluster boundary, the mobile can move into one of four adjacent serving areas, with respect to the mobile's current serving area. The direction of each movement is modeled as a Markov process, as shown in Figure 8.18, where P_{back} is the probability that the mobile will move back to its previous serving area, and P_{same} is the probability that the next move will be in the same direction as the previous move. The probability of the mobile moving in any other direction is P_{other} which is equal to $(1 - P_{\text{same}} - P_{\text{back}})/2$. The probabilities P_{back} and P_{same} allow us to model various degrees of locality in the user's movements.

8.5.3 Handoff Delay with Local Anchor

Handoff delay is defined as the time interval from the instant when the mobile sends the registration request to the new foreign agent, to the instant when the mobile is allowed to send packets to, or receive packets from, the new foreign agent. A handoff will take place when necessary, whether or not a TCP connection exists. If there is no existing TCP connection when the handoff takes place, the handoff will finish when the mobile receives a handoff reply from the new foreign agent. In

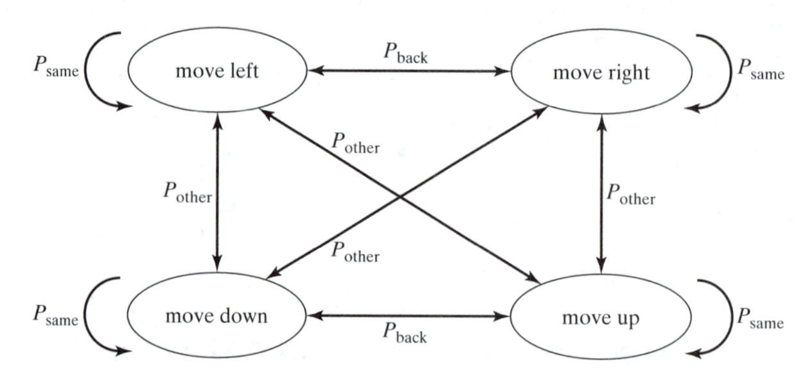

Figure 8.18 Two-dimensional Markov mobility model.

the operation of TCP in a hybrid wireless/IP environment, the air link needs to be considered separately. With M-TCP (see Subsection 8.4.1), the splitting occurs at the supervisor host. With the local anchor scheme in Mobile IP, the split can take place at the anchor agent [165]. This is the focal agent in the anchoring region which maintains the TCP connection state for the mobile host. That is, when the mobile moves within the anchoring region, its new serving agent does not keep track of the TCP connection state for the mobile; it therefore does not need to copy the mobile's TCP connection state from the old foreign agent. This will yield a reduction in handoff delay, in addition to the reduction in delay due to shorter registration path with the local anchor.

In the network model shown in Figure 8.16, if the mobile host has to register with its home agent, the registration request and reply have to go through paths with delays b and d, plus two processing times at the foreign agent and one processing time at the home agent. Let r denote the fixed processing time (cost) at each agent. Assume a split connection scenario, where the split is located at the focal agent of the anchoring region. We will refer to this transport layer connection as an anchor-based indirect TCP (I-TCP). Then, the registration delay, $t_{\text{I-TCP}}$, of this anchor-based I-TCP is given by

$$t_{\text{I-TCP}} = 2(b + d) + 3r. \tag{8.5.1}$$

After a successful registration, the new foreign agent sends a message to the old foreign agent and requests TCP state transfer. This will incur a transfer delay, t_{transfer}, of an amount

$$t_{\text{transfer}} = (n + 2)(f + r), \tag{8.5.2}$$

where n is the TCP buffer size, f is the delay between the anchor and the serving agent (see Figure 8.16 and Problem P8-11). Therefore, when the mobile host has to register with its home agent, the total handoff delay, $t_{\text{h\_I-TCP}}$, for the anchor-based I-TCP scheme is

$$t_{\text{h\_I-TCP}} = 2(b + d) + 3r + (n + 2)(f + r). \tag{8.5.3}$$

On the other hand, if the mobile roams within the anchoring region, it only needs to register with the anchor, as opposed to the home agent. In this case, the registration request and reply will go through the paths with delays d and f, plus two processing times at the new foreign agent and one processing time at the anchor. Thus, the registration delay is

$$t_{\text{h\_anchor}} = 2d + 2f + 3r. \tag{8.5.4}$$

If the new foreign anchor decides that the mobile should register with its home agent, as opposed to the anchor, the handoff delay should be that given by Eq. (8.5.3).

Because of mobility, the mobile will always change its location in the anchor region. Once the mobile moves outside the anchoring region, the new foreign agent will become the focus of the new anchoring region. Thus, the value of f will change when the mobile moves.

Example 8.4 Variable Values of f

In a local anchor-based Mobile IP, the parameter f, the delay between the local anchor and the serving agent, is a variable. Why and how is f a random variable?

Solution The value of f depends on the radius of the anchoring region, in particular, the distance of other agents from the focal agent. An anchoring region with 25 agents is shown in Figure 8.19. The focal agent is located at the center. Any one of the other agents may be the serving agent (see Figure 8.16). Eight of the agents are located at a distance of 1 from the focus and 16 of the agents are located at a distance of 2 from the focus.

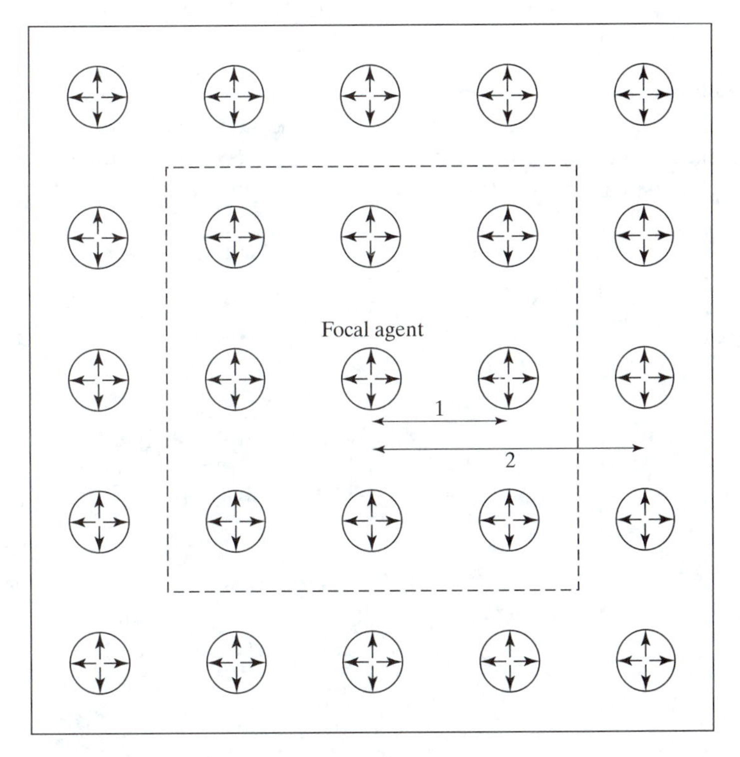

Figure 8.19 Anchoring region with 25 agents.

For the agents with distance 1 from the focus, $f = 2$; for the agents with distance 2 from the focus, $f = 4$. When the mobile moves outside the anchoring region, the value of f to be used to compute the TCP state transferring cost will be greater than 4 (e.g., $f = 6$).

In order to compute the average handoff delay, we need the knowledge of the probability distribution of the mobile staying at each service area. Let π_j be the probability that the mobile stays at service area j with or without the information of its previous area. The average handoff delay is then given by

$$t_{\text{h\_anchor}} = \sum_j \pi_j t_{\text{h\_anchor}}(j). \quad (8.5.5)$$

Example 8.5 Probability Distribution of Mobile's Location

Let $\pi = (\pi_1, \pi_2, \cdots, \pi_J)$ be the probability vector, based on the Mobility Model 3. Consider an anchoring region with 9 agents, as shown in Figure 8.20. Devise a method for calculating the limiting probability vector π.

Solution In Figure 8.20, the focal agent (anchor) is located at the center. Each of the other eight agents is at a distance of 1 from the focus. Consider only the situation of one step movement where a mobile has moved out of its previous area into a new area. Here we have four unique service areas, A, B, C, and the area outside the anchoring region. The mobile's movement and its location can be described by the following six states (i.e., $J = 6$):

S_1: the mobile is located in region A;
S_{21}: the mobile is located in region B and its last location is in region A;
S_{23}: the mobile is located in region B and its last location is in region C;
S_{24}: the mobile is located in region B and its last location is outside the anchoring region;

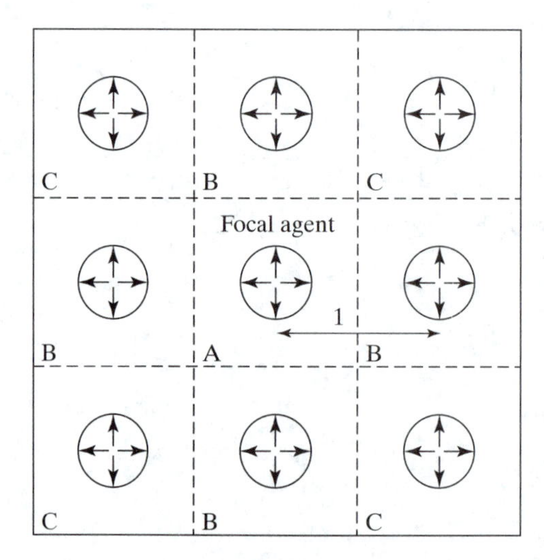

Figure 8.20 Anchoring region with nine agents.

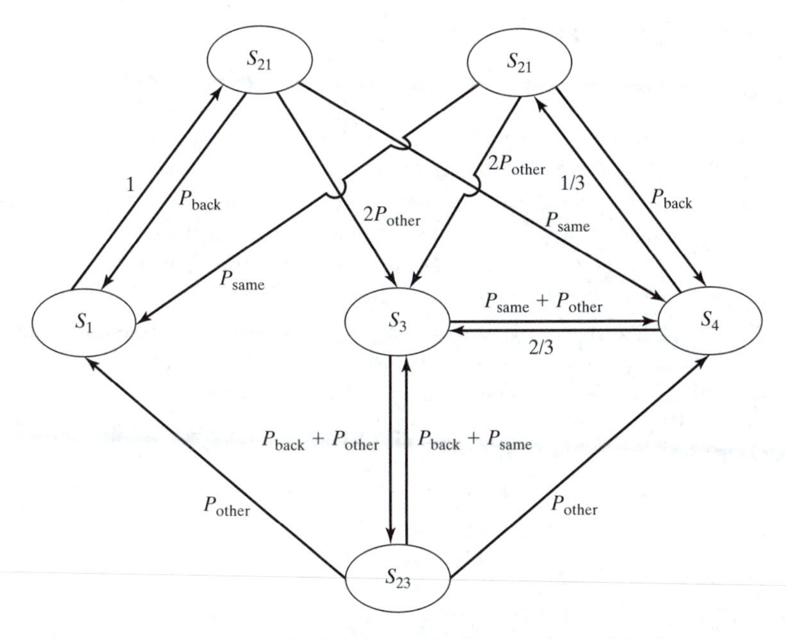

Figure 8.21 Markov chain for the anchoring region with nine agents.

S_3: the mobile is located in region C; and

S_4: the mobile moves out of the current anchoring region and needs to register with the home agent.

The above states characterize the Markov chain shown in Figure 8.21. By assigning $S_1, S_{21}, S_{23}, S_{24}, S_3$ and S_4 the positional values 1, 2, 3, 4, 5, and 6, respectively, we can construct the Markov transition matrix, M, for this Markov chain as

$$M = \begin{pmatrix} 0 & 1 & 0 & 0 & 0 & 0 \\ P_{back} & 0 & 0 & 0 & 2P_{other} & P_{same} \\ P_{other} & 0 & 0 & 0 & P_{back} + P_{same} & P_{other} \\ P_{same} & 0 & 0 & 0 & 2P_{other} & P_{back} \\ 0 & 0 & P_{back} + P_{other} & 0 & 0 & P_{same} + P_{other} \\ 0 & 0 & 0 & 1/3 & 2/3 & 0 \end{pmatrix}.$$

The limiting probability vector, π, can be readily computed from the relationship

$$\pi = \pi M.$$

The Markov chain and the corresponding transition matrix become significantly large when the radius of the anchoring region increases. An alternative approach is to simulate the mobile movements and compute the cost associated with the movements. In the limit, the simulation results should converge to the theoretical value quite well.

8.6 WIRELESS APPLICATION PROTOCOL (WAP)

The Transmission Control Protocol (TCP) has been designed to treat all data losses as being due to network congestion. Thus, conventional TCP erroneously considers data losses due to transmission errors and/or handoff disruptions as the result of network congestion. Taking all losses as being due to network congestion unnecessarily limits the network throughput. Newer versions of TCP (e.g., Tahoe, Reno, NewReno, SACK (Selective Acknowledgment), and other variants [7, 95]) attempt to make TCP aware of transmission errors as well as errors due to network congestion. While TCP/IP can be an effective flow/congestion control protocol for wireless/IP interworking, the large overhead associated with TCP/IP to take care of all sources of data loss is a huge drawback. This is perhaps the main reason behind the formation of the WAP (Wireless Application Protocol) Forum, a consortium of wireless equipment manufacturers and service providers. The WAP Forum introduces a set of WAP specifications for standardization, aiming to facilitate Internet access from wireless terminals such as cellphones, e-commerce portable terminals, and similar devices. The main advance of WAP over TCP is the light overhead that caters to application devices with limited computing power, low power consumption, and similar limitations. Although WAP-enabled wireless terminals and WAP-based Internet services are now available, WAP standards are still evolving.

As discussed in Section 8.4, TCP connections can be end-to-end or split connection (i.e., concatenation of two segments). WAP is based on a split connection principal, where the split is at the base station. The two segments are wireless (from the mobile terminals to the base station) and wireline (from the base station to the backbone network). The wireless segment employs protocols specifically designed for, and fine-tuned to, the wireless propagation channel.

The WAP suite addresses protocols from the transport layer up to the application layer. Specifically, the WAP protocol stack includes the application, session, transaction, and transport layers. To support the WAP protocol stack, the wireless segment of the WAP connection operates as a circuit-switching pipe in handling the transmissions by mobile/portable terminals to the base station. That is, with the WAP suite, the wireless segment is circuit-switched while the wireline segment, from the base station onward, is packet-switched. A request-response procedure is used to establish the WAP connection. The WAP architecture can thus be modeled as a *client-server* scenario, with the mobile terminal as the client and the Internet host as the *original server*. Client-server communication is handled by an interface or *gateway* located at the base station. The gateway interfaces the circuit-switching and the packet-switching modes by performing transmission format translation and conditioning of the data units suitable for forwarding to the client or the server. A functional block diagram of the WAP model is depicted in Figure 8.22.

8.6.1 Wireless Application Environment

Because the WAP standards are still evolving, documentation on the different layers of the WAP protocol stack consists mainly of WAP Forum documents [www.wapforum.org]. The application layer defines the Wireless Application Environment (WAE), which provides for interaction between WAP/Web applications and wireless devices containing a WAP microbrowser. WAE functions include: (*a*) Wireless Markup Language (WML), which accommodates the limitations of wireless devices with limited display capabilities, (*b*) WMLScript, which is WML's accompanying

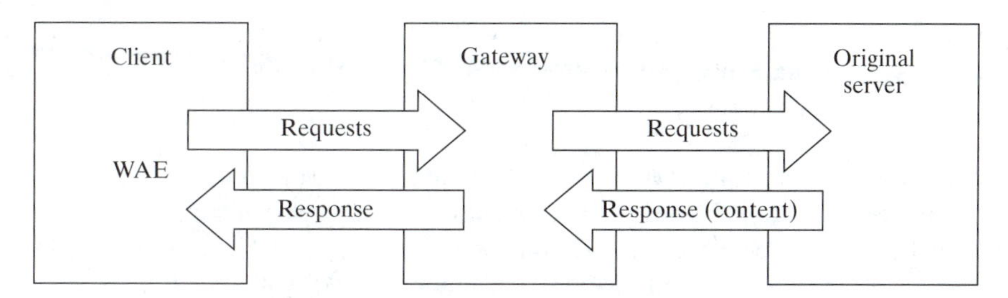

Figure 8.22 WAP programming model.

client-side scripting language that provides for additional intelligence and control over presentation, and (*c*) Wireless Telephony Application. In addition to supporting presentation services, similar to HTML (hypertext markup language), WML adds support for events and variables.

8.6.2 WAP Protocol Stack

The latest release is WAP 2.0 [WAP 2.0 Technical White Paper, Aug. 2001]. These protocols have been optimized for low bandwidth bearer networks with relatively long latency. In addition to the Wireless Application Environment (WAE), the other layers, together with their functionality, are as follows.

Wireless Session Protocol (WSP). WSP provides HTTP/1.1 functionality and incorporates new features, such as long-lived sessions and session suspend/resume. WSP provides the application layer with a consistent interface for two session services: a connection-oriented service that operates above the transaction layer protocol, and a connectionless service that operates above a secure or nonsecure datagram transport service. WSP is optimized for the low bandwidth and long latency inherent in most wireless transmission media. It enables the WAP client to negotiate, open, and maintain a session with the WAP gateway. If a connection closes prematurely, it can go into a sleeping mode and be wakened whenever the connection is reestablished.

Wireless Transaction Protocol (WTP). WTP is a light weight transaction-oriented protocol that is suitable for implementation in "thin" clients (e.g., handsets) and operates efficiently over wireless datagram networks. It is designed to reliably carry out transactions between the client and the server. WTP has the following salient features:

a. Improved reliability over datagram services—WTP relieves the upper layer from retransmissions and acknowledgments that are necessary when datagram services are used;
b. Improved efficiency over connection-oriented services—WTP has no explicit connection setup or tear down phases;
c. Advantages of using a message-oriented protocol, designed for services oriented toward transactions, such as "browsing".

Wireless Transport Layer Security (WTLS). The WTLS layer is designed to provide privacy, data integrity and authentication between two communicating applications. It provides the upper sublayer of WAP with a secure transport service interface that preserves the transport service interface below it. Additionally, WTLS provides an interface for managing (e.g., creating and terminating) secure connections.

Wireless Datagram Protocol (WDP). WDP is a general datagram service which offers a consistent service to the upper layer protocols, and communicates transparently over one of the available underlying bearer services. This consistency is provided by a set of adaptations to specific aspects of these bearers, and provides a common interface with the upper layers to enable operations independent of the services of the wireless network.

8.6.3 WAP Gateway

The WAP gateway provides the interface between the client and the server. The protocol stacks for the client-gateway-server combination are illustrated in Figure 8.23. As can be observed from Figure 8.23, the WAP gateway has two protocol stacks, one for peer communications with the mobile hosts through the wireless propagation channel, and the other for peer communications with the Internet server through wirelines.

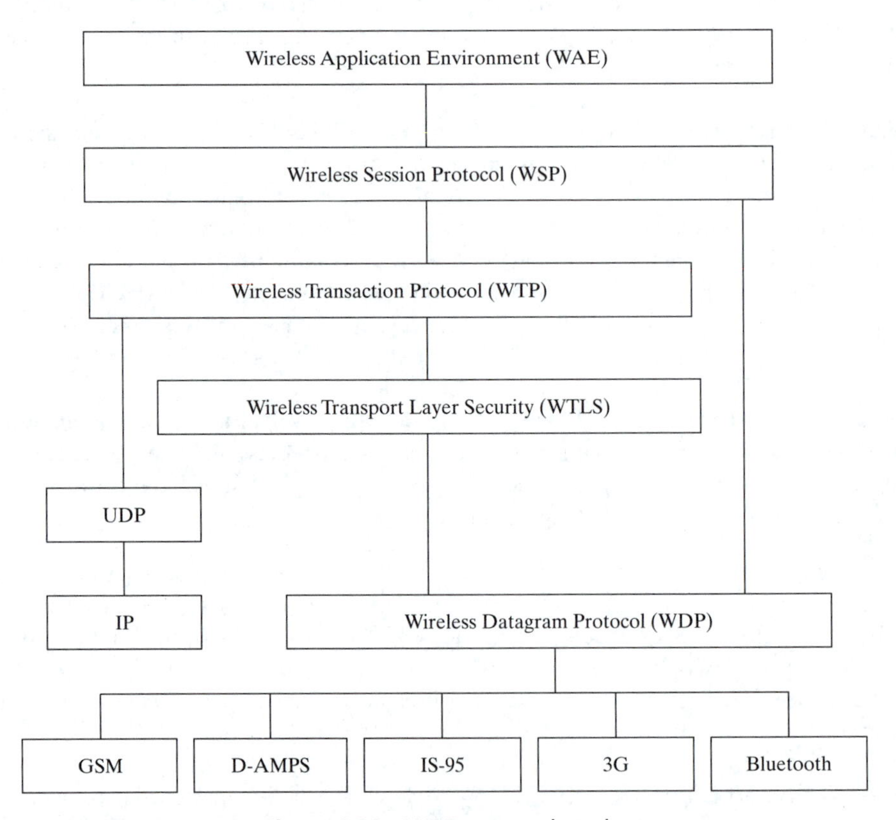

Figure 8.23 WAP protocol stack.

8.7 MOBILE AD HOC NETWORKS

There are two variations of mobile wireless communication networks. The first is known as mobile cellular networks, which is infrastructure-based, and the second is known as mobile ad hoc networks, which is infrastructureless. A mobile cellular network consists of an array of radio cells in which communications in each of the cells is handled by a base station. Thus, the base station is the fixed infrastructure which performs centralized administration. Mobile stations within the footprint of a base station directly communicate with that base station which, in turn, forwards (routes) traffic to designated destinations. Thus, among other tasks, the base station assumes the role of a *router*. A system with a fixed infrastructure is basically a two-hop system. Mobile cellular networks discussed so far in this text have been of the infrastructure-based variety. When a mobile moves outside the footprint of the currently serving base station, its connection with the destination is handed off to the base station with which the mobile station must now communicate.

Infrastructureless mobile networks are commonly known as *ad hoc networks*. Ad hoc networks have no fixed routers; all nodes are capable of movement and can be interconnected dynamically in an arbitrary manner. Nodes of ad hoc networks behave as routers that discover and maintain routes to other nodes in the network. A node in an ad hoc network directly communicates with other nodes in the network if they are within line-of-sight. However, there are also hidden (non-line-of-sight) nodes. Communication between a pair of hidden nodes needs to hop over one or more intermediate nodes. In this sense, ad hoc networks can be thought of as multihop networks.

The connectivity in ad hoc networks is much more complex than that in wireless networks with an infrastructure. This means *routing* in ad hoc networks is a more complex issue than in infrastructure-based networks. It follows that the existence of effective routing protocols is essential for effective and efficient operation of mobile ad hoc networks.

8.7.1 Ad Hoc Routing Protocols

Routing protocols for ad hoc networks can be categorized into two types: (*a*) table-driven and (*b*) source-initiated or demand-driven [17].

Table-Driven Routing Protocols. Table-driven routing protocols try to maintain consistent, up-to-date routing information from each node to every other node in the network. Each node is required to maintain one or more tables to store routing information. An example of table-driven protocols is Destination-Sequenced Distance-Vector (DSDV) Routing. DSDV is a table-driven algorithm based on the classical Bellman-Ford routing mechanism. The table maintained by each and every mobile node in the network contains all of the possible destinations and the number of hops to each destination in the network.

Source-Initiated on-Demand Routing. Source-initiated on-demand routing creates routes only when desired by the source node. When a node requires a *route* to a destination, it initiates a route discovery process within the network. This process is completed once a route is found or all possible route permutations have been explored. Once a route has been established, it is maintained by a route maintenance mechanism until either the destination becomes inaccessible along every path from the source or until the route is no longer needed. An example of on-demand routing is Dynamic Source Routing (DSR) [69]. DSR is based on the concept of source routing. In source routing, mobile nodes are required to maintain route caches that contain the source routes of which the mobile is aware. Entries in the route cache are continually updated as new routes are learned. The source routing protocol consists of two phases: route discovery and route maintenance. When a mobiie node has a packet to send to some destination, it first consults its route cache to determine whether it already has a route to the destination. If there exists a route to the destination, it will use that route to send the packet. On the other hand, if the node does not have a route to the destination, it initiates route discovery.

Route Discovery. The source node initiates route discovery by broadcasting a *route request* packet along all its outgoing links. The route request packet contains the address of the destination, the source node's own address, a unique identification number and a route record field. Each node receiving the packet checks whether it knows a route to the destination. If it does not, it adds its own address to the route record of the packet and then forwards the packet along its outgoing links. Route discovery by broadcasting is tantamount to *flooding*, which can consume a large amount of wireless resources. To limit the number of route requests sent on the outgoing links of a node, a mobile only forwards the route request if the request has not yet been seen by the mobile and if the mobile's address is not already in the route record.

A *route reply* is generated when the route request reaches either the destination node, or an intermediate node that contains in its route cache an unexpired route to the destination. By the time the packet reaches either the destination or such an intermediate node, it contains a route record with the sequence of hops taken. If the node generating the route reply is, in fact, the destination, it places the route record contained in the route request into the route reply. If the responding node is an intermediate node, it will append its cached route to the route record and then generate the route reply. To return the route reply, the responding node must have a route to the initiator. If the responding node does have a route to the initiator in its route cache, it may use that route. Otherwise, if symmetric links are supported, the node may reverse the route in the route record. If symmetric links are not supported, the node may initiate its own route discovery and piggy-back the route reply on the new route request.

Route Maintenance. Route maintenance is accomplished through the use of route error packets and acknowledgments. Route error packets are generated at a node when the data link layer encounters a fatal transmission problem. When a route error packet is received, the hop in error is removed from the node's route cache and all routes containing that hop are truncated at that point. In addition to route error messages, acknowledgments are used to verify the correct operation of the route links. Such acknowledgments include passive acknowledgments, in which a mobile is able to hear the next hop forwarding the packet along the route.

Applications. Ad hoc wireless networks have an important role to play in military applications. There are also commercial scenarios for ad hoc wireless networks, which include:

a. conference/meetings/lectures,
b. emergency services, and
c. law enforcement and similar services.

Approaches reported in the literature for ad hoc routing include paradigms that exploit such features as user demand, user location, power, and association parameters. Adaptivity and self-configuration are key features of these approaches. However, flexibility is also important. A flexible ad hoc routing protocol could responsively invoke table-driven and/or on-demand approaches based on situations and communication requirements. However, the toggling between these two approaches may not be trivial since the nodes involved must be in synchronism with the toggling. Coexistence of both approaches may also exist in spatially clustered ad hoc groups, with intracluster employing the table-driven approach and intercluster employing the demand-driven approach or vice versa.

8.7.2 Comments

In ad hoc networks, dynamic reconfiguration and establishment of routes are the most important features. Two common approaches in route discovery are table-driven and on-demand-driven. There are different schemes available for implementing each of these approaches. Routing protocols are schemes; as such, they are not based on systematic analytical formulation. Any scheme introduced needs to show that it performs well under different network topologies and movement patterns. Algorithmic determination and/or computer simulation are normally used to demonstrate the viability of the routing protocols.

For commercial scenarios such as conference, lectures, and the like, wider geographical coverage than that offered by ad hoc networking would be desirable. In these situations, a backbone network such as the Internet would be used to extend the geographical coverage. There are open challenging problems associated with the interworking of wireless ad hoc networks and IP-based networks.

SUMMARY

While the wireless communication network offers the flexibility for users to roam, the geographical coverage of a wireless system is nevertheless limited. A backbone network, either in the form of a wide area wireline network, or a global satellite network, is needed to provide global communications. The Internet is the most pervasive wireline network that has been enjoying universal acceptance. The interworking of a wireless front-end and an Internet backbone should provide an effective information transfer platform for supporting user roaming on a global basis.

Information transfer across the Internet is in the form of datagrams, and the Internet Protocol is a connectionless datagram transport mechanism that resides in the network layer of the protocol stack. By design, IP has no built-in traffic control capability, and only provides *best effort* services. A connection-oriented and window-based transport layer protocol, referred to as the Transmission Control Protocol, is used to exercise flow control over the Internet. TCP/IP is thus the adopted information transfer mechanism for the Internet. One of the fundamental limitations of the Internet is that each subscriber is only given one IP address, which resides in the subscriber's home network.

The interworking of a wireless segment with the IP-based backbone network requires an effective interface. Mobile IP, introduced by the Internet Engineering Task Force, is the peripheral network that provides the interface between the wireless segment and the IP segment. This chapter has discussed the concepts of, and the entities that make up, Mobile IP. The key elements in Mobile IP are agents, both home agents and foreign agents, that provide the mobile user a care-of address, in addition to its home address. The home address is for *identification* while the care-of address, that changes at each access point, is for *routing*.

The concepts of Mobile IP, and the ramifications of TCP/IP within the context of wireless/IP interworking to provide global information delivery to roaming users, are described and discussed in this chapter.

Although TCP/IP is a viable flow control mechanism for the Internet, when modified forms of TCP are used to enforce flow control for an interworked wireless/IP network, the overhead associated with the application of TCP can be heavy. The WAP (Wireless Application Protocol) Forum has introduced WAP specifications, which provide end-to-end information transmission with QoS provisioning, with lighter overhead compared with TCP, in the wireless Internet access. An overview of WAP is included in this chapter.

Cellular wireless networks use a fixed infrastructure for central administration. In certain applications (e.g., in a military scenario or emergency situations), an infrastructureless network, commonly referred to as an ad hoc network, is preferred. The operational features of ad hoc networks are also described.

ENDNOTES

1. For an overview of IPv6, see the paper by Lee *et al.* [77]. For details of mobility support in IP networks, see the papers by Bhagwat, Perkins, and Tripathi [15], Campbell *et al.* [22], Das *et al.* [37], Perkins and Bhagwat [112], and Manzoni, Ghosal, and Serazzi [93], Ramjee *et al.* [126] [127], and Zhang and Mark [165].
2. For TCP and its performance enhancement, see the papers by Bakre and Badrinath [10], Barakat, Altman, and Dabbous [11], Brown and Singh [18], Caceres and Ifto [20], Ghani and Dixit [53].
3. For details of routing protocols in wireless ad hoc networks, see the papers by Broch, Johnson, and Maltz [17], Johnson and Maltz [69], Perkins and Bhagwat [111], Royer and Toh [131], and the special issue on wireless ad hoc networks of *IEEE Journal on Selected Areas in Communications* [57].
4. For QoS issues in wireless networking, see the papers by Chakrabarti and Mishra [25], Kim and Jamalipour [72], Sobrinho and Krishnakumar [143], Xylomenos and Polyzos [160].
5. For fundamentals of data communication networks, see the book by Bertsekas and Gallager [13]. For wireless network architectures, see the book by Lin and Chlamtac [87].

PROBLEMS

P8-1 The interworking of a wireless network with an IP-based network provides wide geographical coverage for information delivery to mobile users. Describe and discuss the problems created due to user roaming, and the mechanisms needed to ensure seamless information delivery to the mobile user at its current location.

P8-2 When a mobile migrates to a foreign network, certain actions must take place so that messages destined for the mobile in its current location can be delivered.

a. How does the mobile host determine its current location?
b. Describe the steps involved to facilitate delivery of messages from the correspondent host to the mobile host in its current location.
c. Why does Mobile IP need to use two IP addresses?
d. Why is tunneling needed when the home agent forwards messages to the mobile in its current location when it is away from its home network?

P8-3 Mobile IP is a peripheral network (or interface) that bridges a wireless network to an IP-based network for information delivery to a mobile user when it is away from its home network. There are three major processes that must work cooperatively in order for Mobile IP to operate effectively. Name the processes and describe the operational characteristics of each process.

P8-4 In Mobile IP, there are essentially two methods available for routing messages between a mobile host and an Internet correspondent host.

a. What are the two routing methods?
b. Describe and discuss the functions and the operation of the two routing methods.
c. Compare the advantages and disadvantages of the two routing schemes.
d. The simpler method is a popular choice. Under what condition would the more complex routing method be a more appropriate choice?

P8-5
a. Describe and discuss how techniques such as local anchor can reduce the network cost when the mobile user roams away from its home network.
b. If local anchor is the technique used to reduce registration cost, how do you choose the local anchor as the mobile migrates from one foreign network to another?

P8-6 In mobile Internet communications, traffic routing from a source host to a destination host can be performed using flat (one level) routing or hierarchical (multilevel) routing.

a. How does hierarchical routing reduce registration cost?
b. By arranging the foreign agents in a binary tree structure, one can reduce the need for the mobile to register with the home agent. If you have an array of 14 foreign agents, how should the foreign agents be populated in a binary tree to facilitate registration by the mobile as it moves from one leaf node to another leaf node?
c. In hierarchical routing, how are datagrams, which are destined for the mobile in its current location, tunneled to the mobile by the home agent?

P8-7 Internet Protocol version 4 (IPv4) is the basis for the original development of Mobile IP by the IETF (Internet Engineering Task Force). IPv4 has many salient features, but does have certain limitation in supporting Internet administration and operation. As a remedy to the limitation of IPv4, the IETF has introduced Internet Protocol version 6 (IPv6) as the answer.

a. What is the basic limitation of IPv4 that prevents it from being used as full deployment for Internet administration and operation? Explain.
b. Operationally, IPv6 is similar to IPv4, but there are basic differences. Discuss the basic differences that makes IPv6 more suitable for mobile Internet deployment.

c. What are the features that IPv4 lacks, and that make the IETF decide to introduce IPv6?

P8-8 In Mobile IPv6, depending on whether or not the correspondent host has knowledge of the mobile's current care-of address, its messages destined for the mobile in its current location are sent in different ways.

a. Discuss how messages should be sent to the mobile in its current location.
b. What are the basic advantages of Mobile IPv6 compared with Mobile IPv4?

P8-9 As a transport layer protocol, TCP uses a window mechanism to exercise flow control over the best effort IP in the Internet. Flow control is exercised by the edge router, based on congestion status encountered in the core routers between the TCP sender and the TCP receiver.

a. Describe and discuss the operation of the window flow control mechanism.
b. ACKs from the TCP receiver are the basis that the TCP sender uses to adjust the sending window size. Describe and discuss how ACKs are used for this purpose.

P8-10 End-to-end flow control may be designed as a single connection protocol or a split connection protocol.

a. Discuss what is meant by a connection protocol which has end-to-end semantics.
b. Explain why indirect Transmission Control Protocol (I-TCP) has a problem with end-to-end semantics.

P8-11 With the local anchor scheme, when the mobile roams within the anchoring region, it is not necessary to transfer states. However, when the mobile moves outside the anchoring region, it will be necessary to transfer state information from the old anchor to the new anchor, once handoff is successful. Suppose the anchor has a buffer of size n. Since the transmission rate in the wired link is large compared with that in the air link, packets from the correspondent host destined for the mobile have to be buffered at the local anchor.

a. Assuming that the buffer at the local anchor is always full, find an expression for the cost (delay) of state transfer from the old anchor to the new anchor. [Hint: The old anchor does not automatically transfer state information, since only the new anchor knows when handoff is complete.]
b. Assume that (1) the delay between the old anchor and the new anchor is $f = 4$, (2) the buffer size at the anchor is 8 kbytes, (3) the packet length is 1 kbyte, and (4) the processing time is equivalent to a single hop (i.e., $r = 1$). What is the cost (delay) for state transfer?

P8-12 Consider the network model shown in Figure 8.16 with the split for I-TCP at the local anchor.

a. Suppose the mobile only needs to register with the local anchor (i.e., the mobile roams within the anchoring region). Find an expression for registration cost (delay), using the local anchor scheme, that accounts for both transmission and processing costs.
b. When the mobile moves outside the anchoring region, it has to register with its home agent. Suppose that I-TCP uses the same registration procedure as regular Mobile IP. Find an expression for registration cost (delay) using the I-TCP scheme.

P8-13 TCP is a traffic transfer mechanism that relies on feedback from the receiver to the sender in an attempt to protect the integrity of the network and the QoS requirements of the participating users.

a. Describe the salient features of TCP and explain its operation.
b. Explain what can happen to TCP throughput if the messages allowed to enter the IP network themselves contain errors.

P8-14 When the TCP sender window shrinks due to packet losses other than network congestion, the throughput may be quite low. Certain modifications can be used to improve TCP throughput. Describe and discuss what modifications can be made to improve TCP throughput.

P8-15 UDP, as a protocol, has no traffic control capability.

a. Explain why UDP has a role in traffic transfer through the Internet.
b. Explain the ramifications and consequences of using UDP when the traffic entering the IP network contains errors.

P8-16 In wireless/Internet interworking, handoff induced by user mobility is a feature of the wireless segment, while TCP is a backbone Internet mechanism for end-to-end traffic management.

a. How is handoff delay defined?
b. Why is it that a connection may have to be terminated even if handoff is successful?
c. In a hybrid wireless/Internet environment, should both the wireless segment and the wireline segment be considered as a single link, or separate links for traffic control purposes?
d. The handoff delay is a random variable. Explain how the average handoff delay may be computed?

P8-17 The WAP Programming Model is formulated based on three basic elements.

a. Name the three basic elements and describe the function of each, and the functional relationship among the three elements.
b. The Wireless Application Environment (WAE) provides for interaction between WAP/Web applications and wireless devices containing a WAP microbrowser. What are the WAE functions that make WAP a suitable protocol for simple wireless devices?
c. Which layer of the WAP protocol stack facilitates datagram service in WAP? Explain how this datagram service works.

P8-18 In wireless ad hoc networks, routing is one of the most important features. One of the approaches is source-initiated on-demand routing. Describe how routes are discovered in source-initiated routing.

P8-19 User mobility statistics are necessary for performance evaluation of a hybrid wireless/IP network (such as in the handoff delay calculation discussed in Subsection 8.5.3). Here, we are interested in obtaining the probability distribution of mobile locations by computer simulation. Consider Mobility Model 3 as described in Subsection 8.5.2 and an anchoring region with nine agents as illustrated in Figure 8.20. User movements are modeled by boundary crossings between serving areas. The residence time of the mobile in each serving area is an exponentially distributed random variable with mean $\alpha = 3.5$ minutes. Calls to the mobiles are modeled as a Poisson arrival process with a mean rate $\beta = 18$ calls per minute. Calls are generated for randomly selected serving areas. Each call duration is an exponentially distributed random variable with mean value $\gamma = 5$ minutes. Assume that the network has enough resources to accept all the calls. Given that $P_{\text{same}} = 0.5$ and $P_{\text{back}} = 0.1$ (see Fig. 8.18), determine the probabilities that an active mobile stays in the service areas A, B, and C, and the area outside the anchor region, respectively (see Fig. 8.20).

Appendix

A GRAM–SCHMIDT ORTHOGONALIZATION

Given M energy signals $\{v_m(t)\}|_{m=1}^{M}$ defined on $t \in [t_1, t_2]$, we want to find an orthonormal set $\{\tilde{\varphi}_n(t)\}|_{n=1}^{N}$ such that

$$v_m(t) = \sum_{n=1}^{N} v_{mn}\tilde{\varphi}_n(t), \quad t \in [t_1, t_2]; m = 1, 2, \ldots, M, \tag{A.1}$$

where $N \leq M$ and $v_{mn} = \int_{t_1}^{t_2} v_m(t)\tilde{\varphi}_n^*(t)\,dt$ is the correlation between $v_m(t)$ and $\tilde{\varphi}_n(t)$. We can follow the Gram–Schmidt orthogonalization procedure to determine the orthonormal set.

First, let

$$\tilde{\varphi}_1(t) = \frac{1}{\sqrt{\displaystyle\int_{t_1}^{t_2} v_1(t)v_1^*(t)\,dt}} v_1(t), \quad t \in [t_1, t_2], \tag{A.2}$$

where $\int_{t_1}^{t_2} v_1(t)v_1^*(t)\,dt$ is the energy of $v_1(t)$. It can be easily shown that the energy of $\tilde{\varphi}_1(t)$ is equal to 1, due to the normalization to the square root energy of $v_1(t)$ in Eq. (A.2). The correlation between $v_1(t)$ and $\tilde{\varphi}_1(t)$ is

$$v_{11} = \int_{t_1}^{t_2} v_1(t)\tilde{\varphi}_1^*(t)\,dt = \sqrt{\int_{t_1}^{t_2} v_1(t)v_1^*(t)\,dt}.$$

As a result, from Eq. (A.2) we have

$$v_1(t) = v_{11}\tilde{\varphi}_1(t). \tag{A.3}$$

To find the second function in the orthonormal set, we first define an intermediate function

$$u_2(t) = v_2(t) - v_{21}\tilde{\varphi}_1(t),$$

where $v_{21} = \int_{t_1}^{t_2} v_2(t)\tilde{\varphi}_1^*(t)\,dt$ is the correlation between $v_2(t)$ and $\tilde{\varphi}_1(t)$. $u_2(t)$ is the component in $v_2(t)$ which is orthogonal to $\tilde{\varphi}_1(t)$. This can be easily verified as the correlation between $u_2(t)$ and $\tilde{\varphi}_1(t)$ is zero. If the energy of $u_2(t)$ is larger than zero, we can choose the second function as

$$\tilde{\varphi}_2(t) = \frac{1}{\sqrt{\int_{t_1}^{t_2} u_2(t)u_2^*(t)\,dt}} u_2(t). \tag{A.4}$$

It can be easily shown that $\tilde{\varphi}_2(t)$ has unit energy and is orthogonal to $\tilde{\varphi}_1(t)$, and that

$$v_2(t) = v_{21}\tilde{\varphi}_1(t) + v_{22}\tilde{\varphi}_2(t),$$

where $v_{22} = \int_{t_1}^{t_2} v_2(t)\tilde{\varphi}_2^*(t)\,dt$ is the correlation between $v_2(t)$ and $\tilde{\varphi}_2(t)$. If the energy of $u_2(t)$ is zero, it means that $v_2(t) = v_{21}\tilde{\varphi}_1(t)$ and we do not need a new function $\tilde{\varphi}_2(t)$ to accurately describe $v_2(t)$. As a result, we can repeat the procedure by using $v_3(t)$ to find $\tilde{\varphi}_2(t)$.

Similarly, to find other functions in the orthonormal set, $\tilde{\varphi}_n(t)$, $n = 3, 4, \ldots, M$, we first define an intermediate function

$$u_n(t) = v_n(t) - \sum_{i=1}^{n-1} v_{ni}\tilde{\varphi}_i(t),$$

where $v_{ni} = \int_{t_1}^{t_2} v_n(t)\tilde{\varphi}_i^*(t)\,dt$ is the correlation between $v_n(t)$ and $\tilde{\varphi}_i(t)$. By normalizing the intermediate function $u_n(t)$ to the square root of its energy, we obtain the desired unit-energy function $\tilde{\varphi}_n(t)$ for $n = 3, 4, \ldots, M$

$$\tilde{\varphi}_n(t) = \frac{1}{\sqrt{\int_{t_1}^{t_2} u_n(t)u_n^*(t)\,dt}} u_n(t). \tag{A.5}$$

It can be shown that $\tilde{\varphi}_n(t)$ is orthogonal to $\tilde{\varphi}_i(t)$, $i = 1, 2, \ldots, n-1$. The given function $v_n(t)$ can be represented by

$$v_n(t) = \sum_{i=1}^{n} v_{ni}\tilde{\varphi}_i(t),$$

where $v_{nn} = \sqrt{\int_{t_1}^{t_2} u_n(t)u_n^*(t)\,dt}$.

The M given functions, $v_1(t), v_2(t), \ldots, v_M(t)$, are said to be linearly independent if and only if

$$c_1v_1(t) + c_2v_2(t) + \ldots + c_Mv_M(t) = 0, \quad t \in [t_1, t_2]$$

requires $c_1 = c_2 = \ldots = c_M = 0$.

If all the M given functions are linearly independent, then all the intermediate functions, $u_2(t), u_3(t), \ldots, u_M(t)$, have an energy larger than zero and we have $N = M$; otherwise, some of the defined intermediate functions have zero energy and we have $N < M$.

B MAXIMUM LIKELIHOOD DETECTION

Given two random events, A and B, with $P(B) \neq 0$, the conditional probability of event A given that event B has occurred is given by

$$P(A|B) = \frac{P(B|A)P(A)}{P(B)}, \tag{B.1}$$

a result known as *Bayes' theorem*. The term on the right-hand side of Eq. (B.1) is conditioned on A while that on the left is conditioned on B.

Now we want to apply *Bayes' theorem* to determine the optimal decision rule for the receiver in an AWGN channel such that the probability of detection error is minimized. First assume that the elements of the observation vector $\vec{r}$ can take on a finite number of values. Then, given $\vec{r}$, the probability that the symbol s_m was sent is

$$P(s_m|\vec{r}) = \frac{P(\vec{r}|s_m)P(s_m)}{P(\vec{r})}, \quad m = 1, 2, \ldots, M, \tag{B.2}$$

where $P(s_m)$ is the probability of symbol s_m being sent, and $P(\vec{r}|s_m)$ is the conditional probability of obtaining the observation vector $\vec{r}$, given that symbol s_m was sent. Due to the additive Gaussian noise introduced by the channel, the elements of the observation vector $\vec{r}$ are continuous in range. As a result, we modify Eq. (B.2) by using the conditional pdf instead of the conditional probability to describe $\vec{r}$,

$$P(s_m|\vec{r}) = \frac{f(\vec{r}|s_m)P(s_m)}{f(\vec{r})}, \quad m = 1, 2, \ldots, M. \tag{B.3}$$

In the preceding equation, $P(s_m)$ is normally known before a symbol is sent and is therefore called *a priori* probability; on the other hand, $P(s_m|\vec{r})$ is a probability based on the received signal after a symbol is sent, and is therefore called *a posteriori* probability. If we make the decision that $\hat{s} = s_m$ based on the observation vector $\vec{r}$, then the probability of error is

$$P_e(s_m|\vec{r}) = P(s_m \text{ not sent}|\vec{r}) = 1 - P(s_m \text{ sent}|\vec{r}).$$

As a result, a decision rule based on *a posteriori* probability $P(s_m|\vec{r})$ which minimizes the probability of error can be described as follows:

Maximum a posteriori probability (MAP) rule: Set $\hat{s} = s_m$ if $P(s_m \text{ sent}|\vec{r}) \geq P(s_k \text{ sent}|\vec{r})$ for all $k = 1, 2, \ldots, M$ and $k \neq m$.

Using *Bayes' theorem*, we have

$$f(\vec{r}|s_m)P(s_m) \geq f(\vec{r}|s_k)P(s_k).$$

That is, set $\hat{s} = s_m$ if $f(\vec{r}|s_m)P(s_m) \geq f(\vec{r}|s_k)P(s_k)$ for all $k = 1, 2, \ldots, M$ and $k \neq m$.

In the case that all the transmitted symbols are equally likely, that is

$$P(s_m) = 1/M, \quad m = 1, 2, \ldots, M,$$

the decision rule can be simplified as: set $\hat{s} = s_m$ if $f(\vec{r}|s_m) \geq f(\vec{r}|s_k)$ for all $k = 1, 2, \ldots, M$ and $k \neq m$.

The conditional pdf, $f(\vec{r}|s_m)$, or any monotonic function of it, is usually called the *likelihood function*, since it indicates (given that s_m was sent) the likelihood that we obtain the value of the observation vector $\vec{r}$ as compared with other values in the observation space. The preceding decision rule is thus called the *maximum likelihood (ML) decision rule*. Therefore, in the case that the marginal probabilities are known exactly (e.g., the message symbols are equally likely), the MAP criterion is known as the ML criterion. In an AWGN channel with two-sided noise psd $N_0/2$, using the matched filter receiver described in Subsection 3.2.2, we have

$$\ln[f(\vec{r}|s_m)] = -\frac{N}{2}\ln(\pi N_0) - \frac{1}{N_0}\sum_{j=1}^{N}(r_j - x_{mj})^2, \quad m = 1, 2, \ldots, M.$$

Hence, the decision rule is equivalent to: set $\hat{s} = s_m$ if $\sum_{j=1}^{N}(r_j - x_{mj})^2 \leq \sum_{j=1}^{N}(r_j - x_{kj})^2$ for all $k = 1, 2, \ldots, M$ and $k \neq m$.

Maximum-likelihood decision rule: set $\hat{s} = s_m$ if $||\vec{r} - \vec{x}_m|| \leq ||\vec{r} - \vec{x}_k||$ for all $k = 1, 2, \ldots, M$ and $k \neq m$.

In other words, the decision rule is to choose the message point closest to the received signal point in the signal space, which is intuitively satisfying since, in an AWGN channel, a small noise component is more likely than a large noise component.

C MSK SIGNAL REPRESENTATION

Without loss of generality, consider an MSK signal over the time interval $[0,\ T_b]$. From Eq. (3.3.6), an MSK signal can be represented as

$$x(t) = \sqrt{\frac{2E_b}{T_b}}\cos[\phi(t)]\cos(2\pi f_c t) - \sqrt{\frac{2E_b}{T_b}}\sin[\phi(t)]\sin(2\pi f_c t), \tag{C.1}$$

where

$$\begin{aligned}\cos[\phi(t)] &= \cos[\phi(0)]\cos\left(\frac{\pi t}{2T_b}\right) \mp \sin[\phi(0)]\sin\left(\frac{\pi t}{2T_b}\right)\\ &= \cos[\phi(0)]\cos\left(\frac{\pi t}{2T_b}\right), \quad -T_b \leq t \leq T_b.\end{aligned} \tag{C.2}$$

The second equality in Eq. (C.2) arises because $\phi(0) = 0$ or π so that $\sin[\phi(0)] = 0$. Also, the time interval can be extended to $[-T_b,\ T_b]$ as the cosine is an even function. For an expression of

$\sin[\phi(t)]$, we first find an expression for $\phi(t)$ as follows. Let $a_0 \in \{-1, +1\}$ and $a_1 \in \{-1, +1\}$ (+1 for symbol "1" and −1 for symbol "0") denote the information to be sent over the bit intervals $[0,\ T_b]$ and $[T_b,\ 2T_b]$, respectively. Over $[0,\ T_b]$, we have $\phi(t) = \phi(0) + a_0 \frac{\pi t}{2T_b}$, from which we have $\phi(0) = \phi(T_b) - a_0 \frac{\pi T_b}{2T_b}$. Using this relation, $\phi(t)$ can be represented by

$$\phi(t) = \begin{cases} \phi(0) + a_0 \dfrac{\pi t}{2T_b}, & 0 \le t \le T_b \\ \phi(T_b) + a_1 \dfrac{\pi(t - T_b)}{2T_b}, & T_b \le t \le 2T_b \end{cases}$$

$$= \begin{cases} \phi(T_b) + a_0 \dfrac{\pi(t - T_b)}{2T_b}, & 0 \le t \le T_b \\ \phi(T_b) + a_1 \dfrac{\pi(t - T_b)}{2T_b}, & T_b \le t \le 2T_b \end{cases}.$$

As a result,

$$\sin[\phi(t)] = \begin{cases} \sin[\phi(T_b)] \cos\left[a_0 \dfrac{\pi(t - T_b)}{2T_b}\right] + \cos[\phi(T_b)] \sin\left[a_0 \dfrac{\pi(t - T_b)}{2T_b}\right], & 0 \le t \le T_b \\ \sin[\phi(T_b)] \cos\left[a_1 \dfrac{\pi(t - T_b)}{2T_b}\right] + \cos[\phi(T_b)] \sin\left[a_1 \dfrac{\pi(t - T_b)}{2T_b}\right], & T_b \le t \le 2T_b. \end{cases}$$

$$= \sin[\phi(T_b)] \cos\left[\frac{\pi(t - T_b)}{2T_b}\right], \quad 0 \le t \le 2T_b$$

$$= \sin[\phi(T_b)] \sin\left(\frac{\pi t}{2T_b}\right), \quad 0 \le t \le 2T_b, \tag{C.3}$$

where $\cos[\phi(T_b)] = 0$ because $\phi(T_b) = \pm\frac{\pi}{2}$, and $\cos[a_0 \frac{\pi(t-T_b)}{2T_b}] = \cos[a_1 \frac{\pi(t-T_b)}{2T_b}] = \cos[\frac{\pi(t-T_b)}{2T_b}]$. With Eqs. (C.2) and (C.3), from Eq. (C.1) we obtain the MSK signal representation given by Eq. (3.3.7).

D DERIVATION OF THE RICIAN DISTRIBUTION

Consider random variable $Z = X - jY$, where X and Y are independent Gaussian random variables with mean μ_X and μ_Y, respectively, and with the same variance σ^2. We want to derive the pdf of $|Z| = \sqrt{X^2 + Y^2}$.

Let

$$\mu = \sqrt{\mu_X^2 + \mu_Y^2}, \quad \theta_\mu = \arctan(\mu_Y / \mu_X).$$

Then we have

$$\mu_X = \mu \cos\theta_\mu, \quad \mu_Y = \mu \sin\theta_\mu. \tag{D.1}$$

Since $X \sim N(\mu_X, \sigma^2)$, $Y \sim N(\mu_Y, \sigma^2)$, and X and Y are independent, the joint pdf of X and Y is given by

$$\begin{aligned}
f_{XY}(x, y) &= f_X(x) \cdot f_Y(y) \\
&= \frac{1}{\sqrt{2\pi\sigma^2}} \exp\left[-\frac{(x-\mu_X)^2}{2\sigma^2}\right] \cdot \frac{1}{\sqrt{2\pi\sigma^2}} \exp\left[-\frac{(y-\mu_Y)^2}{2\sigma^2}\right] \\
&= \frac{1}{2\pi\sigma^2} \exp\left[-\frac{x^2+y^2+\mu^2-2x\mu\cos\theta_\mu-2y\mu\sin\theta_\mu}{2\sigma^2}\right].
\end{aligned}$$

The cumulative distribution function (cdf) of $|Z|$ is then

$$\begin{aligned}
F_{|Z|}(z) &\stackrel{\triangle}{=} P(|Z| \le z) \\
&= \iint_{\sqrt{x^2+y^2} \le z} f_{XY}(x, y)\, dx\, dy.
\end{aligned}$$

Using the polar coordinates, with the transformations

$$\begin{cases} r = \sqrt{x^2+y^2} \\ \theta = \arctan(y/x) \end{cases} \quad \text{and} \quad \begin{cases} x = r\cos\theta \\ y = r\sin\theta \end{cases},$$

we have

$$\begin{aligned}
F_{|Z|}(z) &= \int_0^z \int_0^{2\pi} r f_{XY}(r\cos\theta, r\sin\theta)\, d\theta\, dr \\
&= \int_0^z \frac{r}{\sigma^2} \exp\left(-\frac{r^2+\mu^2}{2\sigma^2}\right) \left\{\frac{1}{2\pi}\int_0^{2\pi} \exp\left[\frac{r\mu\cos\theta\cos\theta_\mu + r\mu\sin\theta\sin\theta_\mu}{\sigma^2}\right] d\theta\right\} dr \\
&= \int_0^z \frac{r}{\sigma^2} \exp\left(-\frac{r^2+\mu^2}{2\sigma^2}\right) \left\{\frac{1}{2\pi}\int_0^{2\pi} \exp\left[\frac{r\mu\cos(\theta-\theta_\mu)}{\sigma^2}\right] d\theta\right\} dr. \qquad \text{(D.2)}
\end{aligned}$$

Letting $\phi = \theta - \theta_\mu$, the inner integral is

$$\begin{aligned}
&\frac{1}{2\pi}\int_0^{2\pi} \exp\left[\frac{r\mu\cos(\theta-\theta_\mu)}{\sigma^2}\right] d\theta \\
&= \frac{1}{2\pi}\int_{-\theta_\mu}^{2\pi-\theta_\mu} \exp\left[\frac{r\mu\cos\phi}{\sigma^2}\right] d\phi \\
&= \frac{1}{2\pi}\int_0^{2\pi} \exp\left[\frac{r\mu\cos\phi}{\sigma^2}\right] d\phi,
\end{aligned}$$

where the last equality is based on the fact that $\cos\phi$ is periodic with period 2π. The integral does not have a closed-form expression and is defined as the zero-order modified Bessel function of the first kind with argument $r\mu/\sigma^2$. That is

$$\frac{1}{2\pi}\int_0^{2\pi} \exp\left[\frac{r\mu\cos\phi}{\sigma^2}\right] d\phi \triangleq I_0\left(\frac{r\mu}{\sigma^2}\right). \tag{D.3}$$

Substituting the result of the inner integral into Eq. (D.2), we have

$$F_{|Z|}(z) = \int_0^z \frac{r}{\sigma^2}\exp\left(-\frac{r^2+\mu^2}{2\sigma^2}\right) I_0\left(\frac{r\mu}{\sigma^2}\right) dr.$$

As a result, the pdf of $|Z|$, $f_{|Z|}(z) = \frac{dF_{|Z|}(z)}{dz}$, is given by

$$f_{|Z|}(z) = \begin{cases} \frac{z}{\sigma^2}\exp\left(-\frac{z^2+\mu^2}{2\sigma^2}\right) I_0\left(\frac{z\mu}{\sigma^2}\right) & z \geq 0 \\ 0 & z < 0 \end{cases},$$

which is the Rician distribution.

E PSEUDORANDOM NOISE SEQUENCES

Spreading sequences for use in CDMA systems, which exhibit noise-like properties, are generated by shift register generators. A PN sequence (here we consider only m-sequences, where m stands for maximal length) is generated by an autonomous shift register generator in which the feedback connections are specified by a primitive polynomial. In general a primitive polynomial of degree n is represented as

$$F(x) = 1 + \cdots + x^n.$$

A polynomial is said to be primitive if it has no factors other than 1 and itself. Only primitive polynomials can generate PN sequences with maximal lengths. The maximum length of a PN sequence generated by a primitive polynomial of degree n is

$$N = 2^n - 1.$$

That is, the generator has N states, excluding the all 0 state (which is the absorption state).

As an example, consider a PN sequence with the primitive polynomial $F(x) = 1 + x^3 + x^4$. Figure E.1 shows the PN sequence generator, where $\{c_j\}$ denotes the output of the PN sequence generator, with $c_j \in \{0, 1\}$, $j = 1, 2, \ldots, N$.

Before using $\{c_j\}$ as a spreading code, convert it to a ± 1 sequence, $\{a_j\}$, via the transformation

$$a_j = 1 - 2c_j = \begin{cases} +1, & c_j = 0 \\ -1, & c_j = 1 \end{cases}.$$

For $F(x) = 1 + x^3 + x^4$, the maximum length of the generated sequence is $N = 2^4 - 1 = 15$. After 15 shifts, the sequence repeats itself, which implies that the sequence is periodic with period equal to 15.

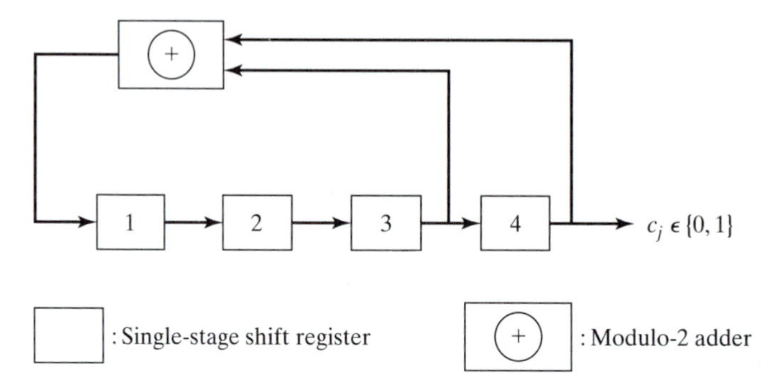

Figure E.1 PN sequence generator with primitive polynomial $F(x) = 1 + x^3 + x^4$.

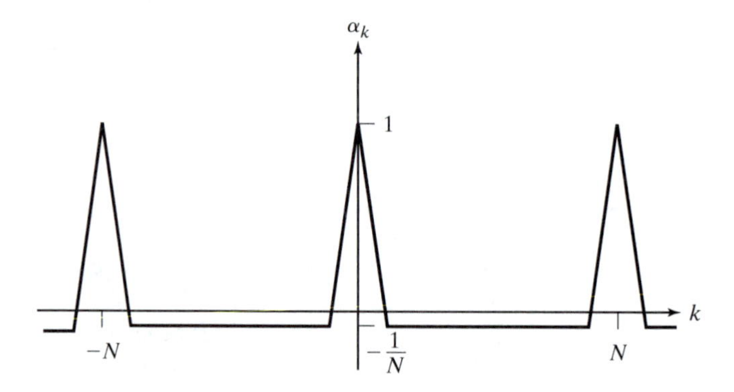

Figure E.2 Autocorrelation function of periodic PN sequences.

The reciprocal of a polynomial is defined as

$$F_{rcp}(x) \triangleq x^n [F(x)]_{rcp},$$

where the symbol $[\cdot]_{rcp}$ denotes component by component inversion of the argument. Thus

$$\begin{aligned} F_{rec}(x) &= x^4 \left[1 + \frac{1}{x^3} + \frac{1}{x^4} \right] \\ &= x^4 + x + 1, \end{aligned}$$

which is also primitive.

In general, the reciprocal of a primitive polynomial is also primitive. For a primitive polynomial $F(x)$ of degree n, the maximum length of the PN sequence is $N = 2^n - 1$.

The autocorrelation function of a periodic PN sequence, $\{a_j\}$, is defined as

$$\alpha_k = \frac{1}{N} \sum_{j=0}^{N} a_j a_{j+k}.$$

α_k is periodic with period N and has the following two values only:

$$\alpha_k = \begin{cases} 1, & k = \pm iN \quad \text{and} \quad i = 0, 1, \cdots \\ -1/N, & \text{elsewhere} \end{cases}.$$

The autocorrelation function of a periodic PN sequence of period $N = 15$ is shown in Figure E.2.

F THE ERLANG-B AND ERLANG-C TABLES

Table F.1 Offered Traffic Load in Erlangs in an Erlang-B System (Number of Channels from 2 to 20)

Channel	Call Blocking Probability							
Number	0.001	0.002	0.005	0.010	0.020	0.050	0.070	0.100
2	0.046	0.065	0.105	0.152	0.223	0.381	0.470	0.595
3	0.194	0.249	0.349	0.455	0.602	0.899	1.057	1.271
4	0.439	0.535	0.701	0.869	1.092	1.525	1.748	2.045
5	0.762	0.900	1.132	1.361	1.657	2.218	2.504	2.881
6	1.146	1.325	1.622	1.909	2.276	2.960	3.305	3.758
7	1.579	1.798	2.157	2.501	2.935	3.738	4.139	4.666
8	2.051	2.311	2.730	3.127	3.627	4.543	4.999	5.597
9	2.557	2.855	3.333	3.783	4.345	5.370	5.879	6.546
10	3.092	3.426	3.961	4.461	5.084	6.216	6.776	7.511
11	3.651	4.021	4.610	5.160	5.841	7.076	7.687	8.487
12	4.231	4.637	5.279	5.876	6.615	7.950	8.610	9.474
13	4.830	5.270	5.964	6.607	7.401	8.835	9.543	10.470
14	5.446	5.919	6.663	7.352	8.200	9.729	10.485	11.473
15	6.077	6.582	7.375	8.108	9.010	10.633	11.434	12.484
16	6.722	7.258	8.100	8.875	9.828	11.543	12.390	13.500
17	7.378	7.946	8.834	9.652	10.656	12.461	13.353	14.522
18	8.046	8.644	9.578	10.437	11.491	13.385	14.321	15.548
19	8.724	9.351	10.331	11.230	12.333	14.315	15.294	16.579
20	9.411	10.068	11.092	12.031	13.182	15.249	16.271	17.613

Table F.2 Offered Traffic Load in Erlangs in an Erlang-B System (Number of Channels from 21 to 55)

Channel	Call Blocking Probability							
Number	0.001	0.002	0.005	0.010	0.020	0.050	0.070	0.100
21	10.108	10.793	11.860	12.838	14.036	16.189	17.253	18.651
22	10.812	11.525	12.635	13.651	14.896	17.132	18.238	19.692
23	11.524	12.265	13.416	14.470	15.761	18.080	19.227	20.737
24	12.243	13.011	14.204	15.295	16.631	19.031	20.219	21.784
25	12.969	13.763	14.997	16.124	17.505	19.985	21.214	22.833
26	13.701	14.522	15.795	16.959	18.383	20.943	22.212	23.885
27	14.439	15.285	16.598	17.797	19.265	21.904	23.213	24.939
28	15.182	16.054	17.406	18.640	20.150	22.867	24.216	25.995
29	15.930	16.828	18.218	19.487	21.039	23.833	25.221	27.053
30	16.684	17.606	19.034	20.337	21.932	24.802	26.228	28.113
31	17.442	18.389	19.854	21.191	22.827	25.773	27.238	29.174
32	18.205	19.175	20.678	22.048	23.725	26.746	28.249	30.237
33	18.972	19.966	21.505	22.909	24.626	27.721	29.262	31.301
34	19.742	20.761	22.336	23.772	25.529	28.698	30.277	32.367
35	20.517	21.559	23.169	24.638	26.435	29.677	31.293	33.434
36	21.296	22.361	24.006	25.507	27.343	30.657	32.311	34.503
37	22.078	23.166	24.846	26.378	28.253	31.640	33.330	35.572
38	22.864	23.974	25.689	27.252	29.166	32.623	34.351	36.643
39	23.652	24.785	26.534	28.129	30.081	33.609	35.373	37.715
40	24.444	25.599	27.382	29.007	30.997	34.596	36.396	38.787
41	25.239	26.416	28.232	29.888	31.916	35.584	37.421	39.861
42	26.037	27.235	29.085	30.771	32.836	36.574	38.446	40.936
43	26.837	28.057	29.940	31.656	33.758	37.565	39.473	42.011
44	27.641	28.881	30.797	32.543	34.682	38.557	40.501	43.088
45	28.447	29.708	31.656	33.432	35.607	39.550	41.529	44.165
46	29.255	30.538	32.517	34.322	36.534	40.545	42.559	45.243
47	30.066	31.369	33.381	35.215	37.462	41.540	43.590	46.322
48	30.879	32.203	34.246	36.108	38.392	42.537	44.621	47.401
49	31.694	33.039	35.113	37.004	39.323	43.534	45.653	48.481
50	32.512	33.876	35.982	37.901	40.255	44.533	46.687	49.562
51	33.331	34.716	36.852	38.800	41.189	45.532	47.721	50.643
52	34.153	35.558	37.724	39.700	42.124	46.533	48.755	51.726
53	34.977	36.401	38.598	40.602	43.060	47.534	49.791	52.808
54	35.803	37.247	39.474	41.505	43.997	48.536	50.827	53.891
55	36.630	38.094	40.351	42.409	44.936	49.539	51.864	54.975

Table F.3 Offered Traffic Load in Erlangs in an Erlang-B System (Number of Channels from 56 to 90)

Channel	Call Blocking Probability							
Number	0.001	0.002	0.005	0.010	0.020	0.050	0.070	0.100
56	37.460	38.942	41.229	43.315	45.875	50.543	52.901	56.059
57	38.291	39.793	42.109	44.222	46.816	51.548	53.940	57.144
58	39.124	40.645	42.990	45.130	47.758	52.553	54.978	58.229
59	39.959	41.498	43.873	46.039	48.700	53.559	56.018	59.315
60	40.795	42.353	44.757	46.950	49.644	54.566	57.058	60.401
61	41.633	43.210	45.642	47.861	50.589	55.573	58.098	61.488
62	42.472	44.068	46.528	48.774	51.534	56.581	59.139	62.575
63	43.313	44.927	47.416	49.688	52.481	57.590	60.181	63.663
64	44.156	45.788	48.305	50.603	53.428	58.599	61.223	64.750
65	44.999	46.650	49.195	51.518	54.376	59.609	62.266	65.839
66	45.845	47.513	50.086	52.435	55.325	60.619	63.309	66.927
67	46.691	48.378	50.978	53.353	56.275	61.630	64.353	68.016
68	47.540	49.243	51.872	54.272	57.226	62.642	65.397	69.106
69	48.389	50.110	52.766	55.191	58.177	63.654	66.442	70.196
70	49.239	50.979	53.662	56.112	59.129	64.667	67.486	71.286
71	50.091	51.848	54.558	57.033	60.082	65.680	68.532	72.376
72	50.944	52.718	55.455	57.956	61.035	66.694	69.578	73.467
73	51.799	53.590	56.354	58.879	61.990	67.708	70.624	74.558
74	52.654	54.463	57.253	59.803	62.945	68.722	71.671	75.649
75	53.511	55.337	58.153	60.727	63.900	69.738	72.718	76.741
76	54.368	56.211	59.054	61.653	64.857	70.753	73.765	77.833
77	55.227	57.087	59.956	62.579	65.814	71.769	74.813	78.925
78	56.087	57.964	60.859	63.506	66.771	72.786	75.861	80.018
79	56.948	58.842	61.763	64.434	67.729	73.803	76.909	81.110
80	57.810	59.720	62.667	65.363	68.688	74.820	77.958	82.203
81	58.673	60.600	63.573	66.292	69.647	75.838	79.007	83.297
82	59.537	61.480	64.479	67.222	70.607	76.856	80.057	84.390
83	60.403	62.362	65.386	68.152	71.568	77.874	81.106	85.484
84	61.268	63.244	66.294	69.084	72.529	78.893	82.156	86.578
85	62.135	64.127	67.202	70.016	73.490	79.912	83.207	87.672
86	63.003	65.011	68.111	70.948	74.452	80.932	84.258	88.766
87	63.872	65.896	69.021	71.881	75.415	81.952	85.309	89.861
88	64.742	66.782	69.932	72.815	76.378	82.972	86.360	90.956
89	65.612	67.668	70.843	73.749	77.342	83.993	87.411	92.051
90	66.484	68.556	71.755	74.684	78.306	85.014	88.463	93.146

Table F.4 Offered Traffic Load in Erlangs in an Erlang-B System (Number of Channels from 91 to 200)

Channel	Call Blocking Probability							
Number	0.001	0.002	0.005	0.010	0.020	0.050	0.070	0.100
91	67.356	69.444	72.668	75.620	79.270	86.035	89.515	94.242
92	68.229	70.333	73.581	76.556	80.236	87.057	90.568	95.338
93	69.103	71.222	74.495	77.493	81.201	88.079	91.620	96.434
94	69.978	72.113	75.410	78.430	82.167	89.101	92.673	97.530
95	70.853	73.004	76.325	79.367	83.133	90.123	93.726	98.626
96	71.729	73.895	77.241	80.306	84.100	91.146	94.779	99.722
97	72.606	74.788	78.157	81.245	85.068	92.169	95.833	100.819
98	73.484	75.681	79.074	82.184	86.035	93.193	96.887	101.916
99	74.363	76.575	79.992	83.124	87.003	94.216	97.941	103.013
100	75.242	77.469	80.910	84.064	87.972	95.240	98.995	104.110
105	79.649	81.951	85.509	88.773	92.821	100.364	104.269	109.598
110	84.072	86.448	90.121	93.493	97.678	105.494	109.549	115.089
115	88.511	90.960	94.746	98.223	102.544	110.630	114.832	120.584
120	92.964	95.484	99.382	102.964	107.419	115.770	120.120	126.082
125	97.431	100.021	104.028	107.713	112.300	120.916	125.412	131.583
130	101.911	104.569	108.684	112.470	117.189	126.066	130.708	137.087
135	106.402	109.128	113.349	117.236	122.084	131.221	136.006	142.592
140	110.904	113.697	118.023	122.009	126.984	136.379	141.308	148.100
145	115.417	118.276	122.706	126.789	131.891	141.541	146.613	153.610
145	115.417	118.276	122.706	126.789	131.891	141.541	146.613	153.610
150	119.940	122.864	127.396	131.575	136.803	146.706	151.920	159.122
155	124.473	127.461	132.093	136.368	141.720	151.874	157.230	164.636
160	129.014	132.065	136.797	141.167	146.641	157.046	162.542	170.151
165	133.564	136.678	141.508	145.972	151.567	162.220	167.856	175.668
170	138.123	141.298	146.225	150.781	156.498	167.397	173.172	181.187
175	142.688	145.924	150.948	155.595	161.432	172.576	178.490	186.706
180	147.262	150.558	155.677	160.415	166.370	177.758	183.810	192.227
185	151.842	155.198	160.411	165.239	171.312	182.942	189.133	197.750
190	156.429	159.844	165.151	170.068	176.257	188.129	194.456	203.273
195	161.023	164.496	169.895	174.901	181.206	193.318	199.781	208.797
200	165.623	169.154	174.644	179.738	186.161	198.508	205.108	214.323

Table F.5 Offered Traffic Load in Erlangs in an Erlang-C System (Number of Channels from 2 to 35)

Channel	Probability of Non-Zero Delay								
Number	0.01	0.02	0.05	0.07	0.10	0.20	0.50	0.70	1.00
2	0.147	0.210	0.342	0.411	0.500	0.740	1.281	1.584	2.000
3	0.429	0.554	0.787	0.900	1.040	1.393	2.116	2.496	3.000
4	0.810	0.994	1.319	1.469	1.653	2.102	2.977	3.422	4.000
5	1.259	1.497	1.905	2.090	2.313	2.847	3.856	4.357	5.000
6	1.758	2.047	2.532	2.748	3.007	3.617	4.747	5.299	6.000
7	2.296	2.633	3.188	3.434	3.725	4.406	5.646	6.245	7.000
8	2.866	3.246	3.869	4.141	4.463	5.210	6.553	7.195	8.000
9	3.460	3.883	4.569	4.867	5.218	6.027	7.466	8.149	9.000
10	4.077	4.540	5.285	5.607	5.986	6.853	8.383	9.104	10.000
11	4.712	5.213	6.015	6.361	6.765	7.688	9.304	10.062	11.000
12	5.362	5.901	6.758	7.125	7.554	8.530	10.229	11.022	12.000
13	6.027	6.601	7.511	7.899	8.352	9.379	11.157	11.983	13.000
14	6.705	7.313	8.273	8.682	9.158	10.233	12.088	12.946	14.000
15	7.394	8.035	9.044	9.473	9.970	11.093	13.021	13.911	15.000
16	8.093	8.766	9.822	10.270	10.789	11.958	13.957	14.876	16.000
17	8.801	9.505	10.607	11.074	11.613	12.826	14.894	15.842	17.000
18	9.517	10.252	11.399	11.883	12.443	13.699	15.834	16.810	18.000
19	10.242	11.006	12.196	12.698	13.277	14.575	16.774	17.778	19.000
20	10.973	11.766	12.998	13.517	14.116	15.454	17.717	18.747	20.000
21	11.711	12.532	13.806	14.341	14.958	16.336	18.661	19.718	21.000
22	12.455	13.304	14.618	15.169	15.805	17.221	19.606	20.688	22.000
23	13.205	14.081	15.434	16.001	16.654	18.109	20.553	21.659	23.000
24	13.960	14.862	16.254	16.837	17.508	18.999	21.500	22.631	24.000
25	14.721	15.648	17.078	17.676	18.364	19.892	22.449	23.604	25.000
26	15.486	16.438	17.905	18.518	19.223	20.786	23.399	24.577	26.000
27	16.255	17.233	18.736	19.364	20.084	21.683	24.350	25.551	27.000
28	17.029	18.031	19.569	20.212	20.948	22.581	25.301	26.525	28.000
29	17.806	18.832	20.406	21.062	21.815	23.481	26.254	27.499	29.000
30	18.588	19.637	21.246	21.916	22.684	24.383	27.207	28.474	30.000
31	19.373	20.446	22.088	22.772	23.555	25.287	28.161	29.450	31.000
32	20.162	21.257	22.932	23.630	24.428	26.192	29.116	30.425	32.000
33	20.953	22.071	23.780	24.490	25.303	27.099	30.071	31.401	33.000
34	21.748	22.888	24.629	25.353	26.180	28.007	31.028	32.378	34.000
35	22.546	23.708	25.481	26.217	27.059	28.916	31.984	33.355	35.000

Table F.6 Offered Traffic Load in Erlangs in an Erlang-C System (Number of Channels from 36 to 70)

Channel	Probability of Non-Zero Delay								
Number	0.01	0.02	0.05	0.07	0.10	0.20	0.50	0.70	1.00
36	23.347	24.530	26.335	27.084	27.940	29.827	32.942	34.332	36.000
37	24.151	25.355	27.190	27.952	28.822	30.739	33.900	35.309	37.000
38	24.957	26.182	28.048	28.822	29.706	31.652	34.859	36.287	38.000
39	25.765	27.011	28.908	29.694	30.591	32.566	35.818	37.265	39.000
40	26.577	27.843	29.769	30.567	31.478	33.482	36.777	38.244	40.000
41	27.390	28.676	30.632	31.442	32.366	34.398	37.738	39.222	41.000
42	28.206	29.512	31.497	32.318	33.256	35.316	38.698	40.201	42.000
43	29.024	30.350	32.363	33.196	34.147	36.234	39.659	41.181	43.000
44	29.844	31.189	33.231	34.076	35.039	37.153	40.621	42.160	44.000
45	30.666	32.030	34.101	34.956	35.932	38.074	41.583	43.140	45.000
46	31.490	32.873	34.972	35.838	36.827	38.995	42.545	44.119	46.000
47	32.316	33.718	35.844	36.722	37.722	39.917	43.508	45.099	47.000
48	33.143	34.564	36.717	37.606	38.619	40.840	44.471	46.080	48.000
49	33.973	35.412	37.592	38.492	39.517	41.763	45.435	47.060	49.000
50	34.804	36.262	38.469	39.379	40.416	42.688	46.399	48.041	50.000
51	35.637	37.113	39.346	40.267	41.316	43.613	47.363	49.022	51.000
52	36.471	37.965	40.225	41.156	42.217	44.539	48.328	50.003	52.000
53	37.308	38.819	41.104	42.046	43.119	45.466	49.293	50.984	53.000
54	38.145	39.674	41.985	42.938	44.021	46.393	50.258	51.965	54.000
55	38.985	40.531	42.867	43.830	44.925	47.321	51.224	52.947	55.000
56	39.825	41.389	43.750	44.723	45.830	48.250	52.190	53.929	56.000
57	40.667	42.248	44.635	45.617	46.735	49.179	53.156	54.911	57.000
58	41.511	43.108	45.520	46.512	47.641	50.109	54.123	55.893	58.000
59	42.355	43.970	46.406	47.408	48.548	51.039	55.090	56.875	59.000
60	43.202	44.833	47.293	48.305	49.456	51.970	56.057	57.858	60.000
61	44.049	45.696	48.181	49.203	50.365	52.902	57.024	58.840	61.000
62	44.897	46.561	49.070	50.102	51.274	53.834	57.992	59.823	62.000
63	45.747	47.427	49.960	51.001	52.184	54.767	58.960	60.805	63.000
64	46.598	48.295	50.851	51.901	53.095	55.700	59.928	61.788	64.000
65	47.450	49.163	51.742	52.802	54.006	56.634	60.897	62.772	65.000
66	48.304	50.032	52.635	53.704	54.918	57.568	61.865	63.755	66.000
67	49.158	50.902	53.528	54.606	55.831	58.503	62.834	64.738	67.000
68	50.014	51.773	54.422	55.510	56.745	59.439	63.804	65.722	68.000
69	50.870	52.645	55.317	56.413	57.659	60.374	64.773	66.705	69.000
70	51.728	53.518	56.212	57.318	58.574	61.311	65.743	67.689	70.000

Table F.7 Offered Traffic Load in Erlangs in an Erlang-C System (Number of Channels from 71 to 100)

Channel	Probability of Non-Zero Delay								
Number	0.01	0.02	0.05	0.07	0.10	0.20	0.50	0.70	1.00
71	52.586	54.392	57.108	58.223	59.489	62.247	66.712	68.673	71.000
72	53.446	55.267	58.006	59.129	60.405	63.184	67.683	69.657	72.000
73	54.306	56.143	58.903	60.036	61.321	64.122	68.653	70.641	73.000
74	55.168	57.019	59.802	60.943	62.238	65.060	69.623	71.625	74.000
75	56.030	57.897	60.701	61.851	63.156	65.998	70.594	72.609	75.000
76	56.894	58.775	61.601	62.759	64.074	66.937	71.565	73.594	76.000
77	57.758	59.654	62.501	63.668	64.993	67.876	72.536	74.578	77.000
78	58.623	60.533	63.402	64.578	65.912	68.816	73.507	75.562	78.000
79	59.489	61.414	64.304	65.488	66.832	69.756	74.478	76.547	79.000
80	60.356	62.295	65.206	66.399	67.752	70.697	75.450	77.532	80.000
81	61.224	63.177	66.109	67.311	68.673	71.637	76.422	78.517	81.000
82	62.092	64.060	67.013	68.223	69.594	72.579	77.394	79.502	82.000
83	62.961	64.943	67.917	69.135	70.516	73.520	78.366	80.487	83.000
84	63.831	65.827	68.822	70.048	71.438	74.462	79.338	81.472	84.000
85	64.702	66.712	69.727	70.961	72.361	75.404	80.311	82.457	85.000
86	65.574	67.598	70.633	71.876	73.284	76.347	81.283	83.443	86.000
87	66.446	68.484	71.539	72.790	74.208	77.290	82.256	84.428	87.000
88	67.319	69.371	72.446	73.705	75.132	78.233	83.229	85.413	88.000
89	68.193	70.258	73.354	74.621	76.056	79.176	84.202	86.399	89.000
90	69.067	71.146	74.262	75.536	76.981	80.120	85.175	87.385	90.000
91	69.943	72.035	75.170	76.453	77.906	81.064	86.149	88.370	91.000
92	70.818	72.924	76.079	77.370	78.832	82.009	87.122	89.356	92.000
93	71.695	73.814	76.989	78.287	79.758	82.954	88.096	90.342	93.000
94	72.572	74.705	77.899	79.205	80.685	83.898	89.070	91.328	94.000
95	73.450	75.596	78.809	80.123	81.612	84.844	90.044	92.314	95.000
96	74.329	76.488	79.720	81.042	82.539	85.790	91.018	93.300	96.000
97	75.208	77.380	80.631	81.961	83.466	86.735	91.992	94.286	97.000
98	76.087	78.273	81.543	82.880	84.394	87.682	92.966	95.273	98.000
99	76.968	79.166	82.455	83.800	85.323	88.628	93.941	96.259	99.000
100	77.849	80.060	83.368	84.720	86.252	89.575	94.915	97.245	100.000

Table F.8 Offered Traffic Load in Erlangs in an Erlang-C System (Number of Channels from 105 to 200)

Channel	Probability of Non-Zero Delay								
Number	0.01	0.02	0.05	0.07	0.10	0.20	0.50	0.70	1.00
105	82.262	84.537	87.938	89.328	90.901	94.313	99.790	102.178	105.000
110	86.690	89.026	92.519	93.944	95.558	99.056	104.668	107.113	110.000
115	91.130	93.527	97.108	98.570	100.223	103.806	109.549	112.049	115.000
120	95.582	98.039	101.707	103.203	104.895	108.561	114.432	116.986	120.000
125	100.046	102.561	106.314	107.844	109.574	113.321	119.318	121.925	125.000
130	104.520	107.092	110.928	112.492	114.259	118.086	124.206	126.865	130.000
135	109.004	111.632	115.550	117.146	118.951	122.855	129.096	131.805	135.000
140	113.498	116.181	120.179	121.808	123.648	127.629	133.988	136.748	140.000
145	118.007	120.751	124.848	126.520	128.413	132.516	139.047	141.827	145.000
150	122.518	125.315	129.490	131.193	133.120	137.297	143.943	146.771	150.000
155	127.038	129.887	134.137	135.871	137.833	142.082	148.840	151.716	155.000
160	131.565	134.465	138.791	140.555	142.550	146.871	153.740	156.662	160.000
165	136.099	139.050	143.449	145.243	147.271	151.663	158.640	161.608	165.000
170	140.640	143.641	148.113	149.936	151.997	156.458	163.543	166.556	170.000
175	145.180	148.223	152.748	154.587	156.663	161.145	168.280	171.368	175.000
180	149.735	152.826	157.421	159.289	161.396	165.945	173.185	176.317	180.000
185	154.295	157.434	162.099	163.994	166.133	170.749	178.091	181.267	185.000
190	158.862	162.048	166.781	168.704	170.874	175.555	182.999	186.217	190.000
195	163.434	166.666	171.468	173.418	175.618	180.364	187.908	191.168	195.000
200	168.011	171.290	176.158	178.135	180.365	185.175	192.818	196.120	200.000

Abbreviations and Acronyms

1G	first generation
2G	second generation
3G	third generation
ACI	adjacent channel interference
ACK	acknowledgment
A/D	analog to digital
AFD	average fade duration
AM	amplitude modulation
AMPS	advanced mobile phone service
ANSI	American National Standards Institute
AP	access point
ARQ	automatic retransmission request
ASK	amplitude shift keying
ATDMA	asynchronous time division multiple access
ATM	asynchronous transfer mode
AWGN	additive white Gaussian noise
BER	bit error rate
BFSK	binary frequency shift keying
BPSK	binary phase shift keying
bps	bits per second
BS	base station
CAC	call admission control
CCI	cochannel interference

CCS	common channel signaling
cdf	cumulative distribution function
CDMA	code division multiple access
CFTP	connection forced termination probability
CMR	call-to-mobility ratio
CPFSK	continuous phase frequency shift keying
CRI	collision resolution interval
CSMA	carrier sense multiple access
D/A	digital-to-analog
dB	decibel
dBm	decibel referenced to 1 milliwatt
dBW	decibel referenced to 1 watt
DBPSK	differential binary phase shift keying
DCA	dynamic channel assignment
DCS	Digital Cellular System
DES-CBC	the US Data Encryption Standard - Cipher Block Chaining
DFE	decision feedback equalizer
DFT	discrete Fourier transform
DHCP	Dynamic Host Configuration Protocol
DQPSK	differential quadrature phase shift keying
DS-CDMA	direct sequence code division multiple access
DSDV	Destination-Sequenced Distance-Vector
DSR	Dynamic Source Routing
EIRP	effective isotropically radiated power
Eq	equation
ETACS	European Total Access Communications System
ESP	Encapsulating Security Payload
exp	exponential
FA	foreign agent
FCA	fixed channel assignment
FDD	frequency division duplexing
FDM	frequency division multiplexing
FDMA	frequency division multiple access
FM	frequency modulation
FFT	fast Fourier transform
FSK	frequency shift keying
G3G	global third generation
GMSK	Gaussian filtered minimum shift keying
GoS	grade of service
GSM	Groupe Special Mobile or Global System for Mobile Communications

HA	home agent
HCDP	handoff call dropping probability
HLR	home location register
hr	hour
HTML	hypertext markup language
HTTP	hypertext transfer protocol
Hz	Hertz
ICMP	Internet Control Message Protocol
IDFT	inverse discrete Fourier transform
IETF	Internet Engineering Task Force
iid	independent and identically distributed
IMT-2000	International Mobile Telecommunications by the year 2000
IP	Internet Protocol
IPv4	Internet Protocol version 4
IPv6	Internet Protocol version 6
IS-41	interim standard 41
IS-54	interim standard 54
IS-95	interim standard 95
IS-136	interim standard 136
ISDN	integrated services digital network
ISI	intersymbol interference
I-TCP	Indirect Transmission Control Protocol
ITU	International Telecommunications Union
kbps	kilobits per second
kHz	kilo hertz
km	kilometer
LA	local anchor
LCR	level crossing rate
LEO	low earth orbit
LMS	least mean square
ln	natural logarithm
$\log_{10}$	logarithm to base 10
LOS	line-of-sight
LPF	low-pass filter
LSTP	local signaling transfer point
LTI	linear time-invariant
LTV	linear time-variant
m	meter
MAC	medium access control
MAHO	mobile assisted handoff

MAI	multiple access interference
MAP	maximum a posteriori probability
MAP	Mobile Application Part
Mbps	megabits per second
MC-CDMA	multicarrier code division multiple access
MCHO	mobile controlled handoff
Mcpc	mega chips per second
MH	mobile host
MHz	megahertz
MIN	mobile identification number
MIRS	Motorola Integrated Radio System
ML	maximum likelihood
MMSE	minimum mean square error
MN	mobile node
MPSK	M-ary phase shift keying
ms	millisecond
MS	mobile station
MSC	mobile switching center
MSE	mean square error
MSK	minimum shift keying
MSR	mobile support router
M-UDP	modified User Datagram Protocol
M-TCP	Transmission Control Protocol for Mobile Cellular Network
m-sequence	maximal-length sequence
μs	microsecond
NCBP	new call blocking probability
NCHO	network controlled handoff
NLOS	non-line-of-sight
NRZ	non-return-to-zero
N-TDMA	narrowband time division multiple access
NTT	Nippon Telephone and Telegraph
OFDM	orthogonal frequency division multiplexing
OQPSK	offset quadrature phase shift keying
pdf	probability density function
psd	power spectral density
PCN	personal communications network
PCS	personal communication service
PDC	Personal Digital Cellular
PLMN	public land mobile network
PN	pseudorandom noise
PSK	phase shift keying
PSTN	public switched telephone network

QoS	quality of service
QPSK	quadrature phase shift keying
RA	registration area
RACH	random access channel
RF	radio frequency
rms	root mean square
RSTP	regional signaling transfer point
RTT	round-trip time
Rx	receiver
s	second
SAC	Stateless Address Configuration
SACK	Selective Acknowledgment
sync	synchronization
SCP	service control point
SDMA	space division multiple access
SH	supervisor host
SINR	signal-to-interference plus noise ratio
SIR	signal-to-interference ratio
SNR	signal-to-noise ratio
SS7	signaling system no. 7
SSP	service switching point
STCM	space-time coded modulation
STDMA	synchronous time division multiple access
STP	signaling transfer point
TCM	trellis coded modulation
TCP	transmission control protocol
TDD	time division duplexing
TDM	time division multiplexing
TDMA	time division multiple access
TLDN	temporary local directory number
Tx	transmitter
UDP	User Datagram Protocol
UHF	ultra high frequency
UMTS	Universal Mobile Telecommunications Service
US	uncorrelated scattering
VHF	very high frequency
VLR	visitor location register
W	watts
WAE	Wireless Application Environment

WAN	wide area network
WAP	Wireless Application Protocol
WATM	wireless asynchronous transfer mode
WDP	Wireless Datagram Protocol
WLAN	wireless local area network
WML	Wireless Markup Language
WSP	Wireless Session Protocol
WSS	wide-sense stationary
WSSUS	wide-sense stationary uncorrelated scattering
W-TDMA	wideband time division multiple access
WTLS	Wireless Transport Layer Security
WTP	Wireless Transaction Protocol
WWAN	wireless wide area network
ZF	zero-forcing

Bibliography

[1] N. Abramson. The ALOHA system – another alternative for computer communications. In *1970 Fall Joint Comput. Conf., AFIPS Conference Proceedings*, pages 37:281–285, 1970.

[2] N. Abramson. The ALOHA system. In *N. Abramson and F.F. Kuo, editors. Computer Communication Networks. Prentice-Hall Inc., Englewood Cliffs, New Jersey*, 1973.

[3] N. Abramson. *Multiple Access Communications: Foundations for Emerging Technologies*. IEEE Press, Piscataway, New Jersey, 1993.

[4] Y. Akaiwa and Y. Nagata. Highly efficient digital mobile communications with a linear modulation method. *IEEE Journal of Selected Areas in Communications*, SAC-5(5):890–895, 1987.

[5] I. F. Akyildiz and S.M. Ho. On location management for personal communications networks. *IEEE Communications Magazine*, 34(9):138–145, 1996.

[6] I.F. Akyildiz, D.A. Levine, and I. Joe. A slotted CDMA protocol with BER scheduling for wireless multimedia networks. *IEEE/ACM Transactions on Networking*, 7(2):146–158, Apr. 1999.

[7] M. Allman, V. Paxson, and W. R. Stevens. TCP congestion control. IETF RFC 2581, 1999.

[8] S. Ariyavisitakul and L. F. Chang. Signal and interference statistics of a CDMA system with feedback power control. *IEEE Transactions on Communications*, 41(11):1626–1634, Nov. 1993.

[9] P.A. Baker. Phase modulation data sets for serial transmission at 2000 and 2400 bits per second. *AIEE Transactions Part 1: Communication and Electronics*, 81:166–171, July 1962.

[10] A. Bakre and B. R. Badrinath. I-TCP: indirect TCP for mobile hosts. In *Proceedings of the 15th IEEE International Conference on Distributed Computing Systems (ICDCS'95)*, pages 136–143, 1995. Vancouver, Canada, May 30-June 2.

[11] C. Barakat, E. Altman, and W. Dabbous. Network layer mobility: an architecture and survey. *IEEE Personal Communications*, 3(3):54–64, 2000.

[12] P. A. Bello. Characterization of randomly time-variant linear channels. *IEEE Transactions on Communication Systems*, CS-11(4):360–393, 1963.

[13] D. Bertsekas and R. Gallager. *Data Networks, 2nd ed.* Prentice-Hall, 1992.

[14] H. H. Beverage and H. O. Peterson. Diversity receiving system of RCA for radiotelegraphy. *Proceedings of the IRE*, 19:531–561, April 1931.

[15] P. Bhagwat, C. E. Perkins, and S. Tripathi. Network layer mobility: an architecture and survey. *IEEE Personal Communications*, 3(3):54–64, 1996.

[16] D. G. Brennan. Linear diversity combining techniques. *Proceedings of the IRE*, 47:1075–1102, June 1959.

[17] J. Broch, D.B. Johnson, and D.A. Maltz. The dynamic source routing protocol for mobile ad hoc networks. Dec. 1992. IETF Internet draft, draft-ietf-manet-dsr-01.txt.

[18] K. Brown and S. Singh. M-TCP: TCP for mobile cellular networks. *ACM Computer Communication Review*, 27(5):19–43, 1997.

[19] BSTJ. Special Issue on Advanced Mobile Phone Service (AMPS). *Bell System Technical Journal*, 58, Jan. 1979.

[20] R. Caceres and L. Ifto. Improving the performance of reliable transport protocol in mobile computing environment. *IEEE Journal on Selected Area in Communications*, 13(5):850–857, 1995.

[21] S. Campanella and D. Schaefer. Time division multiple access systems (TDMA). In *Feher, K., Digital Communications, Satellite/Earth Station Engineering. Prentice-Hall, Inc., Englewood Cliffs, New Jersey*, 1983.

[22] A. T. Campbell, J. Gomez, S. Kim, A. G. Valko, Chieh-Yih Wan, and Z. R. Turanyi. Design, implementation, and evaluation of Cellular IP. *IEEE personal communications*, 7(4):42–49, Aug. 2000.

[23] E. F. Casas and C. Leung. OFDM for data communication over mobile radio FM channels – part I: analysis and experimental results. *IEEE Transactions on Communications*, COM-39(5):783–793, May 1991.

[24] E. F. Casas and C. Leung. OFDM for data communication over mobile radio FM channels – part II: performance improvement. *IEEE Transactions on Communications*, COM-40(4):680–683, April 1992.

[25] S. Chakrabarti and A. Mishra. QoS issues in ad hoc wireless networks. *IEEE Communications Magazine*, 39(2):142–148, Feb. 2001.

[26] G. K. Chan. Effects of sectorization on the spectrum efficiency of cellular radio systems. *IEEE Transactions on Vehicular Technology*, 41(3):217–225, August 1992.

[27] P. Chaudhury, W. Mohr, and S. Onoe. The 3GPP proposal for IMT-2000. *IEEE Communications Magazine*, 37(12):72–81, Dec. 1999.

[28] G. Cherubini, J. M. Cioffi, A. Duel-Hallen, and ed. H. V. Poor. Special issue on multiuser detection techniques with application to wired abd wireless communications systems I. *IEEE Journal on Selected Areas on Communications*, 19(8):1425–1653, Aug. 2001.

[29] A. P. Clark. *Equalizers for Digital Modems*. Pentech Press, London, 1985.

[30] R. H. Clarke. A statistical theory of mobile radio reception. *Bell System Technical Journal*, 47:957–1000, July 1968.

[31] R. B. Cooper. *Introduction to Queueing Theory, 3rd. ed.* CEEPress Books, 1990.

[32] D. C. Cox. Wireless personal communications: what is it? *IEEE Personal Communications*, 2(2):20–35, 1995.

[33] D. C. Cox and D. O. Reudnik. A comparison of some channel assignment strategies in large-scale mobile communications systems. *IEEE Transactions on Communications*, 20:190–195, April 1972.

[34] D. C. Cox and D. O. Reudnik. Dynamic channel assignment in two-dimensional large-scale mobile radio systems. *Bell System Technical Journal*, 51:1611–1630, Sept. 1972.

[35] D. C. Cox and D. O. Reudnik. Increasing channel occupancy in large-scale mobile radio environments: dynamic channel reassignment. *IEEE Transactions on Communications*, 21(11):1302–1306, 1973.

[36] C. E. Dadson, J. Durkin, and E. Martin. Computer prediction of field strength in the planning of radio systems. *IEEE Transactions on Vehicular Technology*, VT-24(1):1–7, February 1975.

[37] S. Das, A. Misra, and P. Agrawal. TeleMIP: telecommunications-enhanced mobile IP architecture for fast intradomain mobility. *IEEE Personal Communications*, 7(4):50–58, Aug. 2000.

[38] G. D. Dill. TDMA, the state-of-the-art. In *Rec. IEEE Electron. Aerosp. Syst. Conv. (EASCON)*, pages 31–5A:31–5I, 1977.

[39] D. Divsalar, M. K. Simon, and D. Raphaeli. Improved parallel interference cancellation for CDMA. *IEEE Transactions on Communications*, 46(2):258–268, Feb. 1998.

[40] R. C. Dixon. *Spread Spectrum Systems, 2nd ed.* Wiley, New York, 1984.

[41] M. L. Doelz and Collins Radio Co. E. H. Heald. Minimum-shift data communication system. 1961. U.S. Patent 2 997 417, March 28.

[42] W. C. Jakes (ed.). *Microwave Mobile Communications.* Wiley, New York, 1974.

[43] R. Edwards and J. Durkin. Computer prediction of service area for VHF mobile radio networks. *Proceedings of the IEE*, 116(9):1493–1500, 1969.

[44] EIA/TIA Interim Standard (IS-41.1-C). Cellular Radio telecommunications Intersystem Operations: Functional Overview. 1995. Electronic Industries Association, Washington, D.C.

[45] S. M. Elnoubi. Analysis of GMSK with discriminator detection in mobile radio channels. *IEEE Transactions on Vehicular Technology*, VT-35(2):71–76, May 1986.

[46] J. D. Parson et al. Diversity techniques for mobile radio reception. *Radio and Electronic Engineer*, 45(7):357–367, 1975.

[47] European Telecommunications Standard Institute (ESTI). Mobile Application Part (MAP) Specification, Version 4.8.0. 1994. Technical Report Recommendation GSM 09.02.

[48] D. E. Everitt. Traffic engineering of the radio interface for cellular mobile networks. *Proceedings of the IEEE*, 82(9):1371–1382, 1994.

[49] D. D. Falconer, F. Adachi, and B. Gudmundson. Time division multiple access methods for wireless personal communications. *IEEE Communications Magazine*, 33(1):50–57, Jan. 1995.

[50] R. C. French. The effects of fading and shadowing on channel reuse in mobile radio. *IEEE Transactions on Vehicular Technology*, 28(3):171–181, 1979.

[51] Jr. G. D. Forney. Maximum-likelihood sequence estimation of digital sequences in the presence of intersymbol interference. *IEEE Transactions on Information Theory*, IT-18:363–378, May 1972.

[52] D. A. George, R. R. Bowen, and J. R. Storey. An adaptive decision-feedback equalizer. *IEEE Transactions on Communications Technology*, COM-19:281–293, June 1971.

[53] N. Ghani and S. Dixit. TCP/IP enhancements for satellite networks. *IEEE Communications Magazine*, 37(7):64–72, 1999.

[54] K. S. Gilhousen, I.M. Jacobs, R. Padovani, A.J. Viterbi, Jr. L.A. Weaver, and C.W. III. On the capacity of a cellular CDMA system. *IEEE Transactions on Vehicular Technology*, 40(2):303–312, 1991.

[55] A. J. Goldsmith and P. P. Varaiya. Capacity of fading channels with channel side information. *IEEE Transactions on Information Theory*, 43(6):1986–1992, Nov. 1997.

[56] J. Griffiths. *Radio Wave Propagation and Antennas: An Introduction.* Prentice Hall, Englewood Cliffs, New Jersey, 1987.

[57] Z. J. Haas, M. Geria, D. B. Johnson, C. E. Perkins, M. B. Pursley, M. Steenstrup, C.-K. Toh, and ed. J. F. Hayes. Special issue on wireless ad hoc networks. *IEEE Journal on Selected Areas in Communications*, 17(8), Aug. 1999.

[58] S. Hara and R. Prasad. Overview of multicarrier CDMA. *IEEE Communications Magazine*, 35(12):126–133, Dec. 1997.

[59] H. Hashemin. Indoor radio propagation channel. *Proceedings of the IEEE*, 81(7):941–968, July 1993.

[60] M. Hata. Empirical formula for propagation loss in land mobile radio services. *ieeevt*, VT-29(3):317–325, August 1980.

[61] S. Haykin. *Adaptive Filter Theory, 2nd ed.* Prentice-Hall, Englewood Cliffs, New Jersey, 1991.

[62] S. Haykin. *Communication Systems, 4th ed.* John Wiley & Sons, New York, 2001.

[63] M. J. Ho and G. L. Stüber. Co-channel interference of microcellular systems on shadowed nakagami fading channels. In *IEEE Vehicular Technology Conference*, pages 568–571, 1993.

[64] M. Honig and W. Xiao. Performance of reduced-rank linear interference suppression for DS-CDMA. *IEEE Transactions on Information Theory*, 47(4):1928–1946, July 2001.

[65] V. Huang and W. Zhuang. Optimal resource management in packet-switching TDD CDMA systems. *IEEE Personal Communications*, 7(6):26–31, Dec. 2000.

[66] V. Huang and W. Zhuang. QoS-oriented access control for 4G mobile multimedia CDMA communications. *IEEE Communications Magazine*, 40(3):118–125, Mar. 2002.

[67] Jr. J. J. Spilker and D. T. Magill. The delay-lock discriminator – an optimum tracking device. *Proceedings of the IRE*, pages 1403–1416, Sept. 1961.

[68] B. Jabbari, G. Colombo, A. Nakajima, and J. Kulkarni. Network issues for wireless communications. *IEEE Communications Magazine*, 33(1):88–98, 1995.

[69] D. B. Johnson and D. A. Maltz. The dynamic source routing in ad-hoc wireless network. pages 153–181, 1996. Mobile Computing, T. Imielinski and H. Korth, Eds., Kluwer.

[70] S. M. Kay. *Moden Spectral Estimation: Theory & Application.* Prentice-Hall, Englewood Cliffs, New Jersey, 1988.

[71] R. S. Kennedy. *Fading Dispersive Communication Channels.* Wiley-Interscience, New York, 1969.

[72] J. Kim and A. Jamalipour. Traffic management and QoS provisioning in future wireless ip networks. *IEEE Personal Communications*, 8(5):46–55, Oct. 2001.

[73] L. Kleinrock and F. A. Tobagi. Packet switching in radio channels: part I - carrier sense multiple access modes and their throughput-delay characteristics. *IEEE Transactions on Communications*, 23(12):1400–1416, Dec. 1975.

[74] S. Knodo and L. B. Milstein. Performance of multicarrier CDMA systems. *IEEE Transactions on Communications*, 44(2):238–246, Feb. 1996.

[75] R. Kohno, R. Meidan, and L.B. Milstein. Spread spectrum access methods for wireless communications. *IEEE Communications Magazine*, 33(1):58–67, Jan. 1995.

[76] Jr. L. J. Cimini. Analysis and simulation of a digital mobile channel using orthogonal frequency division multiplexing. *IEEE Transactions on Communications*, COM-33(7):665–675, July 1985.

[77] D. C. Lee, D. L. Lough, S. F. Midkiff, N. J. Davis, and P. E. Benchoff. The next generation of the internet: aspects of the Internet Protocol Version 6. *IEEE Network*, 14(1):28–33, 1998.

[78] W. C. Y. Lee. Statistical analysis of the level crossings and duration of fades of the signal from an energy density mobile radio antenna. *Bell System Technical Journal*, 46:417–448, Feb. 1967.

[79] W. C. Y. Lee. Spectrum efficiency in cellular. *IEEE Transactions on Vehicular Technology*, 38:69–75, May 1989.

[80] W. C. Y. Lee. Overview of cellular CDMA. *IEEE Transactions on Vehicular Technology*, 40(2):291–302, 1991.

[81] W. C. Y. Lee. *Mobile communications design fundamentals, 2nd ed.* John Wiley & Sons, Inc., 1993.

[82] W. C. Y. Lee. *Mobile Communications Engineering: Theory and Applications, 2nd ed.* McGraw-Hill, New York, 1998.

[83] E. J. Leonardo and M. D. Yacoub. Cell coverage area using statistical methods. In *IEEE Global Telecommunications Conference (Globecom'93)*, pages 1227–1231, 1993. Houston, Texas, Dec.

[84] E. J. Leonardo and M. D. Yacoub. A statistical approach for cell coverage area in land mobile rado systems. In *7th IEE Conference on Mobile and Personal Communications*, pages 16–20, 1993. Brighton, UK, Dec.

[85] V. O. K. Li and X. Qiu. Personal communication system (PCS). *Proceedings of the IEEE*, 83(9):1210–1243, 1995.

[86] Shu Lin and Jr. Daniel J. Costello. *Error Control Coding: Fundamentals and Applications*. Prentice Hall, Inc. Englewood Cliffs, New Jersey, 1983.

[87] Y.-B. Lin and I. Chlamtac. *Wireless and Mobile Network Architectures*. John Wiley & Sons, Inc., New York, 2001.

[88] J-P. M. Linnartz. Exact analysis of the outage probability in multiple-user mobile radio. *IEEE Transactions on Communications*, 40(1):20–23, January 1992.

[89] A. G. Longley and P. L. Rice. Prediction of tropospheric radio transmission loss over irregular terrain: a computer method. *ESSA Technical Report*, ERL 79-ITS 67, 1968.

[90] R. W. Lucky. Automatic equalization for digital communications. *Bell System Technical Journal*, 44:547–588, April 1965.

[91] R. W. Lucky. Techniques for adaptive equalization of digital communications. *Bell System Technical Journal*, 45:255–286, Feb. 1966.

[92] V. H. MacDonald. The cellular concept. *Bell Systems Technical Journal*, 58(1):15–43, 1979.

[93] P. Manzoni, D. Ghosal, and G. Serazzi. Impact of mobility on TCP/IP: an integrated performance study. *IEEE Journal on Selected Area in Communications*, 13(5):858–867, 1995.

[94] J. W. Mark and S. Zhu. Power control allocation in multirate WCDMA systems. In *IEEE Wireless Communications Networking Conference (WCNC'00)*, pages CD–ROM, 2000.

[95] M. Mathis, J. Mahdavi, S. Floyd, and A. Romanow. TCP Selective Acknowledgment option. IETF RFC 2018, 1996.

[96] R. R. Müller. Multiuser receivers for randomly spread signals: fundamental limits with and without decision feedback. *IEEE Transactions on Information Theory*, 47(1):268–283, Jan. 2001.

[97] H. Meyr. Delay-lock tracking of stochastic signals. *IEEE Transactions on Communications*, 24:331–339, March 1976.

[98] A. R. Modarressi and R.A. Skoog. Overview of signaling system no. 7 and its role in the evolving information age network. *Proceedings of the IEEE*, 80(4):590–606, 1992.

[99] P. Monsen. Feedback equalization for fading dispersive channels. *IEEE Transactions on Information Theory*, IT-17:56–64, Jan. 1971.

[100] S. Moshavi. Multi-user detection for DS-CDMA communications. *IEEE Communications Magazine*, 34(10):124–136, Oct. 1996.

[101] S. N. Mukhi and J.W. Mark. Mobile profile information for a feedback-based handoff algorithm. In *IEEE International Conference on Communications (ICC'99) Conf. Record*, 1999. CD-ROM, Vancouver, Canada.

[102] K. Murota and K. Hirade. GMSK modulation for digital mobile radio telephone. *IEEE Transactions on Communications*, COM-29(7):1044–1050, July 1981.

[103] M. Nakagami. The m-distribution – a general formula of intensity distribution of rapid fading. *Statistical Methods in Radio Wave Propagation*, pages 3–36, 1960. W. G. Hoffman, ed.

[104] C. Namislo. Analysis of mobile radio slotted ALOHA networks. *IEEE Journal on Selected Areas in Communications*, 2(4):583–588, July 1984.

[105] A. Noerpel and Y.-B. Lin. Handover management for a PCS network. *IEEE Personal Communications*, 4(6):18–26, 1997.

[106] Yoshihisa Okumura, Eiji Ohmori, Tomihiko Kawano, and Kaneharu Fukuda. Field strength and its variability in VHF and UHF land-mobile radio service. *Review of the Electrical Communication Laboratory*, 16(9-10):825–873, September-October 1968.

[107] J. E. Padgett, C.G. Günther, and T. Hattori. Overview of wireless personal communications. *IEEE Communications Magazine*, 33(1):28–41, Jan. 1995.

[108] A. Papoulis. *Probability, Random Variables, and Stochastic Processes*. McGraw-Hill, New York, 1965.

[109] J. D. Parsons. *The mobile radio propagation channel*. Wiley, New York, NY, 1992.

[110] S. Pasupathy. Minimum shift keying: a spectrally efficient modulation. *IEEE Communications Magazine*, 19:14–22, July 1979.

[111] C. E. Perkins and P. Bhagwat. Highly dynamic destination-sequenced distance-vector routing (DSDV) for mobile computers. *Computer Communication Review*, pages 234–244, Oct. 1994.

[112] C. E. Perkins and P. Bhagwat. A mobile networking system based on the Internet Protocol. *IEEE Personal Communications*, 1(1):32–41, 1994.

[113] H. O. Peterson, H. H. Beverage, and J. B. Moore. Diversity telephone receiving system of RCA. *Proceedings of the IRE*, 19:562–584, April 1931.

[114] R. L. Peterson, R.E. Ziemer, and D.E. Borth. *Introduction to Spread Spectrum Communications*. Prentice-Hall, Englewood Cliffs, New Jersey, 1995.

[115] W. W. Peterson. *Error Correcting Codes*. MIT Press, Cambridge, Massachusetts, 1961.

[116] R. L. Pickholtz, D. L. Schilling, and L. B. Milstein. Theory of spread spectrum communications – a tutorial. *IEEE Transactions on Communications*, 30(5):855–884, May 1982.

[117] G. P. Pollini. Trend in handover design. *IEEE Communications magazine*, 34(2):82–93, 1996.

[118] A. Polydoros and C. L. Weber. A unified approach to serial search spread-spectrum code acquisition – part I & II. *IEEE Transactions on Communications*, 32(5):542–561, May 1984.

[119] A. Polydoros and C. L. Weber. Analysis and optimization of correlative code-tracking loops in spread spectrum systems. *IEEE Transactions on Communications*, 33(1):30–43, Jan. 1985.

[120] R. Prasad and A. Kegel. Effects of Rician faded and log-normal shadowed signals on spectrum efficiency in microcellular radio. *IEEE Transactions on Vehicular Technology*, 42(3):274–281, August 1993.

[121] R. Prasad and T. Ojanperä. An overview of CDMA evolution toward wideband CDMA. *IEEE Communications Survey*, 1(1):http://www.comsoc.org/pubs/surveys, Fourth Quarter 1998.

[122] T. Pratt and C. W. Bostian. *Satellite Communications*. John Wiely & Sons, New York, 1986.

[123] J. G. Proakis. *Digital Communications, 3rd ed.* McGraw-Hill, New Jersey, 1995.

[124] J. G. Proakis and M. Salehi. *Communication System Engineering*. Prentice Hall, New Jersey, 1994.

[125] S. U. H. Qureshi. Adaptive equalization. *Proceedings of the IEEE*, 74(9):1349–1387, 1985.

[126] R. Ramjee, T. F. La Porta, L. Salgarelli, S. Thuel, K. Varadhan, and L. Li. IP-based access network infrastructure for next generation wireless data networks. *IEEE Personal Communications*, 7(4):34–41, Aug. 2000.

[127] R. Ramjee, T. La Porta, S. Thuel, K. Varadhan, and S. Y. Wang. HAWAII: a domain-based approach for supporting mobility in wide-area wireless networks. In *IEEE Proc. 7th Int. Conf. Network Protocols*, pages 283–292, 1999.

[128] T. S. Rappaport. *Wireless Communications: Principle & Practice*. Prentice-Hall PTR, New Jersey, 1996.

[129] P. L. Rice, A. G. Longley, K. A. Norton, and A. P. Barsis. Transmission loss predictions for tropospheric communication circuits. *NBS Tech Note 101*, two volums, 1967. issued May 1965, revised May 1966 and January 1967.

[130] S. Rice. Statistical properties of a sine wave plus noise. *Bell System Technical Journal*, 27:109–157, Jan. 1948.

[131] E. M. Royer and C. K. Toh. A review of current routing protocols for ad hoc mobile wireless networks. *IEEE Personal Communications*, 6(2):46–55, April 1999.

[132] Jr. S. L. Marple. *Digital Spectral Analysis with Applications*. Prentice-Hall, Englewood Cliffs, New Jersey, 1987.

[133] A. M. Saleh and R. A. Valenzula. A statitical model for indoor multipath propagation. *IEEE Journal on Selected Areas in Communications*, SAC-6(2):128–137, 1987.

[134] A. Sampath, P. S. Kumar, and J. M. Holtzman. Power control and resource management for a multimedia CDMA wireless system. In *IEEE International Symposium on Personal, Indoor and Mobile Radio Communications (PIMRC'95)*, pages 21–25, 1995.

[135] R. A. Scholtz. The spread spectrum concept. *IEEE Transactions on Communications*, 25:748–755, Aug. 1977.

[136] M. Schwartz, W.R. Bennett, and S. Stein. *Communication Systems and Techniques*. McGraw-Hill, New York, 1966.

[137] W. Sheen and G. L. Stüber. Effects of multipath fading on delay-locked loops for spread spectrum systems. *IEEE Transactions on Communications*, 42(2/3/4):1947–1956, Feb./Mar./Apr. 1994.

[138] S. Shamai (Shitz) and S. Verdú. The impact of frequency-flat fading on the spectral efficiency of CDMA. *IEEE Transactions on Information Theory*, 47(4):1302–1327, May 2001.

[139] M. K. Simon and C. C. Wang. Differential detection of Gaussian MSK in a mobile radio environment. *IEEE Transactions on Vehicular Technology*, VT-33(4):307–320, November 1984.

[140] B. Sklar. Rayleigh fading channels in mobile digital communications. part I: characterization. *IEEE Communications Magazine*, 35(9):136–146, Sept. 1997.

[141] B. Sklar. Rayleigh fading channels in mobile digital communications. part II: mitigation. *IEEE Communications Magazine*, 35(9):148–155, Sept. 1997.

[142] B. Sklar. *Digital Communications: Fundamental and Applications, 2nd ed.* Prentice Hall PTR, New Jersey, 2001.

[143] J.L. Sobrinho and A.S. Krishnakumar. Quality-of-service in ad hoc carrier sense multiple access wireless networks. *IEEE Journal on Selected Areas in Communications*, 17(8), 1999.

[144] E. A. Sourour and S. C. Gupta. Direct-sequence spread-spectrum parallel acquisition in a fading mobile channel. *IEEE Transactions on Communications*, 38(7):992–998, July 1990.

[145] E. A. Sourour and S. C. Gupta. Direct-sequence spread-spectrum parallel acquisition in nonselective and frequency-selective rician fading channels. *IEEE Journal of Selected Areas in Communications*, 10:535–544, April 1992.

[146] K. W. Sowerby and A. G. Williamson. Outage probability calculations for multiple cochannel interferers in cellular mobile radio systems. *IEE Proceedings Part F*, 135(3):208–215, June 1988.

[147] G. L. Stüber. *Principles of Mobile Communications, 2nd ed.* Kluwer Academic Publishers, Boston, 2001.

[148] R. Steele, J. Whitehead, and W.C. Wong. System aspects of cellular radio. *IEEE Communications Magazine*, 33(1):80–86, 1995.

[149] S. Tabbane. Location management methods for third-generation mobile systems. *IEEE Communications Magazine*, 35(8):72–84, 1997.

[150] V. Tarokh, A. Naguib, N. Seshadri, and A. R. Calderbank. Space-time codes for high data rate wireless communication: Performance criteria in the presence of channel estimation errors, mobility, and multiple paths. *IEEE Transactions on Communications*, 47(2):199–207, Feb. 1999.

[151] V. Tarokh, N. Seshadri, and A. R. Calderbank. Space-time codes for high data rate wireless communication: Performance criterion and code construction. *IEEE Transactions on Information Theory*, 44, March 1998.

[152] D. Tse and S. Hanly. Linear multiuser receivers: effective interference, effective bandwidth and user capacity. *IEEE Transactions on Information Theory*, 45(2):641–657, Mar. 1999.

[153] G. Ungerboeck. Trellis-coded modulation with redundant signal sets. Part I: Introduction; Part II: State of the art. *IEEE Communications Magazine*, 25(2):1987, Feb. 1987.

[154] S. Verdú. *Multiuser Detection.* Cambridge University Press, New York, 1998.

[155] A. J. Viterbi. *CDMA – Principle of Spread Spectrum Communication.* Addison-Wesley Publishing Company, Reading, Massachusetts, 1995.

[156] J. Walfisch and H. L. Bertoni. A theoretical model of UHF propagation in urban environments. *IEEE Transactions on Antennas and Propagation*, AP-36:1788–1796, October 1988.

[157] S. G. Wilson. *Digital Modulation and Coding*. Prentice-Hall, Englewood Cliffs, New Jersey, 1996.

[158] R. W. Wolff. *Stochastic modeling and the theory of queues.* Prentice-Hall, Englewood Cliffs, New Jersey, 1989.

[159] D. Wong and T.J. Lim. Soft handoffs in CDMA mobile system. *IEEE Personal Communications*, 4(6):6–17, 1997.

[160] G. Xylomenos and G.C. Polyzos. Link layer support for quality of service on wireless Internet links. *IEEE Personal Communications*, 6(5):52–60, 1999.

[161] Y. Yeh and S. C. Schwartz. Outage probability in mobile telephone due to multiple log-normal interferers. *IEEE Transactions on Communications*, 32(4):380–388, April 1984.

[162] D. J. Young and N. C. Beaulieu. The generation of correlated Rayleighpart II random variates by inverse discrete Fourier transform. *IEEE Transactions on Communications*, TC-48(7):1114–1127, July 2000.

[163] W. R. Young. Advanced mobile phone service: introduction, background, and objectives. *The Bell System Technical Journal*, 58(1):1–14, 1979.

[164] J. H. Zhang and J.W. Mark. A local VLR cluster approach to location management for PCS networks. In *Proceedings of IEEE Wireless Communications and Networks Conference (WCNC'99)*, pages 311–315, 1999. New Orleans.

[165] J. Z. Zhang and J.W. Mark. A local anchor scheme for Mobile IP. In *Proceedings of International Conference on Performance and QoS of Next Generation Networking (P&QNet'2000)*, pages 137–156, 2000. Nagoya, Japan, Nov. 27-30.

[166] W. Y. Zou and Y. Wu. COFDM: an overview. *IEEE Transactions on Broadcasting*, 41(1):1–8, March 1995.

Answers to Selected Problems

CHAPTER 1

1. a. F, b. T, c. F, d. F, e. F
3. b. smaller cell size to increase frequency reuse, c. mobility management

CHAPTER 2

2. a. $\cos(\Omega t + \theta_0)\left[\frac{1}{j2\pi f} + \frac{1 - \exp(-j2\pi f T)}{(2\pi f)^2 T}\right]$

   b. $$r(t) = \begin{cases} \cos(\Omega t + \theta_0)\left(t - \frac{t^2}{2T}\right), & 0 \le t \le T_s \\ \cos(\Omega t + \theta_0)\left(T_s + \frac{T_s^2}{2T} - \frac{tT_s}{T}\right), & T_s \le t \le T \\ \cos(\Omega t + \theta_0)\left[\frac{t^2 - 2(T+T_s)t + (T+T_s)^2}{2T}\right], & T \le t \le T + T_s \\ 0, & t < 0 \text{ or } t > T + T_s \end{cases}$$

   c. T should not be much smaller than T_s
6. a. $\bar{\tau} = 10$ ns
   b. $\sigma_\tau = 10$ ns
   c. $(\Delta f)_c \approx 10^8$ Hz
7. a. flat fading
   b. $h(\tau, t) = \alpha(t)\exp[j\theta(t)]\delta(\tau)$
   c. $r(t) = \alpha(t)\exp[j\theta(t)]x(t)$; $\bar{\tau} = 0$, $\sigma_\tau = 0$

8. a. (1) $W_s T_m > 1$, (2) $W_s T_m \ll 1$
   b. flat fading with $\bar{\tau} = 0$ and $\sigma_\tau = 0$; the channel introduces Doppler shifts
9. a. $\phi_H(\Delta f) = \dfrac{1 - \exp(-j\pi \Delta f T)}{j4\pi \Delta f} + \dfrac{1 + \exp(-j\pi \Delta f T)}{j4\pi \Delta f + \dfrac{4\pi}{j\Delta f T^2}}$
   b. $\bar{\tau} = T\left(\frac{1}{4} - \frac{1}{\pi^2}\right)$, $\sigma_\tau = T\left[\frac{1}{2}\left(\frac{1}{6} - \frac{1}{\pi^2}\right) - \left(\frac{1}{4} - \frac{1}{\pi^2}\right)^2\right]^{0.5}$, $(\Delta f)_c \approx 1/\sigma_\tau$
   c. frequency-selective fading
10. a. $(\Delta t)_c \approx 50$ s, $(\Delta f)_c \approx 1$ Hz
    b. flat fading
    c. fast fading
13. b. $\Phi_H(\tau, \nu) = \mathcal{F}_{\Delta t}^{-1}[\phi_h(\tau, \Delta t)]$
14. a. $\phi(\tau) = 4/3$ for $0 \le \tau \le 100$ ms and $\phi(\tau) = 0$ otherwise
    b. $\Phi_H(\nu) = \dfrac{0.1}{\nu_m}\left[1 - \left(\dfrac{\nu}{\nu_m}\right)^2\right]$, $|\nu| \le \nu_m$
    c. $\bar{\tau} = 50$ ms, $\sigma_\tau = 28.87$ ms
    d. $\bar{\nu} = 0$, $\sigma_\nu = 4.47$ Hz
    e. $(\Delta f)_c \approx 34.64$ Hz, $(\Delta t)_c \approx 0.224$ s
17. $R = 10^{0.05(\alpha + 147.56 - 20\log_{10} f_c)}$
22. a. 119.45 crossings/s
    b. 0.0033 s
23. a. $\rho_0 = 1/\sqrt{2}$
    b. the pdf of the channel gain is maximal at ρ_0

CHAPTER 3

1. a. 256, b. 8
2. $P_{19200} = 8P_{2400}$
3. a. $\dfrac{A}{T_s}(T_s - t)\sin[2\pi f_c(T_s - t)]$ for $0 \le t \le T_s$ and 0 otherwise
   b. $\dfrac{A^2}{2T_s^2}\left\{\dfrac{T_s^3}{3} - \dfrac{T_s^2}{4\pi f_c}\sin(4\pi f_c T_s) - \dfrac{2T_s}{(4\pi f_c)^2}\cos(4\pi f_c T_s) + \dfrac{2}{(4\pi f_c)^3}\sin(4\pi f_c T_s)\right\}$
   c. the same result as in (*b*)
5. $\Phi_n = \left\{\frac{3\pi}{4}, \pi, \frac{\pi}{4}, 0\right\}$, $I_n = \left\{-\frac{1}{\sqrt{2}}, -1, \frac{1}{\sqrt{2}}, 0\right\}$, $Q_n = \left\{\frac{1}{\sqrt{2}}, 0, \frac{1}{\sqrt{2}}, 1\right\}$
6. $1/T_b$ (corresponding to $h = 1$)
9. $\Phi_x(f) = E_b\{\text{sinc}^2[2(f - f_c)T_b] + \text{sinc}^2[2(f + f_c)T_b]\}$

10. a. $\Phi_v(f) = A^2 T_s \text{sinc}^2(f T_s)$
    b. $\Phi_v(f) = \frac{1}{2} A^2 T_s \text{sinc}^2(f T_s/2)[1 - \cos(\pi f T_s)]$
    c. the first psd null is $1/T_s$ in part (a) and $2/T_s$ in part (b)
11. a. $\Phi_v(f) = T_b \text{sinc}^2(f T_b)[1.25 + \cos(2\pi f T_b)]$
    b. $\Phi_v(f) = \dfrac{9}{8T_b}|G(f)|^2\sigma^2\left[1 - \dfrac{2 - 6\cos(2\pi f T_b)}{10 - 6\cos(2\pi f T_b)}\right]$
14. a. $480\sqrt{N_0}$, b. $1518\sqrt{N_0}$
16. a. 3.87×10^{-6}, b. $A = 0.014$
17. $\dfrac{1}{2a\sqrt{2\pi}} \int_{-a}^{a} \int_{\sqrt{2E_b \cos^2(x)/N_0}}^{\infty} \exp(-z^2/2)\, dz\, dx$
19. a. $P_b = \dfrac{1}{8} + \dfrac{1}{4} Q\left(\sqrt{\dfrac{E_b}{8N_0}}\right) + \dfrac{1}{4} Q\left(\sqrt{\dfrac{E_b}{2N_0}}\right) + \dfrac{1}{4} Q\left(\sqrt{\dfrac{9E_b}{8N_0}}\right)$
    b. $P_b = 1/8$
20. a. exponential distribution with mean $E[\alpha^2]$
    b. the instantaneous SNR/bit is $\alpha^2(t) E_b/N_0$
    c. exponential distribution with mean $E[\alpha^2] E_b/N_0$
21. a. $P_b = Q\left(\sqrt{\dfrac{1}{\sigma_\alpha^2 + 0.5(N_0/E_b)}}\right)$
    b. $P_b = Q(1/\sigma_\alpha)$ as E_b/N_0 approaches infinity
22. b. $P_b = \dfrac{1}{2 + 2\sigma^2(E_b/N_0)}$

CHAPTER 4

4. a. $P_{b1} = 0.4Q\left(\sqrt{\dfrac{4E_b}{N_0}}\right) + 0.4Q\left(\sqrt{\dfrac{0.01E_b}{N_0}}\right) + 0.2Q\left(\sqrt{\dfrac{2.005E_b}{N_0}}\right)$
   b. $P_{b2} = 0.6Q\left(\sqrt{\dfrac{2E_b}{N_0}}\right) + 0.4Q\left(\sqrt{\dfrac{0.005E_b}{N_0}}\right)$
   c. $P_{b1} < P_{b2}$
5. a. $P_{b1} = 0.99Q\left(\sqrt{\dfrac{2E_b}{N_0}}\right) + 0.01Q\left(\sqrt{\dfrac{0.005E_b}{N_0}}\right)$
   b. $P_{b2} = 0.81Q\left(\sqrt{\dfrac{2^2 E_b}{N_0}}\right) + 0.18Q\left(\sqrt{\dfrac{(1.05)^2 E_b}{N_0}}\right) + 0.01Q\left(\sqrt{\dfrac{0.1^2 E_b}{N_0}}\right)$
6. $\Gamma_b = \left[1 + \dfrac{\pi}{4}(L-1)\right]\Gamma_c$

7. a. $f_{\gamma_b}(x) = \frac{2}{\Gamma_c} \exp\left(-\frac{x}{\Gamma_c}\right)\left[1 - \exp\left(-\frac{x}{\Gamma_c}\right)\right], x \geq 0$

   b. $P_b = \frac{1}{2} - \sqrt{\frac{\Gamma_c}{1+\Gamma_c}} + \frac{1}{2}\sqrt{\frac{\Gamma_c}{2+\Gamma_c}}$

8. a. $T_s \gg 0.1$ s

   b. $(\Delta f)_c = 10$ Hz, $(\Delta t)_c = 20$ s

   c. $L = 3$

CHAPTER 5

1. b. $D = \sqrt{3(i^2 + j^2 + ij)}R$

   c. $N = i^2 + j^2 + ij$

   d. system capacity and cochannel interference

4. a. $A_{cell} = \beta R^2$, where $\beta = 3\sqrt{3}/2$ for hexagon, 2 for square, and $3\sqrt{3}/4$ for triangle

   b. closest to a circle and having the largest area given R

5. a. $N = 19$, 7 and 4 for $\kappa = 3$, 4 and 5, respectively

   b. 7.895, 21.43, 37.5

   c. 3947, 10714, 18750

6. a. traffic load = 54.376 Erlangs, trunking efficiency = 0.8366

   b. 2175, 108750

   c. 33.46, 1673

   d. 3250

7. a. 122.98 Erlangs, b. 683, c. 72 s, d. 0.062, e. 0.0031

8. a. 93.44 Erlangs, b. 519, c. 72 s, d. 0.0873, e. 0.004365

9. a. 43 Erlangs, 860, b. 31.3 Erlangs, 626, c. 22.68 Erlangs, 453

   d. $(S/I)_{360^\circ} = 17.3$ dB, $(S/I)_{120^\circ} = 24.6$ dB, $(S/I)_{60^\circ} = 28.97$ dB, $\eta_{t,360^\circ} = 0.77$, $\eta_{t,120^\circ} = 0.58$, $\eta_{t,60^\circ} = 0.42$

10. a. $N_{360^\circ} = 9$, $N_{120^\circ} = 4$, $N_{60^\circ} = 3$

    b. 31.656 Erlangs, 0.7341; 67.866 Erlangs, 0.6925; 68.255 Erlangs, 0.6190

11. a. $N_{360^\circ} = 7$, $N_{120^\circ} = 3$, $N_{60^\circ} = 1$

    b. yes

    c. 60° sectoring

12. b. the transmit power after splitting is 1/8 of that before splitting

CHAPTER 6

5. a. $P_b = \frac{1}{1+\alpha} Q\left(\sqrt{\frac{2E_b}{N_0}}\right) + \frac{\alpha}{1+\alpha} Q\left(\sqrt{\frac{2E_b}{N_0 + \frac{(1+\alpha)P_f}{\alpha W}}}\right)$

b. $\alpha \approx \frac{P_J/W}{2E_b - P_J/W}$

c. $P_{b,\max} \approx \frac{1}{\sqrt{2\pi e}} \frac{P_J/W}{2E_b}$

d. yes

6. a. 27.81, b. 19.60, c. 15.65
7. a. 23, b. 832, c. 116, d. 0.0714 channels/MHz/km^2
8. a. 3132, b. 1566, c. 0.0643 Erlangs/MHz/km^2
9. a. 14 kbps, b. 74 bits, c. 37 bits, d. 5.6 kbps
10. a. 1000, b. 142, c. 0.7424, d. 0.08164 bps/Hz/cell
11. a. 600, b. 1448.4, c. 724.2, d. 0.644 Erlangs/MHz/km^2
12. a. 3.237×10^7 chips/s, b. 1.2948×10^7 chips/s
13. c. 39 users/cell, 0.05054 bps/Hz
14. $N_c = 1 + \frac{W/R_b}{\gamma_r s_f k_f}$
15. $P_b = 3.3458 \times 10^{-5}$

CHAPTER 7

4. b. 0.101 hrs, c. 2.03
11. b. (1) The mobile host has to provide location updates, (2) the network has to send paging messages within some location area, and (3) the mobile host has to acknowledge its current location in response to the paging message. All communications by the mobile hosts are done through the called MSC.
15. a. 8997.52 users/km^2
    b. 181.14 crossings/cluster/s
    c. 8333, 1565
    d. 260835 updates/cluster/hr

CHAPTER 8

3. agent discovery, registration, and tunneling
4. a. triangle routing and optimized routing
11. a. the handoff delay is $t_{ho\_home} = 2(b + d) + 3r + (n + 2)(f + r)$
    b. 50
12. a. $t_{anchor} = 2(b + d) + 3r$
    b. $t_{home} = 2(b + d) + (n + 2)f + (n + 5)r$
17. a. client, gateway and original server
    b. (1) Wireless Markup Language (WML), (2) WMLScript, (3) Wireless Telephone Application
    c. Wireless Datagram Protocol (WDP)

Index

K factor, Rician fading, 50
Q function, 46
$\pi/4$-DQPSK, 79–82
$\pi/4$-QPSK, 79
z transform, 137

A

access point (AP), 162
ACK, 289
ad hoc network, 8, 303
  applications, 304
  route discovery, 304
  route maintenance, 304
  source-initiated on-demand routing, 303
  table-driven routing, 303
adaptive equalizer, *see* equalizer 119
address
  care-of, 275
  home, 162, 274
  link-layer, 275
adjacent channel interference, 175–176
agent
  advertisement, 275
  discovery, 277
  foreign, 275
  home, 162, 275
  solicitation, 275
Aloha, *see* multiple access, Aloha 191
amplitude modulation (AM), 2
AMPS, 5
angle diversity, 121
ANSI, 8
antenna height, 38, 41, 43
antenna sectoring, 182
ARQ, 121
ASK, 66
authentication, 163, 251
autocorrelation, 205, 318
average fade duration (AFD), 54–56
AWGN channel, 70

B

base station (BS), 2
baseband complex signal (envelope), 65
Bayes' theorem, 313
Berkeley Snoop module, 291
binding entry, 275
binomial expansion, 132
bit error rate (BER), *see* probability of transmission error 101
bit rate, 68
blocked call cleared, 179
BPSK, *see* modulation, BPSK 67
broadcasting, 189
bursty errors, 120

C

call
- blocking, 176
- delay, 176
- delivery, 252, 260
- dropping, 237
- duration, 176
- setup, 255, 256

call admission control (CAC), 237
call-to-mobility ratio (CMR), 259
capacity
- cell, *see* cell capacity 223
- Erlang, 226, 230
- soft, 211

CDMA, *see* multiple access, CDMA 190
cdma-2000, 7
cell
- cluster, 164
- coverage, 45
- definition, 3
- residence time, 176
- sectoring, 182
- sojourn time, 245
- splitting, 180

cell capacity
- CDMA, 223–226
- FDMA, 215
- TDMA, 219

channel
- AWGN, 70
- effective, 134
- fading dispersive, 15
- forward, 2
- holding time, 176
- impairments, 3, 12
- interleaving, 176
- LTI model, 16
- LTV model, 16–22
- reverse, 2
- US, 25
- WSS, 25
- WSSUS, 25

channel assignment
- DCA, 185
- FCA, 184

channel coding, 120
channel correlation function, 25
- delay psd, 25–27
- Doppler psd, 29–30
- frequency and time correlation, 27
- frequency correlation, 28
- scattering function, 60
- time correlation, 28

channel encoder/decoder, 64
channel equalization, 119, 134
channel function
- delay-Doppler spread function, 21–22
- Doppler spread function, 21
- impulse response, 16–18
- transfer function, 18–19

chi-square distribution, 128
cluster (of cell)
- frequency reuse, 164
- handoff management, 239

cochannel cell, 165
cochannel interference (CCI), 172–175
coherence bandwidth, 28
coherence time, 28
collision, 189
- resolution interval (CRI), 193

combining, diversity, *see* linear combining 122
common channel signaling (CCS), 254
correlation coefficient, 69
correspondent node, 274
crosscorrelation, 205
CSMA, *see* multiple access, CSMA 195
cyclostationary process, 93

D

decision region, 72
delay
- mean, 27
- moment, 26
- spread, 27

demodulation, *see* detection, of signal 105
detection, of signal, 70–71
- coherent, 77
- noncoherent, 80, 105

DFE, *see* equalizer, DFE 146
DHCP, 277
differential encoding, 80
digital communications
- advantages, 5–6
- system model, 63–64

Dirac delta function, 16

direct sequence spread spectrum, 202
  PN sequence, 201, 317
  power spectral density, 204
  probability of transmission error, 206
  processing gain, 206
  receiver, 202
  transmitter, 202
discrete Fourier transform, 91
dispersion
  frequency, 21
  time, 13–15
diversity, 118
  angle diversity, 121
  frequency diversity, 121
  path diversity, 122
  performance improvement, 126
  space diversity, 121
  time diversity, 121
Doppler shift, 19–21
  mean, 30
  moment, 30
  spread, 30
downlink, 2
DS-CDMA, *see* multiple access, CDMA 7
duplexing
  FDD, 200
  TDD, 200
dynamic channel assignment (DCA), 185

E

effective channel, 134
EIRP, 36
equal-gain combining, 123
equalization error, 144
equalizer
  DFE, 146, 155
    adaptive, 152–155
    MMSE, 150–152
    zero-forcing (ZF), 149–150
  linear equalizer, 136
    adaptive, 145–146
    MMSE, 143–145
    zero-forcing (ZF), 138–143
ergodic process, 51
Erlang, 176
  capacity, 226, 230
  Erlang-B system, 177
  Erlang-C system, 178
  traffic load, 176
error function, 89
error probability, *see* probability of transmission error 100
error propagation, 153
ETACS, 5
Euclidean distance, 70
exponential distribution, 124

F

fading, 14–15
  first-order statistics, 48–50
  flat, 14, 28
  frequency-selective, 14, 28
  long-term, 36, 117
  Rayleigh, 49
  Rician, 50
  second-order statistics, 51–58
  shadowing, 36
  short-term, 36, 117
  small-scale, 48–58
Fast Retransmit, 292
FDD, 200
FDMA, *see* multiple access, FDMA 190
feedback filter, 147
feedforward filter, 147
fixed channel assignment (FCA), 184
flat fading, *see* fading, flat 14
footprint, 162
foreign agent, 275
foreign network, 275
forward channel, 2
forward error correction, 120
frame, 198
free-space path loss model, 36
frequency diversity, 121
frequency modulation (FM), 2
frequency reuse, 161, 211
  efficiency factor, 225
  factor, 164
  ratio, 171
frequency-selective fading, *see* fading, frequency-selective 14
FSK, 67

G

Gaussian distribution, 101
Gaussian filter, 88

generator polynomial, 317
geometry, hexagon, 168
GMSK, 88–90
grade of service (GoS), 177
Gram–Schmidt orthogonalization, 311
Gray encoding, 77
GSM, 6
guard channel, 238

H

handoff
  backward handoff, 240
  delay, 295
  forward handoff, 240
  hard handoff, 239
  intraswitch, 247
  MAHO, 239, 241
  management, 3, 163, 238
  MCHO, 239
  NCHO, 239
  profile, 244
  rate calculation, 263
  soft handoff, 211, 240
  strategies, 239
home agent, 275
home location register (HLR), 250
home network, 162, 275
hybrid multiple access, 230

I

I-TCP, 291
ICMP, 277
IETF, 274
impulse response, 16
IMT-2000, 7
indoor propagation model, 58
interference
  adjacent channel, 175–176
  cochannel (CCI), 172
  factor, 224
  intercell, 172
  intracell, 172
  multiple access (MAI), 204
interference limited environment, 12, 172, 197
intermodulation, 197
Internet Protocol (IP), 284–287
  IPv4, 284
  IPv6, 284
  SAC, 288
intersymbol interference (ISI), 14, 135
inverse filter, 140
IS-136, 252
IS-41, 6, 252
IS-54, 6, 221
IS-95, 6
ISO model, 272
ITU, 8

J

jamming, 208

K

Kronecker delta function, 69

L

least-mean-square (LMS) algorithm, 146
Lee's path loss model, 43
LEO satellite network, 7
level crossing rate (LCR), 51–54
likelihood function, 314
line-of-sight (LOS), 48
linear combining, 122
  equal-gain, 123
  maximal ratio, 122
  selective combining, 123
linear equalizer
  see equalizer, linear equalizer, 136
link, 275
  downlink, 2
  foreign, 275
  home, 275
  uplink, 2
link layer, 4
local anchor, 259, 281, 295
local mean, 36
location management, 3, 163, 250, 257
  local anchor, 259
  overlay approach, 258
location update, 252, 255
log-distance path loss model, 40
log-normal shadowing, 41
long-term fading, 35
LTI channel, 16
LTV channel, 17

M

M-TCP, 292
M-UDP, 293
Marcum's Q function, 102
Markov chain, 299
Markovian process, 176
matched filter, 71
maximal length sequence, 317
maximal ratio combining, 122
maximum a posteriori probability (MAP), 313
maximum likelihood (ML) decision rule, 72, 314
maximum likelihood detection, 313
MC-CDMA, *see* multiple access, CDMA 7
mean squared error (MSE), 143
medium access control (MAC), 275
minimum mean-square error (MMSE), 143, 150
mobile
  host (MH), 162, 274
  node(MN), 274
  station (MS), 2, 274
Mobile Applications Part (MAP), 6, 252
mobile identification number (MIN), 255
Mobile IP, 7, 274–284
  local anchor, 281
  Mobile IPv4, 288
  Mobile IPv6, 287
mobile switching center (MSC), 3, 161
mobility management, 4, 163
mobility model, 245, 246, 263, 295
modified Bessel function, 50
modulation
  $\pi/4$-DQPSK, 79–82
  AM, 2
  ASK, 66
  BPSK, 67, 75
  DBPSK, 131
  efficiency, 221
  FM, 2
  FSK, 67
  GMSK, 88–90
  MPSK, 76
  MSK, 84–88, 314
  OFDM, 8, 90–92
  OQPSK, 6
  QPSK, 76–79
modulation index, 84
MPSK, 76
MSK, 84–88
multipath
  delay spread, 26
  intensity profile, 26
  propagation, 13
multipath intensity profile, 26
multiple access, 189
  Aloha, 191
    delay-throughput, 192
    pure, 191
    slotted, 191
    throughput, 191
  CDMA, 190, 201
    capacity, 223
    direct sequence, 201
    multi-carrier, 230
    near–far problem, 224
    power control, 224
    spectral efficiency, 227
  CSMA, 195
    1-persistent, 195
    non-persistent, 195
    throughput, 195
  FDMA, 190, 197
  hybrid, 230
  SDMA, 190
  TDMA, 190, 198
    ATDMA, 198, 200
    STDMA, 198
multiple access interference (MAI), 204
multiuser detection, 229

N

narrowband
  CDMA, 4
  FDMA, 4, 197
  TDMA, 199
near–far effect, 118, 176
network
  foreign network, 250, 275
  home network, 250, 275
  virtual network, 275
  visitor network, 250
networking layer, 4
noise
  AWGN, 12
  thermal, 12
non line-of-sight (NLOS), 48
norm, 69

NTT, 5
Nyquist criterion, 140

O

OFDM, 8, 90–92
optimum receiver, 73
orthogonality of signals, 190
orthonormal set, 68
out-of-band radiation, 82, 176
outage probability, 185, 234

P

packet-switching, 191
paging, 255, 257
path diversity, 122
path loss exponent, 40
path loss model, 36
  free space, 36–38
  Lee's model, 43–44
  log distance, 40–41
  Okumura–Hata, 41
  smooth plane, 38–39
PCS, 7
PDC, 6
phase distortion, 18
physical layer, 4
PLMN, 7
PN sequence, 317
  m-sequence, 317
  autocorrelation, 318
  generator, 317
  maximal length, 317
power control, 118, 229, 230
  closed-loop, 118
  open-loop, 118
power spectral density (psd), 93–100
  $\pi/4$-DQPSK, 96
  baseband, 94
  BPSK, 94
  GMSK, 97
  MSK, 96
  OFDM, 98
  passband, 94
  QPSK, 95
primitive polynomial, 317
probability density function (pdf)
  chi-square, 128
  exponential, 124
  Gaussian, 101
  log-normal, 41
  Rayleigh, 49
  Rician, 50, 315
  uniform, 49
probability of transmission error
  AWGN, 100–108
    $\pi/4$-DQPSK, 102
    BPSK, 100
    MSK, 103
    QPSK, 101
  Rayleigh fading, 108–112
processing gain, 206
profile, 244
pulse shaping, 82

Q

QPSK, 76–79
quality of service (QoS), 274

R

Rake receiver, 120
random access, 191–197
rate
  bit, 68
  chip, 203
  symbol, 68
Rayleigh fading, 49
registration, 277
registration area (RA), 252
resource management, 4
reverse channel, 2
Rician fading, 50
roaming, 235
roll-off factor, 82
routing
  ad hoc network, 303
  hierarchical, 282
  optimal, 280
  triangle, 279
  tunneling, 277

S

SDMA, *see* multiple access, SDMA 190
sectoring, cell, 182
selective combining, 123
shadowing, 36, 41
shift register, 318

signal space, 68–70
signal-to-interference plus noise ratio (SINR), 206
signal-to-interference ratio (SIR), 173
signal-to-noise ratio (SNR), 109
sinc function, 94
slot, 191
small-scale fading, 48
soft handoff, 211
sojourn time, 245
source encoder/decoder, 64
space diversity, 121
space-time coded modulation (STCM), 121
spectral efficiency, 212
  CDMA, 227–228
  FDMA, 213–218
  TDMA, 218–222
splitting, cell, 180
spread spectrum, direct sequence, 120, 202–208
  despreading process, 205
  spreading process, 203
square root raised cosine filter, 82
SS7, 253
STDMA, 198
subband, 197
symbol rate, 68
synchronization
  carrier, 82
  PN code, 230
  symbol, 80

T

TCP, 273, 289–290
  Berkeley Snoop module, 291
  Fast Retransmit, 292
  flow control, 289
  for Mobile Cellular Network (M-TCP), 292
  indirect TCP (I-TCP), 291
  modified TCP, 290
  sliding window, 289
  slow start, 289
  timeout, 289
TDD, 200
TDM, 255
TDMA, *see* multiple access, TDMA 190
throughput, 191, 195
time correlation function, channel, 28
time diversity, 121
TLDN, 256
traffic intensity, 177
training sequence, 146
transfer function, channel, 18
trellis-coded modulation (TCM), 120
trunking efficiency, 178
tunnel, 275
tunneling, 277, 278

U

UDP, 273
  modified UDP, 293
uniform distribution, 49
uplink, 2
US channel, 25

V

virtual network, 275
visitor location register (VLR), 236
visitor network, 236
vulnerable period, 191

W

WAP, 300–302
  gateway, 302
  WAE, 300
  WDP, 302
  WSP, 301
  WTLS, 301
  WTP, 301
WATM, 7
wideband
  CDMA, 4
  TDMA, 199
Wiener-Hopf equation, 145
Wiener-Khintchine relations, 26
wireless system standards, 4–8
  1G, 4–5
  2G, 4–6
  3G, 4, 7–8
WLAN, 8
WSS channel, 25
WSSUS channel, 25
WWAN, 8

Z

zero-forcing (ZF)
  DFE, 149
  linear equalizer, 138